Ecological and Evolutionary Perspectives on Infections and Morbidity

P.A. Azeez
Salim Ali Centre for Ornithology and Natural History, Coimbatore, India

P.P. Nikhil Raj
Amrita School of Engineering, Amrita Vishwa Vidyapeetham, India

R. Mohanraj
Bharathidasan University, India

A volume in the Advances in Human Services and Public Health (AHSPH) Book Series

IGI Global
PUBLISHER of TIMELY KNOWLEDGE

Published in the United States of America by
IGI Global
Medical Information Science Reference (an imprint of IGI Global)
701 E. Chocolate Avenue
Hershey PA, USA 17033
Tel: 717-533-8845
Fax: 717-533-8661
E-mail: cust@igi-global.com
Web site: http://www.igi-global.com

Library of Congress Cataloging-in-Publication Data

Names: Azeez, P. A., 1957- editor. | Raj, P. P. Nikhil, 1981- editor. | Mohanraj, R. (Rangaswamy), 1973- editor.
Title: Ecological and evolutionary perspectives on infections and morbidity / P.A. Azeez, P.P. Nikhil Raj, and R. Mohanraj, editors.
Description: Hershey, PA : Engineering Science Reference, an imprint of IGI Global, [2022] | Includes bibliographical references and index. | Summary: "This book is an attempt to address human diseases from a wider and to an extent holistic perspective, especially looking at morbidity and mortality (primarily from infections mediated/facilitated by intermediate hosts or vectors) from ecological and evolutionary viewpoints"-- Provided by publisher.
Identifiers: LCCN 2021054729 (print) | LCCN 2021054730 (ebook) | ISBN 9781799894148 (h/c) | ISBN 9781799894155 (s/c) | ISBN 9781799894162 (ebook)
Subjects: LCSH: Communicable diseases. | Human ecology.
Classification: LCC RA643 .E25 2022 (print) | LCC RA643 (ebook) | DDC 362.1969--dc23/eng/20211215
LC record available at https://lccn.loc.gov/2021054729
LC ebook record available at https://lccn.loc.gov/2021054730

This book is published in the IGI Global book series Advances in Human Services and Public Health (AHSPH) (ISSN: 2475-6571; eISSN: 2475-658X)

British Cataloguing in Publication Data
A Cataloguing in Publication record for this book is available from the British Library.

Advances in Human Services and Public Health (AHSPH) Book Series

ISSN:2475-6571
EISSN:2475-658X

MISSION

The well-being of the general public should be a primary concern for any modern civilization. Ongoing research in the field of human services and public healthcare is necessary to evaluate, manage, and respond to the health and social needs of the global population.

The **Advances in Human Services and Public Health (AHSPH)** book series aims to publish high-quality reference publications focused on the latest methodologies, tools, issues, and strategies for managing the health and social welfare of the public. The AHSPH book series will be especially relevant for healthcare professionals, policy makers, government officials, and students seeking the latest research in this field.

COVERAGE

- Healthcare Reform
- Public Funding
- Health Policy
- Public Welfare
- Access to Healthcare Services
- Poverty
- Social Welfare Policy
- Social Work
- Youth Development
- Medicare and Medicaid

IGI Global is currently accepting manuscripts for publication within this series. To submit a proposal for a volume in this series, please contact our Acquisition Editors at Acquisitions@igi-global.com or visit: http://www.igi-global.com/publish/.

The Advances in Human Services and Public Health (AHSPH) Book Series (ISSN 2475-6571) is published by IGI Global, 701 E. Chocolate Avenue, Hershey, PA 17033-1240, USA, www.igi-global.com. This series is composed of titles available for purchase individually; each title is edited to be contextually exclusive from any other title within the series. For pricing and ordering information please visit http://www.igi-global.com/book-series/advances-human-services-public-health/102256. Postmaster: Send all address changes to above address.

Titles in this Series

For a list of additional titles in this series, please visit:
http://www.igi-global.com/book-series/advances-human-services-public-health/102256

The Role of Health Literacy in Major Healthcare Crises
Maria Theodosopoulou (Imperial College London, UK) and Vassilios Papalois (Hammersmith Hospital, UK)
Information Science Reference • © 2023 • 300pp • H/C (ISBN: 9781799896524) • US $235.00

Transformation and Efficiency Enhancement of Public Utilities Systems Multidimensional Aspects and Perspectives
Jordan Gjorchev (International Slavic University, North Macedonia) Samoil Malcheski (International Slavic University, North Macedonia) Tamara Rađenović (University of Niš, Serbia) Dejan Vasović (University of Niš, Serbia) and Snežana Živković (University of Niš, Serbia)
Information Science Reference • © 2023 • 457pp • H/C (ISBN: 9781668477304) • US $215.00

Regulating Human Rights, Social Security, and Socio-Economic Structures in a Global Perspective
Emilia Alaverdov (Georgian Technical University, Georgia) and Muhammad Waseem Bari (Government College University, Faisalabad, Pakistan)
Information Science Reference • © 2022 • 327pp • H/C (ISBN: 9781668446201) • US $215.00

Assessing the Effects of Emerging Plastics on the Environment and Public Health
Sung Hee Joo (University of Seoul, South Korea)
Engineering Science Reference • © 2022 • 441pp • H/C (ISBN: 9781799897231) • US $240.00

Global Dimensions of Democracy and Human Rights Problems and Perspectives
Nika Chitadze (International Black Sea University, Georgia)
Information Science Reference • © 2022 • 345pp • H/C (ISBN: 9781668445433) • US $215.00

For an entire list of titles in this series, please visit:
http://www.igi-global.com/book-series/advances-human-services-public-health/102256

IGI Global
PUBLISHER of TIMELY KNOWLEDGE

701 East Chocolate Avenue, Hershey, PA 17033, USA
Tel: 717-533-8845 x100 • Fax: 717-533-8661
E-Mail: cust@igi-global.com • www.igi-global.com

Table of Contents

Preface xiv

Chapter 1
Changing Climate: Threat to Human Health 1
Abirami Subramanian, Indira Gandhi National Open University, India
Sushmitha Baskar, Indira Gandhi National Open University, India

Chapter 2
Climate Change and Human Infectious Diseases 31
Mercy Oluwatofunmi Olaoluwa, University of Ibadan, Nigeria
Abosede Olufunke Taiwo, Osun State College of Technology, Nigeria
Ayodeji Ojo Oteyola, Federal University of Minas Gerais, Brazil

Chapter 3
Urbanization and Emergence of Infectious Diseases 43
Supriya Ray, Karnatak University, India
Venugopal N. Puluamaghatta, Anthropological Survey of India, India
Suresh Basavaraj Arakera, Karnatak University, India

Chapter 4
Urbanization and the Mosquito Population: Threats to Human Health 68
Debraj Biswal, Government General Degree College at Mangalkote, India

Chapter 5
Men, Mosquito, and Malaria: The Circle of Life in Plasmodium 100
Arunava Mukherjee, Ramakrishna Mission Vivekananda Centenary College, Rahara, India
Pradipta Kumar Ghosh, Ecosystem and Ecology Laboratory, Ramakrishna Mission Vivekananda Centenary College, Kolkata, India

Debojyoty Ghosh, Sundarban Tiger Reserve, India
Banani Mandal, Jogesh Chandra Chaudhuri College, India

Chapter 6
Physical School Environment and Infectious Diseases: A Case of Primary School Context in Bangladesh 126
Afzal Hossain, Institute of Education and Research, University of Rajshahi, Bangladesh
Emaj Uddin, University of Rajshahi, Bangladesh

Chapter 7
Post-Antibiotic Era: A Lurking Global Threat 152
Shafa S. Contractor, Karnatak University, India
Suresh B. Arakera, Karnatak University, India

Chapter 8
The Interplay Between Environmental Pollutants, Gut Microbiota, and Infections: Current Concepts and Therapies 173
Bilal El-Mansoury, Faculty of Science, Chouaib Doukkali University, Morocco
Kamal Smimih, Faculty of Sciences and Techniques, Sultan Moulay Slimane University, Morocco
Youssef Ait Hamdan, Higher Normal School, Cadi Ayyad University, Morocco
Abdelaati El Khiat, Faculty of Medicine and Pharmacy, Cadi Ayyad University, Morocco & ISPITS, Ouarzazate, Morocco.
Samia Elouali, Higher Normal School, Cadi Ayyad University, Morocco
Hanane El Fatimi, Mohammed VI University of Health Sciences, Morocco
Ahmed Draoui, Faculty of Science Semlalia, Cadi Ayyad University, Morocco
Omar El Hiba, Faculty of Sciences, Chouaib Doukkali University, Morocco
Arumugam Radhakrishnan Jayakumar, Miller School of Medicine, University of Miami, USA

Chapter 9
Classification of Melanoma Skin Cancer Based on Transformer Deep Learning Model 208
Nagarjuna Telagam, GITAM University, India
Nehru Kandasamy, Madanapalle Institute of Science and Technology, India

Chapter 10
Clinical Characteristics and Relationship of Acute Kidney Injury With the Severity of Disease and Death in COVID-19 Patients: A Review.....................228
Donovan McGrowder, The University of the West Indies, Jamaica
Fabian Miller, The Mico University College, Jamaica & The University of the West Indies, Jamaica
Magdalene Nwokocha, The University of the West Indies, Jamaica
Melisa Anderson Cross, University of Technology, Jamaica
Cameil Wilson-Clarke, The University of the West Indies, Jamaica
Shelly McFarlane, The University of the West Indies, Jamaica
Lorenzo Gordon, Caribbean School of Medical Sciences, Jamaica

Compilation of References 259

About the Contributors 321

Index 329

Detailed Table of Contents

Preface xiv

Chapter 1
Changing Climate: Threat to Human Health 1
Abirami Subramanian, Indira Gandhi National Open University, India
Sushmitha Baskar, Indira Gandhi National Open University, India

Earth is endowed with sufficient reserves to meet all of our needs. It is the fact that all living organisms use the planet's resources to the extent that they require them. Humans, on the other hand, are the only creatures on the planet who enjoy and even exploit resources beyond their necessity. To be more precise, due to population surge, rapid industrialization, and urbanization, there is a need to meet both the demands and luxuries. The Earth's species are dependent on the ecosystem service functions which are driven by climate stability. Climate-triggered health illnesses are currently showing new trends and are expected to increase in the future if the current climate disturbances persist. In brief, climate instability (warming or cooling) undeniably provides suitable ground for vector and zoonotic life cycles. Climate-related morbidity and mortality are developing public health concerns that, if not addressed properly, will have serious corollaries. This chapter summarizes the various factors responsible for climate variability-induced health impacts.

Chapter 2
Climate Change and Human Infectious Diseases 31
Mercy Oluwatofunmi Olaoluwa, University of Ibadan, Nigeria
Abosede Olufunke Taiwo, Osun State College of Technology, Nigeria
Ayodeji Ojo Oteyola, Federal University of Minas Gerais, Brazil

Climate change is one of the most serious dangers to human health in the 21st century. Climate extremes, air quality, sea-level rise, and water supplies all have a direct impact on human health. Climate change has an influence on infectious diseases as well. In this chapter, the authors explained the influence of climate change on human infectious diseases. The authors also discuss how society may respond to and adapt to climate change. The variables influencing the spread and severity of human diseases were explored as well. These variables include human and animal migration, control methods, medication availability, human behavior, and the quality of public health facilities. The authors also discuss the importance of genome research in climate change and infectious diseases. Finally, as the world faces present and future difficulties in climate change and an increased threat of infectious diseases, the authors emphasize on the importance of scientific networks and collaborations for effective multidisciplinary engagements.

Chapter 3
Urbanization and Emergence of Infectious Diseases 43
Supriya Ray, Karnatak University, India
Venugopal N. Puluamaghatta, Anthropological Survey of India, India
Suresh Basavaraj Arakera, Karnatak University, India

The process of urbanization is ongoing and continuous since the industrial revolution in the 18th century. The United Nations has estimated the growth of world's urban population to be 6.3 billion by 2050. Rapid influx of migrants in cities led to the development of informal urban settlement which is the initial probable source of emerging communicable diseases. Due to overcrowding, closer contacts with wild and domesticated animals increases the risk of emerging zoonotic diseases. Adequate housing, proper city planning, proper hygiene and sanitation, and surveillance can help in preventing transmission of pathogens, emerging infectious diseases, and decline of ecosystem. This chapter considers the role of urbanization that plays in the emergence of zoonotic diseases and cross-species transmission of pathogens into new host population through exploring the dynamics and complexity of ecological system at wildlife-livestock-human interfaces and the factors affecting the epidemiology of disease emergence.

Chapter 4
Urbanization and the Mosquito Population: Threats to Human Health 68
Debraj Biswal, Government General Degree College at Mangalkote, India

Mosquitoes have always been a source of threat to human health because of their ability to transmit deadly diseases. Despite several efforts to curb their population, they have managed to survive and spread havoc. Urbanization has emerged to be one of the major contributors to their reproductive success followed by climate change. Urbanization has stimulated opportunistic breeding behaviour in mosquitoes by generating a vast array of temporary breeding habitats. The effect has been compounded by climate change allowing the mosquitoes to breed throughout the year and spread to regions which were previously inhospitable to them. Further, a study of mosquito breeding behaviour shows that gravid females follow several visual and olfactory cues to select their breeding habitats to ensure the well-being of their young ones. Thus, study of mosquito breeding habitats in urban areas helps to identify their markers to be used in mosquito larval source management. Similar problems can be largely overcome in the future by taking a more inclusive approach during town planning.

Chapter 5
Men, Mosquito, and Malaria: The Circle of Life in Plasmodium......................100
Arunava Mukherjee, Ramakrishna Mission Vivekananda Centenary College, Rahara, India
Pradipta Kumar Ghosh, Ecosystem and Ecology Laboratory, Ramakrishna Mission Vivekananda Centenary College, Kolkata, India
Debojyoty Ghosh, Sundarban Tiger Reserve, India
Banani Mandal, Jogesh Chandra Chaudhuri College, India

The chapter is dedicated to malaria, plasmodium, their ecology, evolution, life cycle, and host interaction. Authors divided the chapter into different subheadings to easily elucidate several aspects of the parasite and disease, which also include disease model evolution of zoonotic malaria of past and present, rise and fall of malaria drugs due to resistant plasmodium strains, and also situations of urban malaria. This may be useful for the reader as the authors also included many scientific anecdotes regarding the history of malaria, plasmodium, and human society.

Chapter 6
Physical School Environment and Infectious Diseases: A Case of Primary School Context in Bangladesh 126
Afzal Hossain, Institute of Education and Research, University of Rajshahi, Bangladesh
Emaj Uddin, University of Rajshahi, Bangladesh

The mechanisms underlying relationships between unhealthy physical environment in schools and infectious disease transmission among students are understudied. Drawing from Bronfenbrenner's ecological systems theory, this chapter analyzes behavioral patterns as mechanisms by which an unhealthy physical school environment transmits infectious diseases to primary students in Bangladesh. For the study, the authors randomly selected 300 students from grade-C primary schools located in disadvantaged areas in Rajshahi City, Bangladesh; over 12 months, they collected data on the physical school environment, children's behavioral patterns, and their infections from infectious diseases in the primary school context. The study found that 20-45% of the students suffered from various infectious diseases during the study period. The results from structural equation modeling suggested that poor qualities of the physical school environment via unhealthy behavioral patterns transmitted infectious diseases, affecting the students' physical health in the primary schools.

Chapter 7
Post-Antibiotic Era: A Lurking Global Threat..152
Shafa S. Contractor, Karnatak University, India
Suresh B. Arakera, Karnatak University, India

The discovery of antibiotics was a breakthrough in the field of medicine in the late 1920s. Antibiotics have saved millions of lives ever since and even increased life expectancy. But due to their irrational use in humans, animals, and agriculture, there is a global concern about antibiotic resistance. The 'magic bullets' are no longer effective against 'superbugs', posing difficulties for the medical practitioners in treatment. The pathogens have evolved different resistance strategies for their survival: enzymatic inactivation of antibiotics, altering the movement of antibiotics across the cell, or modifying target sites that interact with the antibiotics. As the efficacy of the critical drugs wanes and the discovery of new antibiotics stagnates, exploration and use of alternative therapies could offer a much-needed solution. Researchers worldwide are constantly striving to find a solution to combat antibiotic resistance. Some of the alternative therapies and the regulatory measures to deal with antibiotic resistance are discussed in this chapter.

Chapter 8
The Interplay Between Environmental Pollutants, Gut Microbiota, and Infections: Current Concepts and Therapies..173
Bilal El-Mansoury, Faculty of Science, Chouaib Doukkali University, Morocco
Kamal Smimih, Faculty of Sciences and Techniques, Sultan Moulay Slimane University, Morocco
Youssef Ait Hamdan, Higher Normal School, Cadi Ayyad University, Morocco
Abdelaati El Khiat, Faculty of Medicine and Pharmacy, Cadi Ayyad University, Morocco & ISPITS, Ouarzazate, Morocco.
Samia Elouali, Higher Normal School, Cadi Ayyad University, Morocco
Hanane El Fatimi, Mohammed VI University of Health Sciences, Morocco
Ahmed Draoui, Faculty of Science Semlalia, Cadi Ayyad University, Morocco
Omar El Hiba, Faculty of Sciences, Chouaib Doukkali University, Morocco
Arumugam Radhakrishnan Jayakumar, Miller School of Medicine, University of Miami, USA

Environmental pollutants (EPs) have become an increasingly common health hazard. Several studies have sustained the impact of these EPs on human gut microbiota (GM). The human GM is made up of thousands of microbes that play a paramount role to maintain hosts' health by providing many physiological functions, as well as protecting the host from pathogens and aggressions. Considering the crucial role that the host's GM plays in maintaining health, its disruption (gut dysbiosis) by the EPs can increase extraordinary the risk of infections and lead to the development of several conditions. Therefore, interventions aimed to restore the gut microbial population are of great interest. This chapter examines the intricate relationship between EPs and GM, highlighting the profound impact of EPs on human health including the development of infections. Additionally, this chapter discusses the potential role of probiotics, prebiotics, and FMT in mitigating the detrimental effects of EPs on GM.

Chapter 9
Classification of Melanoma Skin Cancer Based on Transformer Deep Learning Model..208
Nagarjuna Telagam, GITAM University, India
Nehru Kandasamy, Madanapalle Institute of Science and Technology, India

An increasing number of genetic and metabolic anomalies have been determined to lead to cancer, which is generally fatal. Cancerous cells may spread to any body part, which can be life-threatening. Skin cancer is significant cancer, and its frequency is increasing worldwide. The main subtypes of skin cancer are squamous and basal cell carcinomas and melanoma. The deep learning methods were used to detect the two primary types of tumours, malignant and benign, by using the MELANOMA dataset. The proposed system utilizes a convolutional neural network (CNN), transformer, and InceptionV3 architecture to learn and extract meaningful features from skin lesion images. The CNN model was trained on a large dataset of dermoscopic images of melanoma and benign lesions. The transformer model in deep learning refers to a neural network architecture based on the transformer architecture specifically designed for image classification tasks. Inception is an image recognition model that has been shown to attain greater than 78.1% accuracy on the ImageNet dataset.

Chapter 10
Clinical Characteristics and Relationship of Acute Kidney Injury With the Severity of Disease and Death in COVID-19 Patients: A Review.....................228
Donovan McGrowder, The University of the West Indies, Jamaica
Fabian Miller, The Mico University College, Jamaica & The University of the West Indies, Jamaica
Magdalene Nwokocha, The University of the West Indies, Jamaica
Melisa Anderson Cross, University of Technology, Jamaica
Cameil Wilson-Clarke, The University of the West Indies, Jamaica
Shelly McFarlane, The University of the West Indies, Jamaica
Lorenzo Gordon, Caribbean School of Medical Sciences, Jamaica

Globally, the coronavirus disease 2019 (COVID-19) pandemic has adversely affected healthcare with significant morbidity and mortality among patients due to the virulent effects of the severe acute respiratory syndrome coronavirus 2 (SARS-CoV-2). The kidneys and other organs are infected and the development of renal disorders such as acute kidney injury (AKI) and subsequently chronic renal disease if there is no clinical intervention. There is increasing evidence of more COVID-19 patients developing AKI, and incidence of mild, moderate and severe stages of the disease. There is also evidence of significant morbidity and mortality, especially in patients with severe AKI admitted to the intensive care units (ICU). We reviewed the recent publications in PubMed, Google Scholar, Embase and Cochrane library relating to AKI in COVID-19 patients. This review examines (i) multifactorial causes of AKI in COVID-19 and the pathophysiology of this renal disorder, (ii) incidence of AKI in COVID-19 as well as the same in mild, moderate and severe disease, (iii) the association between the incidence of AKI and disease severity as well as mortality in patients with confirmed COVID-19, and (iv) incidence of renal replacement therapy among COVID-19 patients with AKI. The evidence points to the need for early therapeutic intervention and multidisciplinary supportive care for COVID-19 patients. This is critical for COVID-19 patients with comorbidities such as diabetes mellitus, hypertension and cardiovascular diseases as their renal function may be compromised.

Compilation of References 259

About the Contributors 321

Index 329

Preface

The world is gradually reviving from the devastating distress of the COVID-19 pandemic. Naturally, there are varied and disparate opinions about the virus's origin, its spread and patient management, and therapeutic, physical, social, and prophylactic strategies to control it. However, threats from many natural and/or artificial virus variants are looming large across the globe. It is well known that disease outbreaks have extensively jeopardized the global socio-economic systems several times in history, and the recent pandemic is one of such grievous episodes. Based on biblical anecdotes and the Greek physician who lived in the 1st Century CE, Rufus of Ephesus, historians of our time have inferred sporadic instances of bubonic plague in the past (Zietz and Dunkelberg, 2004). The bubonic plague, or the Black Death that eliminated more than 25 million people, nearly one-third of the then total human population, during the 14th century, is considered the worst pandemic outbreak documented in human history (Glatter and Finkelman, 2021). Similarly, there are accounts of influenza outbreaks in 412 BC by Hippocrates, who called it an epidemic of his time. It is also believed that an evolved influenza virus caused mass morbidity in Europe during the Middle Ages. The reported mass mortality (25-50 million) and morbidity throughout the world due to the spread of the influenza pandemic, otherwise known as Spanish flu (a well-known misnomer) in 1918, is the first pandemic in the 20th century (Ghendon, 1994).

The primary threat associated with disease outbreaks is the mortality or morbidity of humans and other species, including wild and domestic animals and plants. Such biological catastrophes often lead to large-scale economic and socio-cultural losses (Karesh, 2005) irrespective of the political boundaries, particularly in the modern world of quite open travel and trade, perhaps an important but inadvertent effect of increasing globalization. The trickling effect of such losses is reflected in individuals' psychology and the population's political state (Mikolajkow and Małyszczak, 2020) with widespread repercussions; hence, disease outbreaks are largely significant disasters (Dubb, 2020), mentally, socially and otherwise a grievous 'butterfly defect' (Goldin and Mariathasan 2014) at times stripping the common mass of their wellbeing including life and livelihood-security.

The factors leading to disease outbreaks can be diverse, both intrinsic and extrinsic, to the affected host species. Exposure to pathogens (micro or others) does not always result in infections or ailments in exposed individuals or the spread of diseases across the population since various determinants are involved in creating outbreaks. The intrinsic factors could be genetic, biochemical, physiological or related to an individual's behavioural traits or lifestyle. The extrinsic factors could be those that enhance contact between the pathogen and the hosts or augment the overall susceptibility to infection and essentially could be ecological, environmental, or socio-economic. At times, the external factors trigger internal changes to make the victim/host vulnerable to infections and, thus, the disease. Those external factors also include the quality of the natural environment where the host and the pathogens exist, and they could be related to the population and density of the host and the pathogen and consequent higher host-pathogen contacts. The infectious organisms and the obvious or potential hosts (primary, secondary, paratenic, accidental or reservoir) are a part of the multi-dimensional biodiversity and a quintessential element of the ecological web. Thus, an infectious disease manifests through a chain of events that favour high contact of the pathogenic species with the host species, directly or indirectly, or through intermediate hosts/vectors, in synergy with the host's susceptibility.

The dynamics of the natural systems are panoptic, intricate and intriguing, especially as the multi-dimensional anthropogenic interventions and consequential impacts intensify. The factors influencing vector abundance are significant for the human population density and poor socio-economic conditions (Vikram et al., 2015). It is reported that after World War II, there was a substantial rise in epidemics and zoonotic diseases across the globe (Wilcox and Gubler, 2005). Changes in land use because of population agglomerations associated with urbanization and the growing intrusions to wild habitats resulting in imbalanced co-existence of humans and wildlife are elucidated as one of the primary reasons for the emergence of novel epidemics (Morse, 1995). This context makes it vital to recognize that making widespread in-roads into the wilderness is a threat to human health; the preservation of natural habitats directly bears upon public health since habitats with high biological diversity have the potential for checks and balances of population growth of pathogens and could curtail disease transmission, spread and the manifestation of diseases. Besides, diverse and heterogeneous (and largely anthropogenic) issues such as rapid urbanization, sprawling inroads into natural habitats, simplify the web of life, increasing human-animal conflicts, decreasing environmental quality, passive waste disposal systems, rising contamination/pollution of the natural systems, increasingly unhealthy lifestyle, etc. make the condition much more complex, but conducive for the spread of diseases.

The usual approach of the current medical system is focused on characterizing specific diseases, diagnosing and describing the symptoms of afflicted individuals, and symptomatic management, an evidence-based system where healthcare professionals take care of the symptoms using physical or chemical medications/interventions. As per the system, the management of diseases, essentially its treatment, mainly focuses on intervention after symptoms are evident in the patient. The dominant modern medical-care system is highly technology-oriented, less holistic, and conceivably highly reductionist, concentrating on super-specializations and neglecting the patient's proximate environment. It operates through interventions/techniques, chemical or physical, using diverse gadgets. However, while it helps save a patient, such a strategy is not of much use concerning pre-emptive actions to prevent further incidences of infectious diseases, especially novel unexampled ones. During the last couple of years, as there have been incidences of emerging novel epidemic outbreaks, there has been growing interest in understanding zoonoses from a more ecological perspective (Wilcox and Gubler, 2005).

In that context, disease ecology offers a different approach, is more panoptic and is of high potential for understanding and curtailing potential epidemic outbreaks. It focuses on incidences, spatial distribution and timing of diseases, general processes of inter and intra-specific population interactions in a specific environment, and addresses the issues in a broader context, ecological, social and environmental. Naturally, managing infectious diseases demands a deeper understanding of the environmental and socio-economic drivers and needs a holistic and systematic systems-thinking approach. Issues such as ecological, social and landscape features at the source of disease-causing organisms and how such organisms invade and expand their distribution need to be systematically explored and sufficiently understood. The remedial measures must be handled more holistically, integrating ecology, epidemiology, socio-economics, forestry practices, and agriculture, based on approaches in systems thinking and complex interactions and from an evolutionary viewpoint. It is a paradigm shift from the current relatively reductionist disease management strategy and healthcare system.

This book is an attempt to deliberate upon human diseases from a more comprehensive and, to an extent, holistic perspective, especially looking at morbidity (primarily from infections mediated/facilitated by animals or vectors) from an ecological viewpoint. It also looks at disorders from an evolutionary perspective. Although known much, these perspectives are largely neglected in healthcare systems in vogue. Such a perspective could be an eye-opener for the healthcare providers and the public at the receiving end of increasingly expensive and intrusive healthcare services. The deliberations in the book are significant since books on human health from an ecological perspective are few.

The book's first two chapters explore the impacts of climate change and its implications on human health. Human-induced climate change has been considered one of the major reasons associated with the spread of zoonotic diseases worldwide. Climate change has modified the world's ecosystems along with anthropogenic land use changes, jeopardizing the ecosystems' functioning significantly over the years since the Industrial Revolution, or maybe much before that, but in a less transpicuous and slower mode. The current-day definition of human development characterized by the speedy exploitation and transformation of natural resources significantly increased the ecological footprint along with pollution and greenhouse gas emissions beyond nature's carrying capacity or resilience. Most times, the universally dominant development paradigm, largely obsessed with certain economic indices, goes tangential to egalitarian human welfare and quality of life. Global energy and other resource needs and land transformation associated with agriculture, transportation and other activities are escalating day by day. All these contribute significantly to present-day climate change. Incidentally, the drivers of climate change primarily are the privileged, while the victims are mostly the underprivileged countries or segments of society. The first chapter discusses how human-induced climate modifications significantly affect human health. The chapter examines various epidemic episodes associated with climate events and the spread of various vectors. The chapter also delves into recently reported endemics in India, such as Chikungunya, Malaria, West Nile Fever and Dengue Fever. In the second chapter, the authors explore the effect of climate change on various stages of infectious disease by influencing pathogens, vectors, and host organisms. The chapter also discusses various adaptive strategies to mitigate climate impacts on human health.

In the third chapter, the authors shed light on how urbanisation makes the globe highly vulnerable to diseases. In the recent pandemic episodes, the world has witnessed urban living conditions becoming incredible bottlenecks in managing COVID-19. Rural-to-urban migration, resulting in urban sprawls and urban agglomerations, unhealthy living conditions and deficient amenities, are the primary reasons for increasing the vulnerabilities for the emergence of infectious diseases. Overcrowding with close contact with wild and domesticated animals in urban regions increases the risk of zoonotic disease. This chapter evaluates urbanization in the zoonotic spread and inter-specific transmission of contagious pathogens or parasites by exploring the dynamics of the wildlife-livestock-human interfaces and the factors affecting the epidemiology of diseases.

The following two chapters focus on one of the highly contagious diseases - Malaria. Of these, the first one discusses the ecology, evolution, life cycle and host interactions of Plasmodium. Various aspects of the parasite and the disease, including the disease model, are also discussed. The chapter briefly elucidates the historical aspects of avian malaria, the evolution of other zoonotic malaria and the evolution of malarial drugs. The fifth chapter deliberates upon the evolutionary aspects of malaria, the causative agent and its vectors, anopheline mosquitoes, worldwide, exposing the zoonotic background of the disease and the natural history of its vector. Tropical countries, including India, are considered the hubs of malarial infection. The chapter also sheds light on different anti-malarial drugs and their biomedical evolution. The general biology, lifecycle and survival strategies of the parasite in the host body bypassing the natural immune system are briefly discussed in the chapter with adequate reference to the inherited and genetic immunities of the deadly epidemic. An attempt to codify the diseases' historical context, specifying the cases of urban and rural spread, is also made. The chapters also provide us with a vivid account of how the life cycle of the vector has adapted to the urbanization pattern, especially in tropical countries.

In the sixth chapter, on physical school environment and Infectious Diseases in Bangladesh, the authors provide a quantitative account of how the behavioural patterns of pupils in physically unhealthy primary school environments in the country are vulnerable to the spread of diseases.

The seventh chapter discusses the grievous reality of antibiotic resistance, a severe menace to modern-day humans. Once a magnificent lifesaving remedy, antibiotics turned out to be a global concern over time, probably due to humans' extensive and carefree use of those wonder drugs. Antibiotic resistance due to the irrational use of antibiotics in humans, animals and agriculture is a present-day challenge. The chapter discusses the events that resulted in the discovery and the historical events associated with antibiotic medicines. An in-depth examination and explanation of antibiotic resistance and its cytological and chemical processes are given in this chapter. The authors conclude the chapter by discussing alternative therapies and regulatory measures to deal with antibiotic resistance.

The eighth chapter explores how environmental pollutants alarmingly modify human gut microbiota, resulting in health conditions. The chapter explains how environmental pollutants such as heavy metals, persistent organic Pollutants (POP), pesticides, nanoparticles and microplastics affect the gut microbiota. The chapter also discusses existing remedial measures to mitigate the harmful effects of environmental pollutants on the gut microbiota. The next chapter, the ninth one, discusses melanoma skin cancers. The authors attempt to classify skin cancer using the convolutional neural network (CNN), Transformer, and InceptionV3 architecture to learn and extract meaningful features from skin lesion images taken from the MELANOMA data set.

One would certainly recall the COVID-19 pandemic hitting the world in quite an unprecedented manner in the last couple of years. The healthcare system is one of the sectors that were affected significantly. The ninth chapter of the book explores the impact of COVID-19 on the development of consequent disorders such as acute kidney injury (AKI) and resultant chronic renal failures. It examines multi-factorial causes of AKI in COVID-19 patients, its pathophysiology, the association between AKI and severity of illness in confirmed COVID-19 patients, and other associated issues. The evidence points to the need for early therapeutic intervention and multidisciplinary supportive care for COVID-19 patients.

P.A. Azeez
Salim Ali Centre for Ornithology and Natural History, Coimbatore, India

P.P. Nikhil Raj
Amrita School of Engineering, Amrita Vishwa Vidyapeetham, India

R. Mohanraj
Bharathidasan University, India

REFERENCES

Dubb, S. S. (2020). Coronavirus pandemic: Applying a whole-of-society model for the whole-of-the world. *British Journal of Oral & Maxillofacial Surgery*, *58*(7), 838–842. doi:10.1016/j.bjoms.2020.05.009 PMID:32631753

Ghendon, Y. (1994). Introduction to pandemic influenza through history. *European Journal of Epidemiology*, *10*(4), 451–453. doi:10.1007/BF01719673 PMID:7843353

Glatter, K. A., & Finkelman, P. (2021). History of the plague: An ancient pandemic for the age of COVID-19. *The American Journal of Medicine*, *134*(2), 176–181. doi:10.1016/j.amjmed.2020.08.019 PMID:32979306

Goldin, I., & Mariathasan, M. (2014). *The Butterfly Defect - How Globalization Creates Systemic Risks, and What to Do about It*. Princeton University Press.

Karesh, W. B., & Cook, R. A. (2005). The human-animal link. *Foreign Affairs*, *84*(4), 38. doi:10.2307/20034419

Mikolajkow, A., & Małyszczak, K. (2021). Stress level and general mental state in Polish pregnant women during COVID-19 pandemic. *Journal of Reproductive and Infant Psychology, 2023 Apr;41(2):165-182,* 1–18. doi:10.1080/02646838.2021.1976402 PMID:34555951

Morse, S. S. (1995). Factors in the emergence of infectious diseases. *Emerging Infectious Diseases*, *1*(1), 7–15. doi:10.3201/eid0101.950102 PMID:8903148

Vikram, K., Nagpal, B. N., Pande, V., Srivastava, A., Saxena, R., Singh, H., Gupta, S. K., Tuli, N. R., Yadav, N. K., Olivier, T., & Richard, P. (2015). Detection of dengue virus in individual *Aedes aegypti* mosquitoes in Delhi, India. *Journal of Vector Borne Diseases*, *52*(2), 129. PMID:26119544

Wilcox, B. A., & Gubler, D. J. (2005). Disease ecology and the global emergence of zoonotic pathogens. *Environmental Health and Preventive Medicine*, *10*(5), 263–272. doi:10.1007/BF02897701 PMID:21432130

Zietz, B. P., & Dunkelberg, H. (2004). The history of the plague and the research on the causative agent Yersinia *pestis*. *International Journal of Hygiene and Environmental Health*, *207*(2), 165–178. doi:10.1078/1438-4639-00259 PMID:15031959

Chapter 1
Changing Climate:
Threat to Human Health

Abirami Subramanian
Indira Gandhi National Open University, India

Sushmitha Baskar
Indira Gandhi National Open University, India

ABSTRACT

Earth is endowed with sufficient reserves to meet all of our needs. It is the fact that all living organisms use the planet's resources to the extent that they require them. Humans, on the other hand, are the only creatures on the planet who enjoy and even exploit resources beyond their necessity. To be more precise, due to population surge, rapid industrialization, and urbanization, there is a need to meet both the demands and luxuries. The Earth's species are dependent on the ecosystem service functions which are driven by climate stability. Climate-triggered health illnesses are currently showing new trends and are expected to increase in the future if the current climate disturbances persist. In brief, climate instability (warming or cooling) undeniably provides suitable ground for vector and zoonotic life cycles. Climate-related morbidity and mortality are developing public health concerns that, if not addressed properly, will have serious corollaries. This chapter summarizes the various factors responsible for climate variability-induced health impacts.

DOI: 10.4018/978-1-7998-9414-8.ch001

INTRODUCTION

The economy of any country is dependent on its residents' well health. The underlying health disparities can be reduced by strengthening each country's essential public health services and policies. Specific exposure mechanisms influence climate and health. These pathways can be divided into two categories: (a) direct exposure through extreme weather events such as heat waves or heavy rain; and (b) indirect exposure through ecological changes that alter disease vectors and transmission.

The amount of damage that climate change could inflict and the costs of adaptation are linked to a country's baseline health (World Bank Report, 2010). As a result, climate change's potential dangers and consequences differ from country to country. Droughts, floods, cyclones, heat and cold waves, severe storms, and other natural catastrophes have caused 80-90 per cent of reported disasters worldwide in the last ten years (WHO). The normal functioning of the earth's ecosystems will be affected by even minor changes in global temperature. . In July/August 2021 drought due to the failure of the southwest monsoon has severely affected the state of Gujarat. Several animals including cattle and desert species died due to extreme water scarcity, heat, and drought. Drought elicits a series of health impacts via wildfires, pollution (air, Water and soil), malnutrition, population displacement, psychological stress etc. and also lead to low river flows, accumulating water-borne effluent pathogens (Hofstra, 2011, Semenza and Menne, 2009). According to the India Meteorological Department, Ahmedabad, 30 districts in Gujarat received deficient rainfall in 2021 (Down to Earth, August 2021). The districts Surendranagar, Gandhinagar, and Aravalli received largely deficient rainfall between June and August 2021.

Some alarm bells ringing in the 21st century include increased annual summer monsoon precipitation with enhanced interannual variability over Southeast Asia. Glacier run-off in the Asian mountains will increase up to the mid-21st century and subsequent run-off may decrease due to the loss of glacier storage (The Hindu, 2021). Climate change has the potential to alter the seasonality of some infectious diseases. Vector-borne diseases, such as dengue fever, chikungunya, malaria, and other mosquito-borne infectious virus epidemics, are examples. Because of their inability to respond to variations in climate variability, some florae may die or become extinct. Some species can adapt to changing circumstances. Changes in food production, rural hardship, and socio-economic difficulties are among the other prominent effects of climate change. Table 1 shows some environmental changes, disease outbreaks that resulted from them, and the several mechanisms that generate infectious diseases in people.

Table 1. Environmental changes and infectious diseases in human beings

Environmental changes	Outbreak of Diseases	Pathways Responsible
Dams and irrigation	Malaria, Schistosomiasis	Breeding sites for mosquitoes Snail host habitat – human contact
Agricultural activities	Malaria Skin diseases Helminthiasis	Crop insecticides, vector resistance
Urban crowding	Cholera Dengue Malaria Chagas fever Conjunctivitis Gastrointestinal upset	Sanitation, hygiene, water contamination Water collecting trash Breeding sites for mosquitoes
Deforestation	Malaria Visceral leishmaniasis	Crop insecticides, vector resistance Sand-fly vectors
Reforestation	Lyme disease	Tick hosts, outdoor exposure
Ocean warming	Red tide	Increase in toxic algal blooms
High precipitation	Rift valley fever Hantavirus pulmonary disease	Breeding sites for mosquitoes Increase in rodent food, habitat abundance
Construction activities	Malaria Nipah virus Vector-borne diseases	Water contamination Breeding sites for mosquitoes Bat-human viral transmissions.
Environmental displacement	Yellow fever	The connection between forest corridors and remote areas. Non-human primates act as a carrier.
Mining and Land-use change	Malaria	Deforestation for mining activity alters the rainfall and temperature pattern.
Wildfire and $PM_{2.5}$	Cardiovascular risk and associated mortality	The frequent occurrence of wildfire increases the concentration of $PM_{2.5}$ in the atmosphere
Urbanization	Dengue Chikungunya	The spread of dengue and chikungunya fever is aided by increased urbanization.

Source: Acosta et al. (2021), Chen et al. (2021), Fletcher et al. (2021), Wen et al. (2016), Wilson (2001).

Throughout the year, we are exposed to changing weather patterns worldwide. Changes in water, air, and food quality can also have an influence on agriculture, livelihoods, and infrastructure (Confalonieri et al., 2007). Several health effects have emerged from these direct and indirect exposures. It will be devastating for humanity if current patterns continue. Most of the health consequences of extreme weather events can be avoided using early warning systems and public health preparedness and response.

The latitude, height, terrain, humidity, type of vegetation, and area all influence climatic parameters such as temperature, precipitation, humidity, air pressure, cloudiness, wind, and solar radiation (rural or urban). These variables have been changing continuously since the planet's inception, and it has taken nearly millions of years to make it a suitable habitat for survival. The problem is that depending on how quickly or slowly climate change occurs, and it transforms this phenomenon into a major concern. When climate change occurs at its normal rate, living species have ample time to adapt to the new climatic circumstances. Still when it happens at today's rate, the changes are so rapid that we find it impossible to adapt. Since the beginning of the industrial revolution (the 1800s), the amount of carbon dioxide released by burning fossil fuels has increased dramatically, from 0.930 metric tonnes per second in the 1800s to 756 metric tonnes per second in the 2000s (Conlen, 2021). According to NASA (climate.nasa.gov) estimates, climate change has reversed all anthropogenic emissions into global warming (current CO_2 levels (417ppm) in the atmosphere are the highest in 650,000 years), global temperature rise (nineteen warmest years have been recorded since 2000), sea-level rise (rise of about seven inches over 100 years), and ice sheet melting (polar ice sheets are melting at a rate of 428 billion tonnes per year) (flora and fauna are affected due to climatic shifts). As a result, climate action (SDG 13) has become a top concern for world leaders, as it directly or indirectly impacts the sustainable development objectives.

Health is the most valuable asset we can have during our lives. However, in the current environment, we can extend life spans with the help of modern technologies and medications, but we will barely be able to live a healthy life. Year after year, the number of unknown diseases and the number of persons affected grows regardless of age. Climate change has the potential to directly impact on human health (Figure 1).

According to the Intergovernmental Panel on Climate Change (IPCC), severe weather events will become more common. According to the report, global warming would cause excessive rain in South India. Scientists warn that India will see more heat waves and flooding due to climate change's irreversible effects (IPCC Sixth Assessment Report, Climate Change, 2021). Below is a list of elements that are contributing to the world's poor health conditions as a result of climate change.

Figure 1. Climate variations and impacts on environment and health

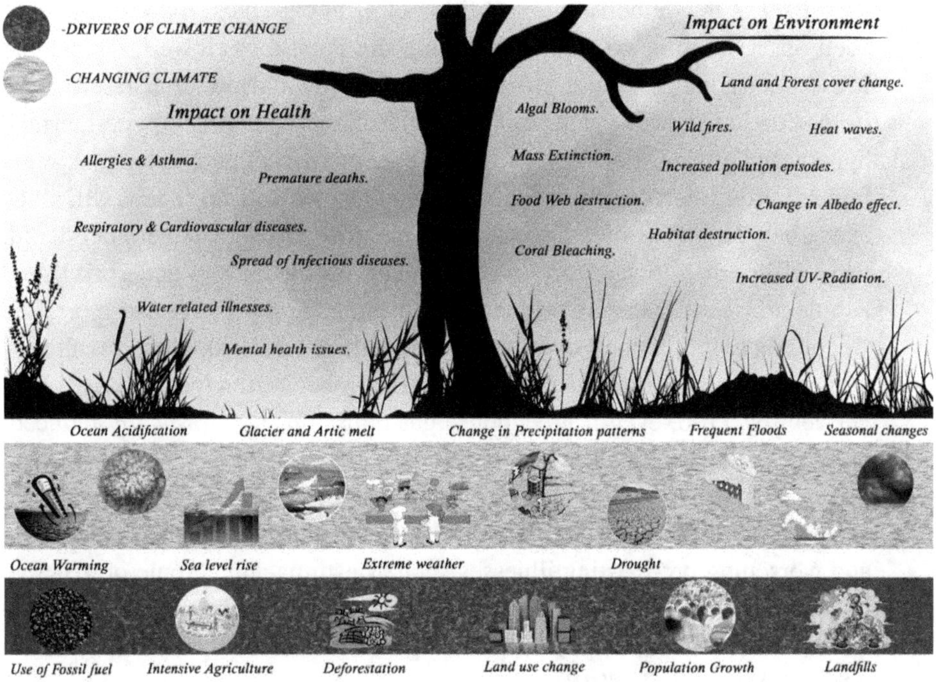

DIRECT EFFECTS OF CLIMATE CHANGE ON HEALTH: DIRECT EXPOSURE

Extreme Weather and Health

(a) ***Heat and cold events:*** The gift of this planet to Homo sapiens is the ambient temperature. The autonomic nervous system's principal job is thermoregulation, which keeps the body's temperature at 37 ± 2°C. This ambient temperature is required for live organisms to maintain proper cellular processes. Humans' thermoregulatory function fails when they are exposed to extremes of heat or cold for an extended period and the normal process (normothermia) is turned into hyperthermia (body temperature over 40°C) or hypothermia (body temperature below 35°C). The global temperatures rise increases the cases of people suffering from heat stroke, heat exhaustion, dehydration, electrolyte imbalance, and other heat-related illnesses (Beker et al., 2018). As far as the human-caused climate change is concerned warmer temperatures have become even warmer, whereas colder temperatures have become even colder. The year 2021 is the demonstration of these climate extremes. Extreme heat has been observed in some places, including the Pacific Northwest, China, and Moscow. In contrast, record-breaking winter storms and low temperatures have been seen in Madrid, Spain, and Texas. Heat and cold stress, hyperthermia, hypothermia, exhaustion, increased hospitalization, respiratory infections, renal failure, diminished cognitive function, and diabetes, to name a few, will all be exacerbated by these high heat and cold circumstances. Whether hot or cold, thermal stress aggravates pre-existing health problems, and underprivileged people are more susceptible to temperature-related morbidity and mortality. According to projections, over 30 per cent of the world's population is already vulnerable to catastrophic heat waves. This number is anticipated to rise by 48 per cent by 2100, even if greenhouse gas emissions are drastically reduced. However, if present emissions levels are maintained, people exposed to heat waves will rise by 74% by 2100 (Mora et al., 2017). Similarly, a sudden cold will leave the individual with no time to acclimate, resulting in respiratory sickness and worsening pre-existing illnesses such as asthma and chronic obstructive pulmonary disease (COPD). In cold-induced hyperpnoea, the airway surface fluid (ASL) evaporates quickly, resulting in dryness and hypertonicity of the ASL (D'Amato et al., 2018).

The global health burden is likewise increasing as the world's population grows. According to a study published in the International Journal of Environmental Research and Public Health, climate-driven high heat events will increase the likelihood of people being affected by heat waves by 500 percent in the most populated areas compared to the 1980s (Li et al., 2021). Because thermoregulation is one of the most critical factors in maintaining blood glucose levels, even brief exposure to excessive heat or cold raises the risk of death by 13.9 percent and morbidity by 1.2 percent in people with diabetes (Song et al., 2021). Air quality is also affected by extreme heat and cold. Heat waves have been linked to a 14.2 percent increase in mortality due to increased ozone and PM10 in the air (Pascal et al., 2021). On the other hand, cold temperatures degrade air quality by encouraging the growth of bacteria, moulds, and spores.

(b) ***Floods and storms:*** A wide range of uncommon weather events, such as flash floods, violent storms, glacial and cloud eruptions, are occurring over the world. Despite the fact that we have ample resources and technologies to anticipate them, the magnitude and timing of these events provide a significant difficulty. Tropical cyclones, hurricanes, and precipitation have become more common and strong as a result of anthropogenic influences on global weather patterns

(USEPA, 2021). To pick a few recent examples from throughout the world in 2021, extreme precipitation in Goa, India, (experienced precipitation that was 122 percent over average), floods in New York, New South Wales, the United Kingdoms, China, and West Germany, as well as cyclone Ana in Fiji, Seroja in Indonesia, Yaas and Taukate in India, have all claimed lives and destroyed property. Incidences of water-borne infectious diseases rise when there is a lot of rain. They have the potential to contaminate water sources with significant levels of faecal bacteria (Jofre et al. 2010). Injuries, drowning, animal attacks, a lack of medical assistance, and respiratory and allergic infections are all direct and immediate effects of floods and storms, while long-term effects include post-traumatic shock, psychological problems, vector-borne diseases, dysentery, malnutrition, and poverty. Several studies have been undertaken and are now being conducted to investigate the impact of floods and storms on human health. Flood catastrophes increase the incidence of childhood atopic dermatitis, exacerbate the conditions of HIV-positive people, degrade general health, lower the birth rate, child nutrition, gastrointestinal health, and increase microbial infections, according to several scientific studies (Anthonj et al., 2015; Chen et al., 2021; Paterson et al., 2018; Zhong et al., 2018). According to the Intergovernmental Panel on Climate Change (IPCC), India will suffer substantial issues as a result of rising sea levels. India has a 7,517-kilometer coastline, and coastal flooding would affect around 28.6 million people if sea levels rose 50 centimetres (IPCC Sixth Assessment Report, Climate Change, 2021).

(c) ***UV-radiation:*** Over the last four decades, climate-induced ozone depletion and ozone depletion's contribution to climate change have increased the amount of surface UV radiation. UVA (315-399 nm), UVB (280-314 nm), and UVC (100-279 nm) are the three main kinds of UV radiation . The majority of UVB and UVC rays are absorbed by the ozone layer, whereas UVA is directly absorbed by the earth's atmosphere. UVB rays are completely absorbed by the epidermis of the skin; however UVA rays can penetrate deeper into the skin. The photoproduct of DNA present in the epidermis, the cyclobutane pyrimidine dimer (CBD), can be formed when exposed to UVB radiation. CBD causes DNA alterations that lead to cutaneous malignant melanomas, keratinocyte malignancies, squamous cell and basal cell carcinomas, and other cancers. Both UVA and UVB have the ability to harm DNA and other biological components through oxidative stress. UV light also has a substantial impact on human health, including immunological modulation, viral infection activation, innate immunity upregulation, and intracellular bacterial and protozoa infections (Lucas et al., 2019). Tropospheric ozone absorbs more UV radiation and has a negative influence on human health. The major component in the generation of tropospheric ozone and secondary organic aerosols is volatile organic compounds (SOAs). Biomass burning, home fuel, coal, cars, industrial emissions, coke oven, glues, paints, cosmetics, varnishes, and detergents are the major anthropogenic contributors to VOCs, while natural sources include plants, animals, forest fires, volcano eruptions, and so on. From 8 Tg NMCOC yr-1 in 1996 to 13.5 Tg NMCOC yr-1 in 2015, non-methane volatile organic compound emissions have increased (Stewart et al., 2021). UV-B radiation is particularly effective in producing skin malignancies, premature skin ageing, skin pigmentation, genomic instability, and oxidative stress, therefore studies on the influence of UV radiation on human health have attracted a lot of attention. According to a study that looked at the global burden of skin cancer from 1990 to 2017, the risk of squamous cell carcinoma increased by 310 percent between those years (Urban et al., 2021).

(d) ***Humidity:*** The transmission of infectious infections and diseases can be influenced by changes in humidity. Variations in humidity levels are linked to the influenza, an airborne viral disease. Researchers have discovered that chilly temperatures and low relative humidity promote the spread of influenza virus (Shaman and Kohn, 2009, Xu et al., 2014). Water-borne viral infections are also caused by changes in humidity. Temperature and humidity fluctuations affect the propagation of the Dengue virus in Yangon and Singapore, according to Patz et al. (2003) and Thu. et al. (1998), resulting in increased disease outbreaks.

(e) ***Air quality and health:*** The air we breathe is composed of 78 percent nitrogen, 21% oxygen, and 0.93 percent argon, accounting for the total of 99.93 per cent. The remaining 0.07 per cent of the air we breathe is made up of Pollen, moulds and spores, water vapour, liquid droplets, carbon dioxide, methane, carbon monoxide, volatile organic compounds (VOCs), nitrogen oxides, sulphur dioxide, black carbon, organic carbon aerosol, and tropospheric ozone.. Depending on environmental and anthropogenic factors, these compounds are found in the atmosphere in various concentrations, sizes (particulate matters – PM10, PM2.5, PM0.1), life spans (long-lived, medium-lived, short-lived, and very short-lived), and combinations (dust, smoke, fumes, ash, mist, spray). It is common knowledge that the chemicals mentioned above trap incoming solar radiation, causing the earth to get warmer. These pollutants are not only bad for the environment, but also for people's health. Climate change is becoming more and more a result of human activity. Similarly, the wind significantly impacts infections that are spread through the air. There is a direct positive relationship between the association of dust particles and virus survival and transmission, according to Chen et al. (2010) and Chung and Sobsey (1993).

Wu and colleagues (2020) did a cohort research to determine the link between PM2.5 exposure and bone strength in pregnant women. The study discovered that exposure to particulate matter (2.5) during pregnancy causes trabecular and cortical bone strength to decline. Increased susceptibility to bone fracture and a longer recovery period of normal bone strength are also linked to high ambient PM2.5 exposure. Ischemic and hemorrhagic strokes are also increased by long-term exposure to air pollution. Exposure to air pollution in the environment is considered to be the cause of around 14% of deaths attributed to strokes (Verhoeven et al., 2021). Table 2 lists the infectious diseases that could emerge from harsh weather.

Table 2. Potential infectious diseases after extreme weather events

Type of Extreme Event	Place	Type of Disease Transmission	Findings	Reference
El Niño	World wide	Chikungunya Hantavirus Rift Valley fever Dengue Cholera Plague Zika virus	Plague and cholera are associated with abnormal precipitation and dengue is associated with abnormal land surface temperature.	(Anyamba et al., 2019)
Extreme precipitation	Canada	Gastrointestinal illness due to giardiasis cryptosporidiosis	The rate of transmission is higher during extreme precipitation following prolonged dry periods.	(Chhetri et al., 2017)
Drought	UK, France, Germany, Italy, and Spain	Plague	Drought followed by rainfall elevated the number of plague-carrying fleas.	(Yue & Lee, 2020)
Floods	South-Eastern Australia	Ross River Virus (RRV)	Frequent flood increases the spread of arboviral disease.	(Tall & Gatton, 2020)
	Kelantan, Malaysia Thailand	Leptospirosis	Severe floods increase the incidence of *Leptospira* infection.	(Chadsuthi et al., 2021; Radi et al., 2018)
Hurricanes	United States	Viral, bacterial, and fungal infections	Hurricane facilitates the spread of infectious diseases and also exaggerates pre-existing conditions like HIV, tuberculosis, gastrointestinal and respiratory illnesses.	(Shukla et al., 2018)
Warming temperatures	Sub-Saharan Africa	Arbovirus	The spread of arboviral infections such as dengue fever is likely to be aided by rising temperatures.	(Mordecai et al., 2020)
	Brazil	Dengue		(Lee et al., 2021)
	UK	West Nile virus	Warming temperature increases the mosquito-biting periods.	(Ewing et al., 2021)
	Ethiopia	Malaria	Temperature, rainfall, and altitude are the major factors for the spread of the infection.	(Dabaro et al., 2021)

Some Diseases Caused Due to Extreme Climate Events

(a) ***Asthma:*** Climate change feedbacks differ depending on the habitat, such as rural or urban, mountains or plains, forests or deserts, and so on. The climate feedback that causes asthma in humans is a worldwide concern. Increased greenhouse gas emissions, monsoon fluctuations, unexpected floods and storms, shifts in plant reproductive cycles, indoor air quality, fungal spores, excessive heat, and cold all contribute to the rise in asthma cases. Asthma is characterised by inflammation of the airways, allergic reactions, and breathing difficulties. Increased CO_2 levels will enhance photosynthesis in plants, resulting in improved reproductive cycles and greater pollen production. The allergenicity potential of pollens is likewise increased as CO_2 levels rise. In asthma patients, exposure to fungal spores, pollutants from automobile emissions, and volatile organic compounds from both indoor and outdoor environments trigger the immunological response, leading to toxic consequences and infection progression (G. D'Amato et al., 2020; Eguiluz-Gracia et al., 2020). According to the most recent global burden of illness data published in Lancet (2019), asthma caused 21.6 million Disability-adjusted life years (DALYs), with low and middle-income nations having the highest death rate.

(b) ***Cardiopulmonary diseases:*** Air contaminants affect human respiratory and heart health differently. In conclusion, sulphur dioxide is harmful to the upper airway, whereas nitrous oxide and ozone can reach deep into the lungs. Carbon monoxide is a toxic gas that causes tissue hypoxia when inhaled. The upper airways will be harmed by large particulate matter (PM10) in the atmosphere, which will cause acute coughing and endothelial dysfunction. Fine particles can penetrate the lungs, damaging the membranes and inducing self-apoptosis in the cells. The immune system is stimulated by pollutants in the airways and on the skin, which might contribute to atherosclerosis. Lung cancer is caused by long-term pollutant accumulation (Duffin et al., 2007; Oberdörster, 1995; Samoli et al., 2007; Schraufnagel et al., 2019; Tamagawa et al., 2008). Upper respiratory infections include cough, acute nasopharyngitis, sinusitis, pharyngitis, tonsillitis, laryngitis, tracheitis, epiglottitis, rhinitis, rhinosinusitis, rhinopharyngitis, and supraglottis, whereas lower respiratory infections include pneumonia, bronchiolitis, respiratory syncytial virus (RSV), and influenza-like illness. The global burden of respiratory and cardiac disorders induced by various pollutants is shown in Table 3.

Table 3. Global burden of deaths caused by pollutants

Pollutants	Deaths (in Millions)
Particulate Matter	4.41
House hold air pollutants	4.36
Ozone	0.365
Residential radon	0.0837
Lead exposure	0.902
Occupational particulate matter, gases, and fumes	0.524

Source: *Global Burden of Disease: GBD Cause and Risk Summaries* (2019)

DIRECT EFFECTS OF CLIMATE CHANGE ON HEALTH: INDIRECT EXPOSURE

Impacts of Climate Change on Vector-Borne and Zoonotic Diseases

Climate change has a significant impact on host-pathogen interactions. The spread of vector-borne and zoonotic infections is aided by unusual precipitation, drought, high temperatures, sea-level rise, habitat degradation, and displacement. Because the vectors are expected to adapt to changing weather patterns, the infection is becoming more widespread over the planet. Infectious diseases that had previously been endemic to a specific geographic area have now expanded to non-endemic places.

Cholera epidemics erupted in South Sudan between 2014 and 2017 because of flooding, drought, population displacement, and civil fighting. Hundreds of individuals died because of the outbreak, which infected thousands of people. According to scientists cholera was most likely transported into South Sudan from Uganda between 2013 and 2014 because of population displacement accompanied by harsh occurrences such as drought and flooding (Jones et al., 2020).

(a) **Leishmaniasis:** Leishmaniasis is a sandfly-borne infection caused by the protozoan parasite Leishmania (Phlebotomus and Lutzomyia species). Environmental and climatic conditions have a big impact on how this virus spreads. The three most frequent kinds of leishmaniasis in humans are visceral leishmaniasis (VL), cutaneous leishmaniasis (CL), and mucocutaneous leishmaniasis (World Health Organization, 2021b). Humans and other mammals such as dogs, cats, rats, marsupials, anteaters, foxes, monkeys, coatis, sloths, armadillos, porcupines, kinkajous, raccoons, and red squirrels serve as reservoirs for human leishmaniasis (Alemayehu & Alemayehu, 2017). The transmission trend of the leishmaniasis infection is given in Figure 2.

Figure 2. Transmission trend of the leishmaniasis infection

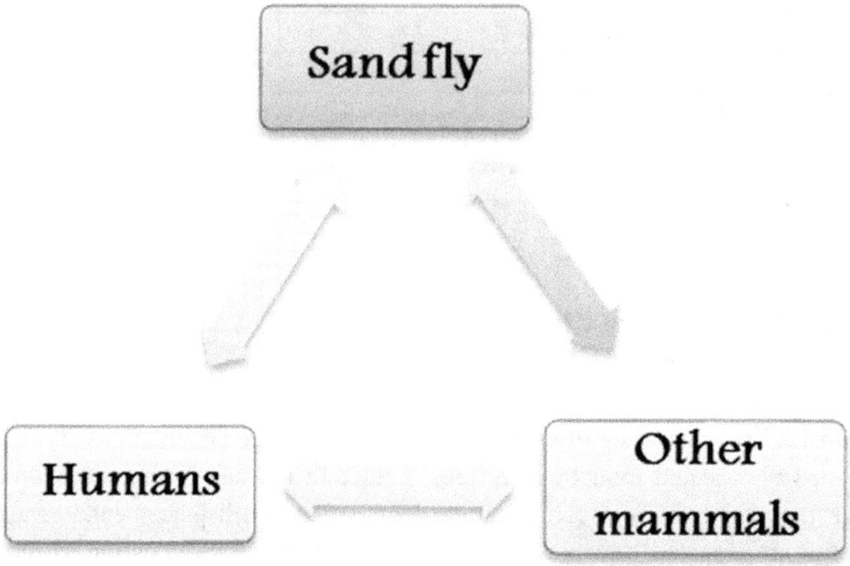

Human leishmaniasis is caused by more than 20 types of Leishmania protozoa, which are transmitted by about 90 species of sandflies (World Health Organization, 2021b). Visceral leishmaniasis (VL) is a parasite infection that kills more people than malaria (Wamai et al., 2020). VL is caused by liver and spleen enlargement and is responsible for 5710 fatalities in 2019. (Global Burden of Disease: GBD Cause and Risk Summaries, 2019). Climate change, according to several future projection models, will lead to the coexistence of vector and human populations in the future (the 2050s). In this scenario, leishmaniasis infections will spread faster than they are now in both endemic and non-endemic countries (Charrahy et al., 2021).

(b) **Hemorrhagic fever with renal syndrome (HFRS) and Hantavirus cardiopulmonary syndrome (HCPS):** HRFS and HCPS are developing zoonotic diseases spread by rodents and caused by the orthohantavirus genus, which belongs to the Bunyaviridae family. The principal method of viral transmission to humans is the inhalation of the rodents' aerosolized excreta or direct contact with the droppings. HFRS is native to Asia and Europe, while HCPS is only found in North America. The prevalence and transmission of illness are affected by climate change (He et al., 2017). Temperature and precipitation can encourage rodent populations, leading to increased viral infections. HRFS incidences rise by 1.6 percent and 0.2 percent, respectively, with a 1°C increase in temperature and a 1mm increase in rainfall (Xiang et al., 2018). The transmission of hantavirus from rodents to humans is depicted in Figure 3.

Figure 3. Flowchart showing the transmission of Hantavirus from rodents to humans

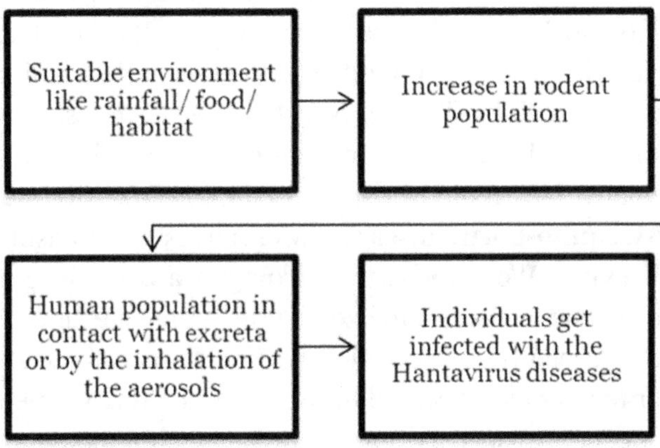

(c) **Tick-borne encephalitis:** Tick-borne encephalitis is a neurological infection caused by the tick-borne encephalitis virus (TBEV), a member of the Flaviviridae family and is transmitted by the tick *Ixodes ricinus. I.ricinus* ticks are climate-sensitive, requiring at least 80% relative humidity and moderate to high rainfall (European Centre for Disease Control and Prevention, 2014). If left untreated, the TBEV is known to produce severe inflammation in the brain, which can be fatal. Although there is no specific treatment for tickborne encephalitis, vaccination can help avoid the condition (Bogovic & Strle, 2015). According to a study conducted in Hungary, climate change would provide the best feeding conditions for ticks, increasing the likelihood of TBEV infection in the future (Nah et al., 2020). In 2019, 89,900 people died globally as a result of encephalitis illness.

(d) **Lyme disease:** Erythema migrans (skin rash) is a symptom of Lyme disease, which is caused by the bacteria *Borrelia burgdorferi* and, in rare circumstances,

Borrelia mayonii. The bacterial infection is transmitted to humans by the black-legged tick *Ixodes scapularis.* In the United States and Canada, the disease is more common. According to estimates, roughly 476,000 cases were recorded annually in the United States between 2010 and 2018, and around 329,000 cases between 2005 and 2010 (Kugeler et al., 2021). Ticks carrying B.burgdorferi need a warm environment to lay their eggs, and the larval phases occur throughout the summer. The larva remains dormant in cold temperatures, and the nymph stage of development occurs in the spring (12°C-22°C), then transfers to the other host during the summer of the second year (Shapiro & Gerber, 2000).

(e) **Nipah virus:** The virus is discovered in the Pteropus bat species, which serves as a natural reservoir. It spreads through contaminated food and through direct contact between people. Livestock is also affected by the infection.. Bats' feeding requirements are influenced by climate variability and the loss of their native habitat owing to anthropogenic activity. The places with the most deforestation are the ones that suffer the most. Bats are growing closer to being able to live in close proximity to humans. This could be a factor in the spread of disease.

(f) **Climate change and Mosquito-borne diseases:** Mosquitoes can transmit various diseases. The Zika virus, West Nile virus, Chikungunya virus, dengue virus, and the malaria-causing parasite protozoa Plasmodium are among them. Mosquitoes are significantly influenced by temperature parameters such as humidity and air temperature because they are ectothermic species. As a result, climate change has provided a suitable environment for the vector population to multiply. The sudden change in precipitation patterns and warmer temperatures provide the vector with a breeding environment as well as the necessary incubation times for the virus to complete its life cycle. Migration, behavioural changes, ecological imbalance, socio-economic situations, and vector control activities are indirect climate influences on mosquito-borne diseases.

(i) **Malaria:** *Plasmodium vivax, Plasmodium falciparum, Plasmodium malariae, Plasmodium ovale, and Plasmodium nowlesi* are parasitic protozoa that cause malaria. Compared to other species, the parasites *Plasmodium falciparum* and *Plasmodium vivax* are the most common and pose a major threat to humans. The female Anopheles mosquito transmits Plasmodium malaria. According to the World Health Organization, malaria affects roughly half of the world's population (World Malaria Report 2021).The transmission of illness is largely climate-dependent, as the mosquito population grows in response to rainfall and warmer temperatures. According to the World Malaria Report (2020), the number of people infected with malaria has increased from 217 million in 2014 to 229 million in 2019. Climate variability is closely linked to this growth.

(ii) **Dengue:** Dengue fever is a mosquito-borne disease caused by four strains of the dengue virus (DENV1-DENV4) and disseminated by two mosquito genera, Aedes aegypti and Aedes albopictus. Because there are four strains of DENV, a person can become infected four times in a row. People infected with the virus may show no symptoms, or develop significant illnesses such as high fever, severe bleeding, organ dysfunction, and plasma leakage in severe cases. The infection was the sole epidemic in nine countries until the 1970s, but it has since expanded rapidly over the world, infecting more than 100 countries (World Health Organization, 2021a). Various scientists' experimental research supports the idea that global warming is one of the primary drivers of disease spread in endemic and non-endemic nations (Gurevitz et al., 2021; Liu et al., 2017; Pea-Garca et al., 2017). The global incidence of dengue infection has climbed from 11164524 individual cases in 1990 to 52414653 person cases in 2017, according to figures published in the Lancet, while mortality cases have increased from 9409 in 1990 to 20616 in 2017 (Zeng et al., 2021). Dengue fever is primarily seen in tropical and sub-tropical climates and is heavily impacted by climatic conditions. Flooding boosts the number of eggs and the adult population of Aedes aegypti, according to a recent study conducted in Kenya (Nosrat et al., 2021).

(iii) **Chikungunya:** Chikungunya virus (CHIKV) is the virus that causes the chikungunya infection, which is spread by Aedes mosquitos. Fever, severe joint and muscular pain, arthralgia, joint swelling, nausea, headache, and exhaustion are some of the symptoms. According to the most recent estimate released by the European Centre for Disease Prevention and Control (2021), about 26,543 cases were reported across American and Caribbean countries from January to May 2021. In Asia (India, Cambodia, Malaysia, and Thailand), there have been 10,166 cases reported, with 8,170 suspected and 1,116 confirmed cases in India. The study, which was published in Nature, shows how climate change affects the transmission of disease to areas where no or few cases had previously been reported. Climate change is predicted to establish the right transmission ground in Northern and Southern Italy, Southwest France, Northeast Spain, Sub-Saharan Africa, Northern Australia, and the southernmost part of Florida in the United States (Tjaden et al., 2017). Meanwhile, Pakistan claimed in 2017 that the chikungunya virus had resurfaced in its largest metropolis, Karachi, for the first time since the 1980s. Over 3,000 people have been affected by the virus (Mallhi et al., 2017).

(iv) **Zika virus:** A flavivirus is what this is. It was first discovered in 1947 among monkeys on the African continent. In Uganda and the United Republic of Tanzania, the disease was first discovered in people in 1952. In 2013, an epidemic was detected in French Polynesia and other Pacific countries. The abrupt recurrence of this disease has been attributed to a number of environmental causes, including severe storms and rising temperatures.

Indirect Effects of Climate Change on Human Health

Climate change has an indirect impact on the human population, particularly children and the vulnerable, through diarrhoea, cholera, and other infectious diseases. Food supplements and health care, epidemic surveillance and early warning systems, and water safety strategies that analyse and manage the health hazards posed by climate variability and change should all be part of the response. The indirect consequences of climatic fluctuations on human health are depicted in Figure 4.

Figure 4. Chart showing indirect effects of climate variations on human health

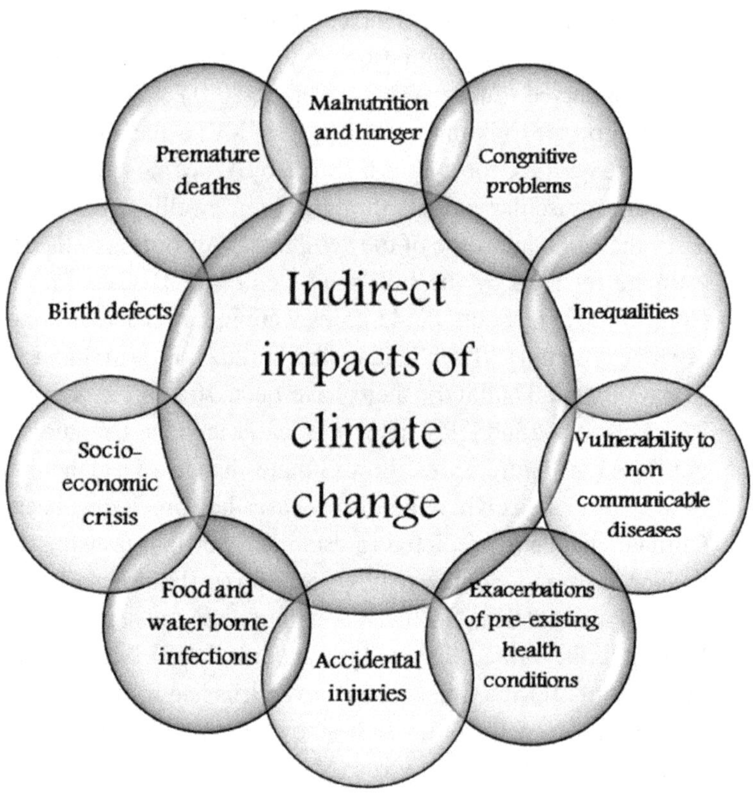

Malnutrition and Hunger

Reduced production of staple foods will be influenced by climate variability. As a result, the risk of malnutrition will rise (IPCC, 2014). Crop yields in Central and South Asia may drop by the mid-twentieth century. Increased greenhouse gas emissions impact plant development and nutrient levels. Malnutrition and hunger hazards will rise as a result of rapid population growth and urbanisation. There is a huge imbalance in the world when it comes to underweight children under the age of five. In many developing countries, population growth, undernutrition, and poor health will directly impact on productivity and economic development. Around 700 million people worldwide do not have access to better drinking water, and 2.5 billion do not have access to improved sanitation (WHO, 2014). Children under three years in underdeveloped nations experience at least three episodes of diarrhoea each year. Malnutrition, stunted growth, vitamin D deficiency, Beri-beri, Kwashiorkar, and hypoproteinemia are all consequences of this.

Food Insecurity

Agriculture is being intensified in the modern era to meet rising food demands. This has influenced food security. Many countries' economies are based on agriculture. Climate change directly impact on the agricultural, aquaculture, and forestry industries. The growth of flora and animals, their metabolic activity, and nutrient cycling in food chains are all influenced by climate. Pests that damage plants and cattle may emerge because of climate change, posing a threat to food security. Similarly, water-borne infections caused by weather extremes can have a negative impact on human health. Several vectors and illnesses thrive during certain humidity-temperature circumstances and irrigation management regimes. Crops, cattle, fish, and eventually humans will be exposed to new dangers as a result of these changes. Climate change will have an impact on food availability, accessibility, consumption, and food system stability. Supply networks will be disrupted, resulting in higher market pricing. Those whose livelihoods are depending on agriculture are the most susceptible. People who work in agriculture in mountainous and coastal locations will be affected as well (Raj et al., 2022). We can mitigate these effects to some extent by reducing greenhouse gas emissions and choosing sustainable agriculture techniques.

Climate Change Impacts on Human Settlements

Natural calamities have wreaked havoc on human communities. Thousands of people were displaced as a result of the tsunami in 2004. Natural disasters, such as earthquakes, have uprooted millions of people and wreaked havoc on infrastructure. Climate change and disasters compel people to look for work and housing in surrounding places in order to maintain their livelihoods. Floods, such as the Kosi floods of 2008, affected a large number of people since they are unpredictable events. Several million people have been compelled to relocate due to sudden onset risks. Drought, for example, is a slow-onset hazard that gives individuals adequate time to adjust to the situation and take appropriate action. Human migration is influenced by social, environmental, cultural, economic, and political factors. Migration and displacement are mentioned in the Paris Agreement. It mentions migrants' vulnerability in the preamble and urges nations to respect, promote, and consider their particular human rights commitments' when addressing climate change (UNFCCC, 2015: 1). It also mentions the need for a task force to "create suggestions for integrated methods to avoid, mitigate, and remedy displacement connected to climate change's detrimental effects" (UNFCCC, 2015). Displacement was also identified as a potential generator of vulnerability in the Hyogo Framework for Action (UNISDR, 2005). The Sendai Framework is concerned with displacement as a result of extreme occurrences (UNISDR, 2015).

The 2015 Chennai floods are a well-known cause of human settlements being threatened. It was an unexpected occurrence that leveled nearly half of Chennai city, killed several people, and wreaked havoc on human settlements. Approximately 500 people died, while nearly 1.8 million people were rescued and sent to safer areas. During this time, the city as a whole has suffered from a lack of essential home supplies, proper food, and access to clean water and sanitation. Extreme weather occurrences, such as flash floods, droughts, and other natural and man-made calamities, pose a major threat to human life and civilization. Human settlement disruptions, compounded with climate change, wreak havoc on the population's well-being.

Climate Change Impacts on Livelihoods

The various forms of assets, abilities, and activities that enable a person or household to subsist are referred to as livelihoods (FAO, 2003). Physical assets such as infrastructure and domestic assets; financial assets such as currency, jewels, bank assets, land assets, and agricultural fields; and human assets such as education and survival skills are all examples of these assets.

Different groups exist in society, and not all can adapt to the effects of climate change. People from marginalized groups have fewer resources and are more exposed to the effects. Climate change can make some foods and agricultural products unavailable at times. Such seasonal fluctuations might have a negative influence on people's livelihoods. Agriculture, livestock husbandry, and fishing are the main sources of income for marginalized people. In most cases, the obstacles that rural livelihoods face force people to migrate to cities. Fishermen living around beaches, for example, may be subjected to natural disasters, which have wreaked havoc on their homes, fishing nets, and infrastructure. Only one extreme catastrophe has wreaked havoc on their lives.

In 2021 Gujarat in India has faced the worst drought. According to a report from Down to Earth, (August 2021) it has been bad produce for farmers, especially those growing cash crops such as cotton and groundnut in Gujarat. The farmers have a hand-to-mouth living. In Banaskantha district in North Gujarat, there was insufficient enough water even for livestock to drink. Due to drought, the crops were infested with worms and the farmers spent more on pesticides. From this recent case study, we can see the impacts of drought on farmer's livelihoods. Therefore climate variabilities have serious consequences and affect the livelihoods of a million people globally. Therefore, such studies show that adaptation to such events is very much the need of the hour. We need to prepare ourselves with heatwave early warning systems, urban planning to reduce the urban heat island effect, better land-use planning, community relocation, reducing the vulnerability of essential services such as water, energy, and food, and measures to assist vulnerable sectors and households.

Climate Change and Mental Health

Physical and mental health is inextricably linked in life, as physical health has a profound influence on mental state and vice versa. Tough life has become much simpler in our modern period as we buzz around with technology to replace our mental and physical labour, but psychological problems have gotten a lot worse. The most challenging part is the mental illnesses related to the climate change crisis and even psychologists are new to this dimension of mental state. Thus, the new field called "ecopsychology" is booming to treat the environment and climate-associated threats to mental well-being like eco-anxiety, eco guilt, ecological grief, solastalgia, biospheric concern, etc.

The association between climate change and mental health is an important public health concern. Mental health issues before and after extreme climatic events are assessed by various researchers. The psychological effects of post-flood events have increased the number of psychiatric cases in the hospital for PTSD, depression, anxiety, psychiatric disorders, sleep disorder, and suicide (Dar & Lone, 2021; Zhong et al., 2018). The poll conducted by American Psychological Association in 2020 shows that two-thirds of the respondents (n=1000) feel extremely or moderately anxious about the impacts of climate change (APA, 2020). Climate change-related events like extreme weather events, wildfires, drought, floods, have caused several devastating effects on both life and properties. Individuals who have lost their close ones and homes or seeing their younger generation suffer from illnesses like asthma, chest infections (Caused by poor air quality), etc. develops fear or anxiety which eventually leads to depression, sleep disturbances, aggression, obsessive thoughts, post-traumatic stress disorder, and so on (Usher et al., 2019).

Climate change has evolved into an inescapable global issue that must be addressed in every way imaginable. Unpredictable climatic phenomena, such as excessive heat, floods, storms, sea-level rise, drought, and wildfire, have had and continue to have serious negative consequences on not just human health, but also exaggerate environmental issues and societal crises. Climate change and environmental health are inextricably related processes that cannot be separated. If there is any imbalance between the two components, unanticipated climatic events will occur, as mentioned above. Pre-existing physical and mental health disorders, poverty, age, sex, indigenous peoples, population residing in places prone to extreme weather events, and geopolitical variables are among the sensitivity factors to consider in terms of climate change and human health. As a result, in order to adapt to and prevent climate change, scientists throughout the world are attempting to develop environmentally acceptable alternatives to fossil fuels, plastics, and other materials in order to reduce carbon emissions, forest damage, and other issues. On the other hand, the government sector collaborates with policymakers and non-governmental organisations to create a climate-resilient society. As a result, it is the responsibility of each and every citizen to take the required steps to counteract climate change.

CONCLUSION

Climate change is accelerating, bringing with it new pests, microbiological agents, the recurrence of infections, and the advent of diseases. Climate change also puts reproductive, bone, and immunological health at risk. As a result of this exposure to a variety of climatic conditions and associated factors, persons' general health and life span will suffer. Changes in climate variability lead to more frequent extreme weather events, ecosystem changes, and disease spread. Land-use change, population growth and urbanisation, migration, and economic development are among socioeconomic elements that contribute to the changing climate and the host-pathogen relationship. Climate fluctuation also has the potential to destroy food webs and cause global extinction. This has an impact on not only the human population, but also on animal health and, eventually, the entire ecosystem.

ACKNOWLEDGEMENT

Sushmitha Baskar - Corresponding Author

ORCID ID - 0000-0003-3839-5557

REFERENCES

Acosta, A. L., Pinter, A., Faria, N., Sabino, E., & Mureb Sallum, M. A. (2021). Landscape connectivity for yellow fever: A proxy approach to detect displacement and circulation through environmental corridors and interfaces. *The Lancet. Planetary Health*, *5*, S22. doi:10.1016/S2542-5196(21)00106-6

Alemayehu, B., & Alemayehu, M. (2017). Leishmaniasis: A Review on Parasite, Vector and Reservoir Host. *Health Science Journal*, *11*(4), 519. doi:10.21767/1791-809X.1000519

Anthonj, C., Nkongolo, O. T., Schmitz, P., Hango, J. N., & Kistemann, T. (2015). The impact of flooding on people living with HIV: A case study from the Ohangwena Region, Namibia. *Global Health Action*, *8*(1), 26441. Advance online publication. doi:10.3402/gha.v8.26441 PMID:25813771

Anyamba, A., Chretien, J.-P., Britch, S. C., Soebiyanto, R. P., Small, J. L., Jepsen, R., Forshey, B. M., Sanchez, J. L., Smith, R. D., Harris, R., Tucker, C. J., Karesh, W. B., & Linthicum, K. J. (2019). Global Disease Outbreaks Associated with the 2015–2016 El Niño Event. *Scientific Reports*, *9*(1), 1–14. doi:10.1038/s41598-018-38034-z

APA. (2020). *New APA Poll Reveals That Americans are Increasingly Anxious About Climate Change's Impact on Planet, Mental Health*. https://www.psychiatry.org/newsroom/news-releases/climate-poll-2020

Beker, B. M., Cervellera, C., De Vito, A., & Musso, C. G. (2018). Human physiology in extreme heat and cold. *International Archieves of Clinical Physiology*, *1*(1). Advance online publication. doi:10.23937/iacph-2017/1710001

Bogovic, P., & Strle, F. (2015). Tick-borne encephalitis: A review of epidemiology, clinical characteristics, and management. *World Journal of Clinical Cases*, *3*(5), 430. doi:10.12998/wjcc.v3.i5.430 PMID:25984517

Chadsuthi, S., Chalvet-Monfray, K., Wiratsudakul, A., & Modchang, C. (2021). The effects of flooding and weather conditions on leptospirosis transmission in Thailand. *Scientific Reports*, *11*(1), 1486. doi:10.103841598-020-79546-x PMID:33452273

Charrahy, Z., Yaghoobi-Ershadi, M. R., Shirzadi, M. R., Akhavan, A. A., Rassi, Y., Hosseini, S. Z., Webb, N. J., Haque, U., Omid, F. B., & Hanafi-Bojd, A. A. (2021). Climate change and its effect on the vulnerability to zoonotic cutaneous leishmaniasis in Iran. *Transboundary and Emerging Diseases*. Advance online publication. doi:10.1111/tbed.14115 PMID:33876891

Chen, G., Guo, Y., Yue, X., Tong, S., Gasparrini, A., Bell, M. L., Armstrong, B., Schwartz, J., Jaakkola, J. J. K., Zanobetti, A., Lavigne, E., Nascimento Saldiva, P. H., Kan, H., Royé, D., Milojevic, A., Overcenco, A., Urban, A., Schneider, A., Entezari, A., ... Li, S. (2021). Mortality risk attributable to wildfire-related PM2·5 pollution: A global time series study in 749 locations. *The Lancet. Planetary Health, 5*(9), e579–e587. doi:10.1016/S2542-5196(21)00200-X PMID:34508679

Chen, N. T., Chen, M. J., Da Wu, C., & Guo, Y. L. (2021). Emergency room visits for childhood atopic dermatitis are associated with floods? *The Science of the Total Environment, 773*, 145435. doi:10.1016/j.scitotenv.2021.145435 PMID:33940726

Chhetri, B. K., Takaro, T. K., Balshaw, R., Otterstatter, M., Mak, S., Lem, M., Zubel, M., Lysyshyn, M., Clarkson, L., Edwards, J., Fleury, M. D., Henderson, S. B., & Galanis, E. (2017). Associations between extreme precipitation and acute gastro-intestinal illness due to cryptosporidiosis and giardiasis in an urban Canadian drinking water system (1997–2009). *Journal of Water and Health, 15*(6), 898–907. doi:10.2166/wh.2017.100 PMID:29215354

Confalonieri, U., Menne, B., Abeku, T., Alam, M., Beggs, P., Clot, B., Furgal, C., Curto, S., McMichael, A., Parry, M., Canziani, O., Palutikof, J., van der Linden, P., & Hanson, C. (2007). *Impacts, Adaptation and Vulnerability. Contribution of Working Group II to the Fourth Assessment Report of the Intergovernmental Panel on Climate Change*. Academic Press.

Conlen, M. (2021). *How Much Carbon Dioxide Are We Emitting?* NASA's Vital Signs of the Planet. https://climate.nasa.gov/news/3020/how-much-carbon-dioxide-are-we-emitting/

Control European Centre for Disease Prevention and Control. (2021). *Chikungunya worldwide overview*. https://www.ecdc.europa.eu/en/chikungunya-monthly

D'Amato, G., Chong-Neto, H. J., Ortega, O. P. M., Vitale, C., Ansotegui, I., Rosario, N., Haahtela, T., Galan, C., Pawankar, R., Murrieta-Aguttes, M., Cecchi, L., Bergmann, C., Ridolo, E., Ramon, G., Diaz, S. G., D'Amato, M., & Annesi-Maesano, I. (2020). The effects of climate change on respiratory allergy and asthma induced by pollen and mold allergens. *Allergy, 75*(9), 2219–2228. doi:10.1111/all.14476 PMID:32589303

D'Amato, M., Molino, A., Calabrese, G., Cecchi, L., Annesi-Maesano, I., & D'Amato, G. (2018). The impact of cold on the respiratory tract and its consequences to respiratory health. *Clinical and Translational Allergy, 8*(1), 1–8. doi:10.1186/s13601-018-0208-9

Dabaro, D., Birhanu, Z., Negash, A., Hawaria, D., & Yewhalaw, D. (2021). Effects of rainfall, temperature and topography on malaria incidence in elimination targeted district of Ethiopia. *Malaria Journal, 20*(1), 104. Advance online publication. doi:10.118612936-021-03641-1 PMID:33608004

Dar, B., & Lone, S. A. (2021). Pre and post Vulnerability of Floods to Mental Health Among the Residents of Srinagar City, J&K-India. *Sustainability, Agri. Food and Environmental Research, 11*(1), 1–12. doi:10.7770afer-V11N1-art2393

Duffin, R., Mills, N. L., & Donaldson, K. (2007). Nanoparticles-a thoracic toxicology perspective. *Yonsei Medical Journal, 48*(4), 561–572. doi:10.3349/ymj.2007.48.4.561 PMID:17722227

Eguiluz-Gracia, I., Mathioudakis, A. G., Bartel, S., Vijverberg, S. J. H., Fuertes, E., Comberiati, P., Cai, Y. S., Tomazic, P. V., Diamant, Z., Vestbo, J., Galan, C., & Hoffmann, B. (2020). The need for clean air: The way air pollution and climate change affect allergic rhinitis and asthma. *Allergy, 75*(9), 2170–2184. doi:10.1111/all.14177 PMID:31916265

European Centre for Disease Control and Prevention. (n.d.). *Tick-borne encephalitis in Europe*. Fact Sheet.

Ewing, D., Purse, B., Cobbold, C., & White, S. (2021). A novel approach for predicting risk of vector-borne disease establishment in marginal temperate environments under climate change: West Nile virus in the UK. *Journal of the Royal Society, Interface, 18*(178), 20210049. Advance online publication. doi:10.1098/rsif.2021.0049 PMID:34034529

Fletcher, I. K., Hernández-Villena, J., Moreno, J. E., Drakeley, C., Jones, K., Grillet, M. E., & Lowe, R. (2021). The effect of environmental degradation and land use change on malaria re-emergence in south Venezuela: A spatiotemporal modelling study. *The Lancet. Planetary Health, 5*, S13. doi:10.1016/S2542-5196(21)00097-8

Global Burden of Disease: GBD cause and risk summaries. (2019). https://www.thelancet.com/gbd/summaries

Gurevitz, J. M., Antman, J. G., Laneri, K., & Morales, J. M. (2021). Temperature, traveling, slums, and housing drive dengue transmission in a non-endemic metropolis. *PLoS Neglected Tropical Diseases, 15*(6), e0009465. doi:10.1371/journal.pntd.0009465 PMID:34115753

He, J., Christakos, G., Zhang, W., & Wang, Y. (2017). A Space-Time Study of Hemorrhagic Fever with Renal Syndrome (HFRS) and Its Climatic Associations in Heilongjiang Province, China. *Frontiers in Applied Mathematics and Statistics, 0*, 16. doi:10.3389/fams.2017.00016

Hugo, G. (2010). *Climate change-induced mobility and the existing migration regime in Asia and the Pacific*. Hart Publishing.

IPCC. (2007). Fourth Assessment Report. *Climatic Change.*

IPCC. (2012). *Managing the Risks of Extreme Events and Disasters to Advance Climate Change Adaptation. Special Report of the Intergovernmental Panel on Climate Change*. Cambridge University Press.

IPCC. (2014a). Regional Context. In *Climate Change 2014: Impacts, Adaptation, and Vulnerability. Part B: Regional Aspects. Contribution of Working Group II to the Fifth Assessment Report of the Intergovernmental Panel on Climate Change.* Cambridge University Press.

IPCC. (2014b). Adaptation Needs and Options. In *Climate Change 2014: Impacts, Adaptation, and Vulnerability. Part A: Global and Sectoral Aspects. Contribution of Working Group II to the Fifth Assessment Report of the Intergovernmental Panel on Climate Change*. Cambridge University Press.

IPCC. (2014c). Summary for Policymakers. In *Climate Change 2014: Impacts, Adaptation, and Vulnerability. Part A: Global and Sectoral Aspects. Contribution of Working Group II to the Fifth Assessment Report of the Intergovernmental Panel on Climate Change*. Cambridge University Press.

Jones, F. K., Wamala, J. F., Rumunu, J., Mawien, P. N., Kol, M. T., Wohl, S., Deng, L., Pezzoli, L., Omar, L. H., Lessler, J., Quilici, M. L., Luquero, F. J., & Azman, A. S. (2020). Successive epidemic waves of cholera in South Sudan between 2014 and 2017: A descriptive epidemiological study. *The Lancet. Planetary Health, 4*(12), e577–e587. doi:10.1016/S2542-5196(20)30255-2 PMID:33278375

Kugeler, K. J., Schwartz, A. M., Delorey, M. J., Mead, P. S., & Hinckley, A. F. (2021). Estimating the Frequency of Lyme Disease Diagnoses, United States, 2010–2018 - Volume 27, Number 2—February 2021 - Emerging Infectious Diseases journal - CDC. *Emerging Infectious Diseases*, *27*(2), 616–619. doi:10.3201/eid2702.202731 PMID:33496229

Lee, S. A., Economou, T., Barcellos, C., Catão, R., Carvalho, M. S., & Lowe, R. (2021). Effect of climate change, connectivity, and socioeconomic factors on the expansion of the dengue virus transmission zone in 21st century Brazil: An ecological modelling study. *The Lancet. Planetary Health*, *5*, S14. doi:10.1016/S2542-5196(21)00098-X

Li, L., Jiang, C., Murtugudde, R., Liang, X.-Z., & Sapkota, A. (2021). Global Population Exposed to Extreme Events in the 150 Most Populated Cities of the World: Implications for Public Health. *International Journal of Environmental Research and Public Health, 18*(3), 1293. doi:10.3390/ijerph18031293

Liu, Z., Zhang, Z., Lai, Z., Zhou, T., Jia, Z., Gu, J., Wu, K., & Chen, X.-G. (2017). Temperature Increase Enhances Aedes albopictus Competence to Transmit Dengue Virus. *Frontiers in Microbiology*, *0*(DEC), 2337. doi:10.3389/fmicb.2017.02337 PMID:29250045

Lucas, R. M., Yazar, S., Young, A. R., Norval, M., De Gruijl, F. R., Takizawa, Y., Rhodes, L. E., Sinclair, C. A., & Neale, R. E. (2019). Human health in relation to exposure to solar ultraviolet radiation under changing stratospheric ozone and climate. *Photochemical & Photobiological Sciences*, *18*(3), 641–680. doi:10.1039/c8pp90060d PMID:30810559

Mallhi, T. H., Khan, Y. H., Khan, A. H., Tanveer, N., & Qadir, M. I. (2017). First chikungunya outbreak in Pakistan: A trail of viral attacks. *New Microbes and New Infections*, *19*, 13–14. doi:10.1016/j.nmni.2017.05.008 PMID:28663798

Martens, P., & McMichael, A. J. (Eds.). (2009). *Environmental change, climate and health: issues and research methods*. Cambridge University Press.

Mora, C., Dousset, B., Caldwell, I. R., Powell, F. E., Geronimo, R. C., Bielecki, C. R., Counsell, C. W. W., Dietrich, B. S., Johnston, E. T., Louis, L. V., Lucas, M. P., McKenzie, M. M., Shea, A. G., Tseng, H., Giambelluca, T. W., Leon, L. R., Hawkins, E., & Trauernicht, C. (2017). Global risk of deadly heat. *Nature Climate Change 7, 7*(7), 501–506. doi:10.1038/nclimate3322

Mordecai, E. A., Ryan, S. J., Caldwell, J. M., Shah, M. M., & LaBeaud, A. D. (2020). Climate change could shift disease burden from malaria to arboviruses in Africa. *The Lancet. Planetary Health, 4*(9), e416–e423. doi:10.1016/S2542-5196(20)30178-9 PMID:32918887

Nah, K., Bede-Fazekas, Á., Trájer, A. J., & Wu, J. (2020). The potential impact of climate change on the transmission risk of tick-borne encephalitis in Hungary. *BMC Infectious Diseases, 20*(1), 1–10. doi:10.1186/s12879-019-4734-4

Nosrat, C., Altamirano, J., Anyamba, A., Caldwell, J. M., Damoah, R., Mutuku, F., Ndenga, B., & LaBeaud, A. D. (2021). Impact of recent climate extremes on mosquito-borne disease transmission in Kenya. *PLoS Neglected Tropical Diseases, 15*(3), e0009182. doi:10.1371/journal.pntd.0009182 PMID:33735293

Oberdörster, G. (1995). Lung particle overload: Implications for occupational exposures to particles. *Regulatory Toxicology and Pharmacology, 21*(1), 123–135. doi:10.1006/rtph.1995.1017 PMID:7784625

Paterson, D. L., Wright, H., & Harris, P. N. A. (2018). Health Risks of Flood Disasters. *Clinical Infectious Diseases, 67*(9), 1450–1454. doi:10.1093/cid/ciy227 PMID:30986298

Patil, S. S., Patil, S. R., Durgawale, P. M., & Patil, A. G. (2013). A study of the outbreak of Chikungunya fever. *Journal of Clinical and Diagnostic Research : JCDR, 7*(6), 1059. PMID:23905103

Patz, J. A., Graczyk, T. K., Geller, N., & Vittor, A. Y. (2000). Effects of environmental change on emerging parasitic diseases. *International Journal for Parasitology, 30*(12-13), 1395–1405. doi:10.1016/S0020-7519(00)00141-7 PMID:11113264

Paz, S. (2015). Climate change impacts on West Nile virus transmission in a global context. *Philosophical Transactions of the Royal Society B: Biological Sciences, 370*(1665), 20130561.

Peña-García, V. H., Triana-Chávez, O., & Arboleda-Sánchez, S. (2017). Estimating Effects of Temperature on Dengue Transmission in Colombian Cities. *Annals of Global Health, 83*(3–4), 509–518. doi:10.1016/j.aogh.2017.10.011 PMID:29221523

Powers, A. M., & Logue, C. H. (2007). Changing patterns of chikungunya virus: Re-emergence of a zoonotic arbovirus. *The Journal of General Virology, 88*(9), 2363–2377. doi:10.1099/vir.0.82858-0 PMID:17698645

Qin, D., Manning, M., Chen, Z., Marquis, M., Averyt, K. B., & Tignor, M. (Eds.). (2007). *IPCC Fourth Assessment Report: Climate Change 2007: The Physical Science Basis*. Cambridge University Press.

Radi, M. F. M., Hashim, J. H., Jaafar, M. H., Hod, R., Ahmad, N., Nawi, A. M., Baloch, G. M., Ismail, R., & Ayub, N. I. F. (2018). Leptospirosis Outbreak After the 2014 Major Flooding Event in Kelantan, Malaysia: A Spatial-Temporal Analysis. *The American Journal of Tropical Medicine and Hygiene*, *98*(5), 1281–1295. doi:10.4269/ajtmh.16-0922 PMID:29532771

Raj, S., Roodbar, S., Brinkley, C., & Wolfe, D. W. (2022). Food Security and Climate Change: Differences in impacts and adaptation strategies for rural communities in the Global South and North. *Frontiers in Sustainable Food Systems*, *5*(1), 691191. Advance online publication. doi:10.3389/fsufs.2021.691191

Samoli, E., Touloumi, G., Schwartz, J., Anderson, H. R., Schindler, C., Forsberg, B., Vigotti, M. A., Vonk, J., Košnik, M., Skorkovsky, J., & Katsouyanni, K. (2007). Short-Term Effects of Carbon Monoxide on Mortality: An Analysis within the APHEA Project. *Environmental Health Perspectives*, *115*(11), 1578–1583. doi:10.1289/ehp.10375 PMID:18007988

Schraufnagel, D. E., Balmes, J. R., Cowl, C. T., De Matteis, S., Jung, S. H., Mortimer, K., Perez-Padilla, R., Rice, M. B., Riojas-Rodriguez, H., Sood, A., Thurston, G. D., To, T., Vanker, A., & Wuebbles, D. J. (2019). Air Pollution and Noncommunicable Diseases: A Review by the Forum of International Respiratory Societies' Environmental Committee, Part 1: The Damaging Effects of Air Pollution. *Chest*, *155*(2), 409–416. doi:10.1016/j.chest.2018.10.042 PMID:30419235

Shapiro, E. D., & Gerber, M. A. (2000). Lyme Disease. *Clinical Infectious Diseases*, *31*(2), 533–542. doi:10.1086/313982 PMID:10987718

Shukla, M. A., Woc-Colburn, L., & Weatherhead, J. E. (2018). Infectious Diseases in the Aftermath of Hurricanes in the United States. *Current Tropical Medicine Reports*, *5*(4), 217–223. doi:10.100740475-018-0162-6

Song, X., Jiang, L., Zhang, D., Wang, X., Ma, Y., Hu, Y., Tang, J., Li, X., Huang, W., Meng, Y., Shi, A., Feng, Y., & Zhang, Y. (2021). Impact of short-term exposure to extreme temperatures on diabetes mellitus morbidity and mortality? A systematic review and meta-analysis. *Environmental Science and Pollution Research International*, *2021*(41), 1–15. doi:10.100711356-021-14568-0 PMID:34105073

Stewart, G. J., Nelson, B. S., Acton, W. J. F., Vaughan, A. R., Hopkins, J. R., Yunus, S. S. M., Hewitt, C. N., Wild, O., Nemitz, E., Gadi, R., Sahu, L. K., Mandal, T. K., Gurjar, B. R., Rickard, A. R., Lee, J. D., & Hamilton, J. F. (2021). Emission estimates and inventories of non-methane volatile organic compounds from anthropogenic burning sources in India. *Atmospheric Environment: X, 11*, 100115. doi:10.1016/j.aeaoa.2021.100115

Tall, J. A., & Gatton, M. L. (2020). Flooding and Arboviral Disease: Predicting Ross River Virus Disease Outbreaks Across Inland Regions of South-Eastern Australia. *Journal of Medical Entomology, 57*(1), 241–251. doi:10.1093/jme/tjz120 PMID:31310648

Tamagawa, E., Bai, N., Morimoto, K., Gray, C., Mui, T., Yatera, K., Zhang, X., Xing, L., Li, Y., Laher, I., Sin, D. D., Man, S. F. P., & van Eeden, S. F. (2008). Particulate matter exposure induces persistent lung inflammation and endothelial dysfunction. *American Journal of Physiology. Lung Cellular and Molecular Physiology, 295*(1), L79–L85. Advance online publication. doi:10.1152/ajplung.00048.2007 PMID:18469117

The Potsdam Institute for Climate Impact Research and Climate Analytics. (2012). Turn-down the Heat – Why a 4 degree Warmer World Must be Avoided. International Bank for Reconstruction and Development and World Bank.

Tjaden, N. B., Suk, J. E., Fischer, D., Thomas, S. M., Beierkuhnlein, C., & Semenza, J. C. (2017). Modelling the effects of global climate change on Chikungunya transmission in the 21 st century. *Scientific Reports, 7*(1), 1–11. doi:10.1038/s41598-017-03566-3

UNFCCC. (2011). *Report of the Conference of the Parties: The Cancun Agreements, Outcome of the work of the Ad Hoc Working Group on the Long-term Cooperative Action under the Convention*. FCCC/CP/2010/7/Add.1, Part Two: Action taken by the Conference of the Parties at its sixteenth session. Bonn: UNFCCC, http://unfccc.int/resource/ docs/2010/cop16/eng/07a01.pdf#page=4

UNFCCC (United Nations Framework Convention on Climate Change). (2016). *Task Force on Displacement*. UNFCCC.

Urban, K., Mehrmal, S., Uppal, P., Giesey, R. L., & Delost, G. R. (2021). The global burden of skin cancer: A longitudinal analysis from the Global Burden of Disease Study, 1990–2017. *JAAD International, 2*, 98–108. doi:10.1016/j.jdin.2020.10.013 PMID:34409358

Usher, K., Durkin, J., & Bhullar, N. (2019). Eco-anxiety: How thinking about climate change-related environmental decline is affecting our mental health. International Journal of Mental Health Nursing, 28(6), 1233–1234. doi:10.1111/inm.12673

Verhoeven, J. I., Allach, Y., Vaartjes, I. C. H., Klijn, C. J. M., & de Leeuw, F. E. (2021). Ambient air pollution and the risk of ischaemic and haemorrhagic stroke. *The Lancet. Planetary Health*, *5*(8), e542–e552. doi:10.1016/S2542-5196(21)00145-5 PMID:34390672

Wamai, R. G., Kahn, J., McGloin, J., & Ziaggi, G. (2020). Visceral leishmaniasis: A global overview. *Journal of Global Health Science*, *2*(1), e3. Advance online publication. doi:10.35500/jghs.2020.2.e3

Wen, T. H., Tsai, C. T., & Chin, W. C. B. (2016). Evaluating the role of disease importation in the spatiotemporal transmission of indigenous dengue outbreak. *Applied Geography (Sevenoaks, England)*, *76*, 137–146. doi:10.1016/j.apgeog.2016.09.020

Wilson, M. L. (2001). Ecology and infectious disease. In J. L. Aron & J. A. Patz (Eds.), *Ecosystem Change and Public Health: A Global Perspective* (pp. 283–324). Johns Hopkins University Press.

World Health Organization. (2020). *World malaria report 2020*. https://www.who.int/publications/i/item/9789240015791

World Health Organization. (2021a). *Dengue and severe dengue*. Fact Sheet. https://www.who.int/news-room/fact-sheets/detail/dengue-and-severe-dengue

World Health Organization. (2021b). *Leishmaniasis*. Fact Sheet. https://www.who.int/news-room/fact-sheets/detail/leishmaniasis

Wu, H., Kioumourtzoglou, M. A., Just, A. C., Kloog, I., Sanders, A., Svensson, K., McRae, N., Tamayo-Ortiz, M., Solano-González, M., Wright, R. O., Téllez-Rojo, M. M., & Baccarelli, A. A. (2020). Association of ambient PM2·5 exposure with maternal bone strength in pregnant women from Mexico City: A longitudinal cohort study. *The Lancet. Planetary Health*, *4*(11), e530–e537. doi:10.1016/S2542-5196(20)30220-5 PMID:33159880

Xiang, J., Hansen, A., Liu, Q., Tong, M. X., Liu, X., Sun, Y., Cameron, S., Hanson-Easey, S., Han, G. S., Williams, C., Weinstein, P., & Bi, P. (2018). Impact of meteorological factors on hemorrhagic fever with renal syndrome in 19 cities in China, 2005–2014. *The Science of the Total Environment*, *636*, 1249–1256. doi:10.1016/j.scitotenv.2018.04.407 PMID:29913587

Yue, R. P. H., & Lee, H. F. (2020). Drought-induced spatio-temporal synchrony of plague outbreak in Europe. *The Science of the Total Environment*, *698*, 134138. doi:10.1016/j.scitotenv.2019.134138 PMID:31505345

Zeng, Z., Zhan, J., Chen, L., Chen, H., & Cheng, S. (2021). Global, regional, and national dengue burden from 1990 to 2017: A systematic analysis based on the global burden of disease study 2017. *EClinicalMedicine*, *32*, 100712. doi:10.1016/j.eclinm.2020.100712 PMID:33681736

Zhong, S., Yang, L., Toloo, S., Wang, Z., Tong, S., Sun, X., Crompton, D., FitzGerald, G., & Huang, C. (2018). The long-term physical and psychological health impacts of flooding: A systematic mapping. *The Science of the Total Environment*, *626*, 165–194. doi:10.1016/j.scitotenv.2018.01.041 PMID:29339262

KEY TERMS AND DEFINITIONS

Ecopsychology: Climate-associated threats to mental well-being like eco-anxiety, eco guilt, ecological grief, solastalgia, biospheric concern, etc.

Health: Health can be defined as a complete state of physical, mental, and social well-being and not merely the absence of disease or infirmity (WHO).

Livelihood: Livelihoods can be defined as the different types of assets, abilities, and activities that enable a person or household to survive (FAO, 2003).

Morbidity: A disease or condition that reduces health and the quality of life. The morbidity rate is a measure of the frequency of disease among a defined population during a specified time period.

Mortality: It is the number of deaths in a defined population during a specified time period.

Chapter 2
Climate Change and Human Infectious Diseases

Mercy Oluwatofunmi Olaoluwa
University of Ibadan, Nigeria

Abosede Olufunke Taiwo
Osun State College of Technology, Nigeria

Ayodeji Ojo Oteyola
https://orcid.org/0000-0002-4319-8868
Federal University of Minas Gerais, Brazil

ABSTRACT

Climate change is one of the most serious dangers to human health in the 21st century. Climate extremes, air quality, sea-level rise, and water supplies all have a direct impact on human health. Climate change has an influence on infectious diseases as well. In this chapter, the authors explained the influence of climate change on human infectious diseases. The authors also discuss how society may respond to and adapt to climate change. The variables influencing the spread and severity of human diseases were explored as well. These variables include human and animal migration, control methods, medication availability, human behavior, and the quality of public health facilities. The authors also discuss the importance of genome research in climate change and infectious diseases. Finally, as the world faces present and future difficulties in climate change and an increased threat of infectious diseases, the authors emphasize on the importance of scientific networks and collaborations for effective multidisciplinary engagements.

DOI: 10.4018/978-1-7998-9414-8.ch002

INTRODUCTION

According to the World Health Organization, among the greatest dangers to human health is climate change. Climate change directly affects health (Epstein 1999; Kovats *et al.,* 2000; Costello *et al.,* 2009; Willox *et al.,* 2015). Ways through which climate change could affect human health include the following:

(1) Changes in temperature and rainfall,
(2) Climatic extremes,
(3) Reduction in air quality, and
(4) Influences on food production and water quality (Field *et al.,* 2014).

As reported by the European Environment Agency, in the 20th century, there is an increase in the global average temperature by 0.74⁰C the global average surface temperature. In addition, since 1961, there has been a 1.8 mm increase every year in the sea level. Furthermore, mountain icebergs are contrasting. The acidic nature of the water in the ocean needs urgent attention (EEA 2008). In the wake of all these developments, it was forecasted that in the 21st century, there would be an increase in temperature of about 1.5 – 5.8⁰C. The temperature rise would be associated with irregularities in weather events which may in turn lead to floods, and drought (IPCC 2001).

Three constituents are crucial for most infectious diseases: (1) pathogen, (2) vector/host, and (3) the environment for the disease transmission (Epstein 2001). The lifecycle of many pathogens is so dynamic that they require the intervention of a vector or an intermediate host for completion. In fact, the distribution, breeding, and pathogen transmission need a suitable climatic condition. Hence, irregularities in the climate could affect pathogens and vectors which may in turn have influence on infectious diseases (Epstein 2001). Over time, studies have confirmed that the increase in infectious diseases could be linked to climatic variations extremity (Epstein *et al.,* 1999, Ostfeld and Brunner, 2015), while extreme weather conditions could lead to more outbreak of diseases in a place (Epstein 2000). Generally, the geographical and periodic distribution of infectious diseases is controlled by climate change while weather affects the timeline and severity of the disease epidemic (Kuhn *et al.,* 2005; Wu *et al.,* 2014).

Studies have shown that many of the infectious diseases in which insects act as the vector are very nociceptive to climate change (Kuhn *et al.,* 2005; Tian *et al.,* 2015). Examples of such infectious diseases include malaria, dengue, hantavirus, and cholera. Increase in temperature and flooding may also increase the outbreaks of giardiasis, salmonellosis, and cholera. Watts and colleagues reported that the effect of climate change on health has been considered in the implementation of the Paris climate agreement (Watts *et al.,* 2017; Watts *et al.,* 2018).

In this chapter, the authors focus on the impacts of climate change on human infectious diseases, by explaining the influence that climate change has on pathogens, vectors, or hosts, and its effect on infectious diseases transmission. The authors discuss the adaptive measures that can help to control the impact of climatic conditions on infectious diseases. The authors also explain the importance of genome research in climate change and infectious diseases as well as the importance of effective collaboration between health experts, professionals and researchers in climate and infectious diseases research.

The Impact of Climate Change on Pathogens

Pathogen refers to organisms or agents that can cause disease, this includes viruses, bacterium, and fungal agents. Changes in climate could either affect pathogens directly or indirectly. It could be direct by affecting survival, life cycle, and reproduction of pathogens. When the effect of climate change is evident on the environment or habitat of pathogens, this can be referred to as indirect. Temperature could influence disease by altering the life cycle of the pathogens. The survival and development of pathogen requires a temperature range. For instance, maximum temperature of 22 ^{0}C – 23 ^{0}C is needed for the development of mosquitoes (Mellor and Leake, 2000). Gerba (1999) reported that the death rate of several pathogens increases when heat increases. Patz and colleagues reported that when the temperature exceeds 33^{0}C-39^{0}C, it affects the development of malaria parasite. Temperature range between 7^{0}C and 37^{0}C increases the reproduction of Salmonella (IWGCCH 2010). Development and survival of pathogens related to water-borne diseases is largely influenced by rainfall. Jofre and group (2012) reported that there is increase in fecal pathogens during the season of rain. This is because of settled deposits in water which is caused by rainfall which in turn can lead to increase in fecal pathogens. Change in humidity can also influence the pathogens of infectious diseases. Scientific research by Shaman and Kohn in 2009 has shown that the transmission of virus associated with influenza could also be affected by humidity and temperature. This was also reported by Xu *et al.* 2014. Humidity change can also affect virus of vector-borne diseases. Another important climatic variable that influences pathogens of infectious diseases is sunshine. Islam *et al.,* 2009 reported that the action of sunshine together with temperature in the course of cholera time increase the multiplication of *Vibrio cholerae* in aquatic environments. In addition, pathogens associated with air-borne diseases are affected by wind. Reports have shown that during the time of the Asian Dust Storms (ADSs), dust presence in the environment is connected to an increase in bacteria and fungi.

Impact of Climate Change on Vector

The living organisms where pathogens live are referred to as the host. The intermediate hosts that transmit pathogens to living organisms which eventually become hosts are referred to as vectors. The influence of climate change on vectors can influence the span, duration, and severity of infectious diseases. Temperature greatly influences the geographical distribution of vectors. Changes in temperature may lead to geographical shift of diseases because increase in temperature leads to insects (vectors) in low latitude regions migrating to high latitude regions. Harvell together with colleagues reported about the spread of some vector-borne infectious diseases, such as malaria, yellow fever, trypanosomiasis, and dengue (Harvell *et al.,* 2002). Zhou and group reported that when temperature began to increase in China, *Oncomelania hupensis,* which is *Schistosoma japonicum* intermediate host, spread to new habitat which includes the northern China (Zhou *et al.,* 2010). Not only that temperature can cause geographical shift in vector's habitat, but it could also restrain the spread of vectors. For instance, Epstein (2001) reported that change in temperature could affect the distribution of *Aedes aegypti* which is the intermediate host for the virus of yellow fever and dengue fever. Precipitation changes could also affect vectors or hosts. Rainfall is linked to many vector-borne infectious diseases. Hoshen and Morse reported that increase in rainfall and temperature hasten the larval development of mosquito vectors (Hoshen and Morse, 2004). Zell reported that in Mexico, cocoliztli outbreak confirmed that rainfall can influence outbreak of rodent-borne diseases as a result of its effect on rodent ecosystem (Zell 2004). Besides, it may also have a devastating effect on the population of vectors. For instance, Kuhn and colleagues (2005) reported that too much rainfall can affect the population of mosquitoes by flooding away their breeding site. Relative humidity influence mosquito survival and activity which in turn influences the transmission of malaria. Pampana reported that when the humidity is low, the mosquito lifetime is reduced, thereby reducing the rate of malaria transmission (Pampana 1969). Wind effect on vectors or host can either be favorable or unfavorable. Persistent wind may affect the biting ability of a mosquito, but it could also extend the distance of their flight.

Impact of Climate Change on Infectious Disease Transmission

Disease transmission as it relates to climate change could either be primary or secondary. Primary transmission is the transmission from one individual to another individual via physical contact, droplets contact, air transmission, and fecal-oral transmission. Secondary transmission is disease transmission to humans via an intermediate host. Climate change may affect disease transmission by influencing the variability of pathogens. The transmission of diseases can be influenced by either temperature alone or in conjunction with other climatic factors. Wind also affects the transmission of infectious disease via acting as a means of transportation for pathogens and viruses linked to air-borne infections. Climate change plays an important role in human and host behaviors, which include migration, lifestyles, physical exercise, and occupation (Viboud *et al.,* 2004). The following factors can therefore influence the transmission of infectious disease. In Europe, studies have revealed that the periodic incidence pattern of influenza infection is associated with humans staying more indoors in the winter (Halstead 1996; Lofgren *et al.,* 2007). Changes in climate can also affect human resistance and vulnerability to disease. This can in turn influence disease transmission. Climate change could also cause soil degradation, which may in turn lead to crop failure, malnutrition, and starvation. These factors could also increase vulnerabilities to infectious diseases. Reduction in access to clean water because of climate change may lead to the incidence of water-borne diseases (CDC, 2010).

Adaptive Measures to Climate Change

Society and human response to climate change may reduce or increase the prevalence of infectious diseases. Some communities are more susceptible to infectious diseases because of the inability to respond adequately to the danger that is pose by extreme conditions of the climate (Wei *et al.,* 2012; Li *et al.,* 2014). However, the difference in susceptibility to infectious diseases across communities, regions, and countries is as a result of measures and programs that are put in place to check and control the threat from climate change and extreme weather conditions. These measures include:

1. Availability of public health facilities,
2. Availability of safe water and good sanitation,
3. Surveillance program to respond to outbreaks of infectious diseases (Bai *et al.,* 2014; Li *et al.,* 2014).

Lloyd and colleagues (2007) reported that scarcity of water often leads to an increase in diarrhea morbidity, this implies that people living in a community with less access to clean water would be susceptible to diarrhea. In addition, Guill and Shandera (2001) reported that many infectious diseases occur more in the developing countries than the developed countries. Examples of these infectious diseases include Balantidiosis spread in Turkey (Toole 1997), acute respiratory infection in Puerto Rico (Sanders *et al.,* 1997) and in Dominican Republic (CDC 1999). The minimal occurrence of infectious diseases in the developed countries is as a result of adequate health education, effective communication, and financial strength and resources that are abundant in developed countries. Infectious diseases are prevalent in developing countries as a result of the no- or inadequate public health facilities. The facilities in the public health centers in the developing countries are minimal, which makes it difficult to control or check the spread of infectious diseases in these regions. For instance, PAHO (1998) reported that the death rate from dengue fever shoot up following the occurrence of Hurricane Mitch in Guatemala and Honduras. In the year 1979 in Dominican Republic, about five (5) months following Hurricane David and Fredrick, there was increase in infectious diseases which includes infectious hepatitis, typhoid and paratyphoid, and measles. The delay in the outbreak of this infectious disease was as a result of insufficient sanitary facilities, contamination of food and water, and low immunization rates (Ebi *et al.*, 2001). Appropriate adaptation measures put in place by private individuals, government can help to reduce the risk of infectious diseases. As reported by UNEP, these adaptation measures include good drainage system, building sea walls, and reforestation. These measures could help to reduce the influence of climate change on disease transmission.

Significance of Genome Research in Climate Change and Infectious Disease

Recent panel discussion using 'One Health and approach' emphasized the importance of genetic sequencing as a means to address climate change and infectious illness (Yeh *et al.,* 2020). Yeh and colleagues emphasize that one specific focus was on how organisms move and develop in response to environmental change and the difficulties that accompany it (Yeh *et al.,* 2020). Chown and colleagues reported that when genome sequencing techniques are integrated with risk and surveillance control initiatives, an efficient structure for illustrating priorities at several levels of impact may be established (Chown *et al.,* 2015). This combination of multidisciplinary approach and developments in predictive analytics employing artificial intelligence and machine learning (AI/ML) techniques, provides exceptional prospects for successful methods in addressing infectious diseases transmission. Parekh et al. (2020) reported that environmental disasters from man-made infrastructures may impact climatic variables and infectious disease epidemics resulting because of ecosystem interruption, flooding, pollution, and human population migration. The following elements may either increase or decrease transmission of disease. Next generation research, particularly those that uncover quick health risk and mitigation techniques in conjunction with the identification of gene of pathogen and vector or host, will become very important in the future.

Collaborations: Essential for Multidisciplinary Engagements

Research partnership network could either be formal or informal. Networking and partnership play an important role in stopping and reducing infectious diseases outbreaks. Establishing effective adaptive measures among countries requires connecting researchers and experts together to enable expertise and information sharing (Fair *et al.,* 2016; Yeh *et al.,* 2021). Research aimed at addressing more global and complicated concerns necessitates interdisciplinary teams that may span many domains in a single project and assemble numerous machineries and data sources. Because of the complexity of research teams, it is necessary to reconsider how to collaborate, interact, harmonize, and communicate (Bennett and Gadlin, 2012). Scientists and researchers usually use informal collaborations across their professional career. These interactions include associate researchers, friends, and mentors/mentees relationship. As a result of the multidisciplinary nature of science, it becomes clearer that collaborations and networking between researchers are necessary for effective and efficient results. For instance, infectious diseases research demands the having the knowledge of etiology of an emerging infectious disease. This may require epidemiologists working together with meteorologists, biological scientists, and geographers. The internet has made connections easier between researchers and professionals. Yeh and colleagues reported that quick and effective response during the outbreaks of infectious diseases requires collaboration and teamwork, be it formal or informal (Yeh *et al.,* 2023). As the world faces present and future difficulties in climate change and an increased danger of infectious diseases, teamwork and collaborations are necessary for effective multidisciplinary engagements.

CONCLUSION

The severity of the impact of climate change on infectious diseases depends on its influence on pathogens and vectors. Humans are at the receiving end of climate change through susceptibility to infectious diseases. Society and the government play a vital role in alleviating the effects of climate change by putting in place adaptive measures which can in turn reduce the negative impact of climate change. It would be interesting to see the developed countries partnering with the developing and underdeveloped countries to control the negative impact of climate change. State-of-the-art research, with emphasis on those that unravel quick health risk in conjunction with fast genome identification should be encouraged. Finally, researchers and professionals should engage in collaborations and partnerships in order to have a rapid and coordinated response to infectious diseases outbreaks.

REFERENCES

Bai, L., Woodward, A., & Liu, Q. (2014). Temperature and mortality on the roof of the world: A time-series analysis in three Tibetan counties. *The Science of the Total Environment*, *485*, 41–48. doi:10.1016/j.scitotenv.2014.02.094 PMID:24704955

Bennett, L. M., & Gadlin, H. (2012). Collaboration and team science: From theory to practice. *Journal of Investigative Medicine*, *60*(5), 768–775. doi:10.2310/JIM.0b013e318250871d PMID:22525233

CDC. (1999). Needs assessment following Hurricane Georges—Dominican Republic, 1998. *Morbidity and Mortality Weekly Report*, *48*, 93–95. PMID:10072265

CDC. (2010). *Waterborne diseases. Climate and Health Program*. Centers for Disease Control and Prevention.

Chown, S. L., Hodgins, K. A., Griffin, P. C., Oakeshott, J. G., Byrne, M., & Hoffmann, A. (2015). Biological invasions, climate change and genomics. *Evolutionary Applications*, *8*(1), 23–46. doi:10.1111/eva.12234 PMID:25667601

Costello, A., Abbas, M., Allen, A., Ball, S., Bell, S., Bellamy, R., Friel, S., Groce, N., Johnson, A., Kett, M., Lee, M., Maslin, M., McCoy, D., McGuire, B., Montgomery, H., Napier, D., Pagel, C., Patel, J., de Oliveira, J. A. P., ... Patterson, C. (2009). Managing the health effects of climate change. *Lancet*, *373*(9676), 1773–1964. doi:10.1016/S0140-6736(09)60935-1 PMID:19447250

Ebi, K. L., Exuzides, K. A., Lau, E., Kelsh, M., & Barnston, A. (2001). Association of normal weather periods and El Niño events with hospitalization for viral pneumonia in females: California, 1983–1998. *American Journal of Public Health*, *91*(8), 1200–1208. doi:10.2105/AJPH.91.8.1200 PMID:11499104

EEA. (2008). *Impact of Europe's Changing Climate—2008 Indicator-based Assessment. Joint EEA-JRC-WHO report*. European Environment Agency.

Epstein, P. R. (1999). Climate and health. *Science*, *285*(5426), 347–348. doi:10.1126cience.285.5426.347 PMID:10438299

Epstein, P. R. (2000). Is global warming harmful to health? *Scientific American*, *283*(2), 50–57. doi:10.1038cientificamerican0800-50 PMID:10914399

Epstein, P. R. (2001). Climate change and emerging infectious diseases. *Microbes and Infection*, *3*(9), 747–754. doi:10.1016/S1286-4579(01)01429-0 PMID:11489423

Epstein, P. R., Diaz, H. F., Elias, S., Grabherr, G., Graham, N. E., Martens, W. J., Mosley Thompson, E., & Susskind, J. (1998). Biological and physical signs of climate change: Focus on mosquito-borne diseases. *Bulletin of the American Meteorological Society, 79*(3), 409–417. doi:10.1175/1520-0477(1998)079<0409:BAPSOC>2.0.CO;2

Fair, J. M., Stokes, M. M., Pennington, D., & Mendenhall, I. H. (2016). Scientific collaborations: How do we measure the return on relationships? *Frontiers in Public Health, 4*, 9. doi:10.3389/fpubh.2016.00009 PMID:26913278

Field, C. B., Barros, V. R., & Dokken, D. J. (2014). *Climate Change 2014: Impacts, Adaptation, and Vulnerability. Part A: Global and Sectoral Aspects. Working Group II Contribution to the Fifth Assessment Report of the Intergovernmental Panel on Climate Change*. Cambridge University Press.

Gerba, C. P. (1999). Virus survival and transplant in groundwater. *Journal of Industrial Microbiology & Biotechnology, 22*(4-5), 535–539. doi:10.1038j.jim.2900663

Guill, C. K., & Shandera, W. X. (2001). The effects of Hurricane Mitch on a community in northern Honduras. *Prehospital and Disaster Medicine, 16*(3), 166–171. doi:10.1017/S1049023X00025929 PMID:11875801

Halstead, S. B. (1996). Human factors in emerging infectious disease. *Eastern Mediterranean Health Journal, 2*(1), 21–29. doi:10.26719/1996.2.1.21

Harvell, C. D., Mitchell, C. E., Ward, J. R., Altizer, S., Dobson, A. P., Ostfeld, R. S., & Samuel, M. D. (2002). Climate warming and disease risks for terrestrial and marine biota. *Science, 296*(5576), 2158–2162. doi:10.1126cience.1063699 PMID:12077394

Hoshen, M. B., & Morse, A. P. (2004). A weather-driven model of malaria transmission. *Malaria Journal, 3*(1), 32. doi:10.1186/1475-2875-3-32 PMID:15350206

IPCC. (2001). Climate Change 2001: Synthesis Report. In A Contribution of Working Groups I, II, and III to the Third Assessment Report of the Integovernmental Panel on Climate Change. Cambridge University Press.

Islam, M. S., Sharker, M. A. Y., Rheman, S., Hossain, S., Mahmud, Z. H., Islam, M. S., Uddin, A. M. K., Yunus, M., Osman, M. S., Ernst, R., Rector, I., Larson, C. P., Luby, S. P., Endtz, H. P., & Cravioto, A. (2009). Effects of local climate variability on transmission dynamics of cholera in Matlab, Bangladesh. *Transactions of the Royal Society of Tropical Medicine and Hygiene, 103*(11), 1165–1170. doi:10.1016/j.trstmh.2009.04.016 PMID:19477477

IWGCCH. (2010). A human health perspective on climate change. In K. T. Tart (Ed.), *A Report Outlining the Research Needs on the Human Health Effects of Climate Change: Environmental Health Perspectives and the National Institute of Environmental Health Sciences*. Author.

Jofre, J., Blanch, A. R., & Lucena, F. (2010). Water-borne infectious disease outbreaks associated with water scarcity and rainfall events. In S. Sabater & D. Barcelo (Eds.), *Water Scarcity in the Mediterranean: Perspectives Under Global Change*. Springer.

Kovats, R. S., Menne, B., McMichael, A. J., Corvalan, C., & Bertollini, R. (2000). *Climate Change and Human Health: Impact and Adaptation*. World Health Organization.

Kuhn, K., Campbell-Lendrum, D., Haines, A., & Cox, J. (2005). *Using Climate to Predict Infectious Disease Epidemics*. World Health Organization.

Li, R., Jiang, Z. B., & Xu, B. (2014). Global spatiotemporal and genetic footprint of the H5N1 avian influenza virus. *International Journal of Health Geographics, 13*(1), 14. doi:10.1186/1476-072X-13-14 PMID:24885233

Lloyd, S. J., Kovats, R. S., & Armstrong, B. G. (2007). Global diarrhoea morbidity, weather and climate. *Climate Research, 34*, 119–127. doi:10.3354/cr034119

Lofgren, E., Fefferman, N. H., Naumov, Y. N., Gorski, J., & Naumova, E. N. (2007). Influenza seasonality: Underlying causes and modeling theories. *Journal of Virology, 81*(11), 5429–5436. doi:10.1128/JVI.01680-06 PMID:17182688

Mellor, P. S., & Leake, C. J. (2000). Climatic and geographic influences on arboviral infections and vectors. *Revue Scientifique et Technique (International Office of Epizootics), 19*(1), 41–54. doi:10.20506/rst.19.1.1211 PMID:11189725

Ostfeld, R. S., & Brunner, J. L. (2015). Climate change and Ixodes tick-borne diseases of humans. *Philosophical Transactions of the Royal Society of London. Series B: Biological Science, 370*(20).

PAHO. (1998). Impact of Hurricane Mitch on Central America. *Epidemiological Bulletin, 19*, 1–14. PMID:10330785

Pampana, E. (1969). *A Textbook of Malaria Eradication* (2nd ed.). Oxford University Press.

Parekh, F. K., Yeh, K. B., Olinger, G., & Ribeiro, F. A. (2020). Infectious disease risks and vulnerabilities in the aftermath of an environmental disaster in Minas Gerais, Brazil. *Vector Borne and Zoonotic Diseases (Larchmont, N.Y.), 20*(5), 387–389. doi:10.1089/vbz.2019.2501 PMID:31944914

Patz, J. A., Epstein, P. R., Burke, T. A., & Balbus, J. M. (1996). Global climate change and emerging infectious diseases. *Journal of the American Medical Association, 275*(3), 217–223. doi:10.1001/jama.1996.03530270057032 PMID:8604175

Sanders, E. J., Rigau-Perez, J. G., Smits, H. L., Deseda, C. C., Vorndam, V. A., Aye, T., Spiegel, R. A., Weyant, R. S., & Bragg, S. L. (1999). Increase of leptospirosis in dengue-negative patients after a hurricane in Puerto Rico in 1996. *The American Journal of Tropical Medicine and Hygiene, 61*(3), 399–404. doi:10.4269/ajtmh.1999.61.399 PMID:10497979

Shaman, J., & Kohn, M. (2009). Absolute humidity modulates influenza survival, transmission, and seasonality. *Proceedings of the National Academy of Sciences of the United States of America, 106*(9), 3243–3248. doi:10.1073/pnas.0806852106 PMID:19204283

Tian, H. Y., Zhou, S., Dong, L., Van Boeckel, T. P., Cui, Y. J., Wu, Y. R., Cazelles, B., Huang, S. Q., Yang, R. F., Grenfell, B. T., & Xu, B. (2015). Avian influenza H5N1 viral and bird migration networks in Asia. *Proceedings of the National Academy of Sciences of the United States of America, 112*(1), 172–177. doi:10.1073/pnas.1405216112 PMID:25535385

Toole, M. J. (1997). Communicable Diseases and Disease Control. In E. Noji (Ed.), *The Public Health Consequences of Disasters*. Oxford University Press.

Viboud, C., Pakdaman, K., Boelle, P.-Y., Wilson, M. L., Myers, M. F., Valleron, A.-J., & Flahault, A. (2004). Association of influenza epidemics with global climate variability. *European Journal of Epidemiology, 19*, 1055–1059. doi:10.100710654-004-2450-9 PMID:15648600

Watts, N., Adger, W. N., Ayeb-Karlsson, S., Bai, Y., Byass, P., Campbell-Lendrum, D., Colbourn, T., Cox, P., Davies, M., Depledge, M., Depoux, A., Dominguez-Salas, P., Drummond, P., Ekins, P., Flahault, A., Grace, D., Graham, H., Haines, A., Hamilton, I., ... Costello, A. (2017). The Lancet Countdown: Tracking progress on health and climate change. *Lancet, 389*(10074), 1151–1164. doi:10.1016/S0140-6736(16)32124-9 PMID:27856085

Watts, N., Amann, M., Ayeb-Karlsson, S., Belesova, K., Bouley, T., Boykoff, M., Byass, P., Cai, W., Campbell-Lendrum, D., Chambers, J., Cox, P. M., Daly, M., Dasandi, N., Davies, M., Depledge, M., Depoux, A., Dominguez-Salas, P., Drummond, P., Ekins, P., ... Costello, A. (2018). The Lancet Countdown on health and climate change: From 25 years of inaction to a global transformation for public health. *Lancet, 391*(10120), 581–630. doi:10.1016/S0140-6736(17)32464-9 PMID:29096948

Wei, T., Yang, S. L., Moore, J. C., Shi, P. J., Cui, X. F., Duan, Q. Y., Xu, B., Dai, Y. J., Yuan, W. P., Wei, X., Yang, Z., Wen, T., Teng, F., Gao, Y., Chou, J., Yan, X., Wei, Z., Guo, Y., Jiang, Y., ... Dong, W. (2012). Developed and developing world responsibilities for historical climate change and CO_2 mitigation. *Proceedings of the National Academy of Sciences of the United States of America*, *109*(32), 12911–12915. doi:10.1073/pnas.1203282109 PMID:22826257

Willox, A. C., Stephenson, E., Allen, J., Bourque, F., Drossos, A., Elgarøy, S., Kral, M. J., Mauro, I., Moses, J., & Pearce, T. (2015). Examining relationships between climate change and mental health in the Circumpolar North. *Regional Environmental Change*, *15*(1), 169–182. doi:10.100710113-014-0630-z

Wu, X. X., Tian, H. Y., Zhou, S., Chen, L. F., & Xu, B. (2014). Impact of global change on transmission of human infectious diseases. *Science China. Earth Sciences*, *57*(2), 189–203. doi:10.100711430-013-4635-0 PMID:32288763

Xu, B., Jin, Z. Y., Jiang, Z. B., Guo, J. P., Timberlake, M., & Ma, X. L. (2014). Climatological and geographical impacts on global pandemic of influenza A (H1N1). In Q. Weng (Ed.), *Global Urban Monitoring and Assessment through Earth Observation*. Taylor & Francis/CRC Press.

Yeh, K., Fair, M., Smith, J., Martinez Torres, W., Lucas, T., Monagin, J., & .... (2020). Assessing climate change impact on ecosystems and infectious disease: Important roles for genomic sequencing and a one health perspective. *Tropical Medicine and Infectious Disease*, *5*, 90. doi:10.3390/tropicalmed5020090 PMID:32503239

Yeh, K. B., Parekh, F. K., Mombo, I., Leimer, J., Hewson, R., Olinger, G., Fair, J. M., Sun, Y., & Hay, J. (2023). Climate change and infectious disease: A prologue on multidisciplinary cooperation and predictive analytics. *Frontiers in Public Health*, *11*, 1018293. doi:10.3389/fpubh.2023.1018293 PMID:36741948

Yeh, K. B., Parekh, F. K., Tabynov, K., Tabynov, K., Hewson, R., Fair, J. M., Essbauer, S., & Hay, J. (2021). Operationalizing cooperative research for infectious disease surveillance: Lessons learned and ways forward. *Frontiers in Public Health*, *9*, 659695. doi:10.3389/fpubh.2021.659695 PMID:34568249

Zell, R. (2004). Global climate change and the emergence/re-emergence of infectious diseases. *International Journal of Medical Microbiology NLM*, *293*, 16–26. doi:10.1016/S1433-1128(04)80005-6 PMID:15146981

Zhou, Y. B., Zhuang, J. L., Yang, M. X., Zhang, Z. J., Wei, J. G., Peng, W. X., Zhao, G. M., Zhang, S. M., & Jiang, Q. W. (2010). Effects of low temperature on the schistosome-transmitting snail *Oncomelania hupensis* and the implications of global climate change. *Molluscan Research*, *30*, 102–108.

Chapter 3
Urbanization and Emergence of Infectious Diseases

Supriya Ray
Karnatak University, India

Venugopal N. Puluamaghatta
Anthropological Survey of India, India

Suresh Basavaraj Arakera
https://orcid.org/0000-0002-6234-666X
Karnatak University, India

ABSTRACT

The process of urbanization is ongoing and continuous since the industrial revolution in the 18th century. The United Nations has estimated the growth of world's urban population to be 6.3 billion by 2050. Rapid influx of migrants in cities led to the development of informal urban settlement which is the initial probable source of emerging communicable diseases. Due to overcrowding, closer contacts with wild and domesticated animals increases the risk of emerging zoonotic diseases. Adequate housing, proper city planning, proper hygiene and sanitation, and surveillance can help in preventing transmission of pathogens, emerging infectious diseases, and decline of ecosystem. This chapter considers the role of urbanization that plays in the emergence of zoonotic diseases and cross-species transmission of pathogens into new host population through exploring the dynamics and complexity of ecological system at wildlife-livestock-human interfaces and the factors affecting the epidemiology of disease emergence.

DOI: 10.4018/978-1-7998-9414-8.ch003

INTRODUCTION

Urbanization is an ongoing process of increased and continuous movement and settlement of people around specific surroundings (Phillips, 1993). Economic growth, higher education, job employment, better health care, and safety of social services around cities are the reasons that alter the demographic and socioeconomic structures of both urban and rural areas (United Nations, Department of Economic and Social Affairs, 2014). The growth of the world's urban population, mainly in Africa and Asia, is expected to be the addition of another 2.5 billion by 2050. The most significant urban growth will likely occur in India, Nigeria, and China (Cohen, 2015). According to the UN development agenda, India, China, and Nigeria are expected to account for 37% of the total growth in urban areas of the world's total population by 2050. India, China, and Nigeria will have an additional 404 million, 292 million, and 212 million urban residents respectively by 2050. With the increase in urban population globally, the world is experiencing an increase in the growth of large cities to 'megacities' with around 453 million inhabitants, covering almost 12% of the world's urban population. Over recent years, megacities like Delhi and Shanghai, with 25 million and 23 million residents respectively, are growing at 3% per annum (Cohen, 2015).

United Nations Development Program (UNDP) reported a continual upsurge in the unequal wealth and resource distribution in both developing and developed countries, causing hindrance to economic development, threatening social harmony, and leading to recessions. Due to income inequality, corruption leads to a decline in environmental regulations, (Wang et al., 2021) causing various types of pollution, including greenhouse gas emissions. Income inequality has varied effects on the societal classes. Due to unequal distribution of resources, the poorest people of society are most affected, leading to the poorest migrating to cities and forcefully adapting an unhygienic lifestyle that not only affects environmental health but also surges the spread of infectious diseases. While protecting environmental degradation and attaining sustainable goals worldwide, climate uncertainty and income inequality are repeatedly featured as obstacles (Kyte, 2014). Besides globalization, climate change, lifestyle changes, unemployment, and psychology are some of the whys behind the mass migration of people from villages to cities. Because of migration, trade, and lifestyle of the people changes. Changes in eating patterns, living in unhygienic conditions, and increased demand for buying and selling products associated with urbanization also may help spread infectious diseases (Li et al., 2022). Urbanization also imparts adverse effects that increase the transmission of pathogens due to various human and social factors such as population size and density, migration, transport hubs, pollution, hygiene and sanitation, climatic factors, socioeconomic inequalities, and urban inequity (Boyce et al., 2019). The rise of new modern cities acts as a catalyst that creates danger and threats leading to more infectious diseases (World Health Organization, WHO, 2013). The risk factors include poor housing that offers favourable conditions for the multiplication of rodents and insects, leading to more vector-borne diseases and helminth infections. The lack of ventilation systems in buildings can spread respiratory tract infections. Poor storage facilities for food also can spread infection due to zoonoses and microbial toxins (WHO, 2010). The rise of the severe acute respiratory syndrome (SARS), an infectious disease, was spreading fast in densely populated urban areas. (WHO, 2008)

International trade and tourism are the other reasons behind the rise and spread of infectious diseases. The travel and tourism industry for many urban-centric countries with favourable climate conditions holds around 10.2% of the World's Gross Domestic Product, which is high compared to less developed countries. Tourism creates a chief foreign income and exchange for these countries. However, favourable climate conditions allow pathogens to grow and spread infectious diseases. The rise of international tourist flows echoes the fast movement of the population, which may create an increased risk of infectious diseases such as SARS, COVID-19, Yellow fever, and Ebola virus. Poor hygiene and sanitation facilities, poor socioeconomic conditions, and different cultural practices contribute to spreading diseases. Recent outbreaks of the severe acute respiratory syndrome (SARS) and the COVID-19 pandemic have significantly impacted health policy debates. Besides, SARS and COVID-19, Malaria, Dengue, Yellow fever, and Ebola are the most dominant pathogens among ill-returned tourism passengers. Many large populations are associated with the international trading of urban-centric goods. The risk of catching any infectious diseases among travellers and traders depends upon several factors such as season, immunization status, travel location, rate of virus transmission locally at the time of travel, length of virus exposure, several occupational activities, trading of product activities and recreational activities (Rosselló et al., 2017).

To understand the condition where urbanization can lead to emerging zoonotic diseases spreading across species, it is crucial to reduce the convolution of the metropolitan structure by understanding the network interfaces like physical interfaces where both human and animal interaction leads to the exchange of pathogens (Alberti, 2005; Pickett et al., 2011). Although it can be considered that the original sources of emerging infectious diseases are reservoirs in wildlife and domesticated animals, but it is the anthropogenic effect on ecological systems that determine risk levels at the human and animal interface in zoonotic disease emergence (Hassell et al., 2017). Synanthropic species are referred to as urban adapted wildlife species, sometimes acting as a reservoir host for pathogens and leading to the development of zoonotic diseases (Deplazes et al., 2004; Wright & Gompper, 2005).

Studies have estimated that 80% of lately infectious emerging diseases are zoonotic in ancestry, i.e., their survival depends on animal reservoir (Jones et al., 2008;

Woolhouse & Gowtage-Sequeria, 2005), and 70% of these have a wildlife ancestry (Karesh et al., 2012; Jones et al., 2008). The major risk factors for transmission in urban environments are wet markets that sell freshly slaughtered wild animals (e.g. bats, rodents, wild as well as domesticated birds and other mammal species that are hosts for zoonotic diseases). That introduces novel pathogens, creating animal/human interfaces that help develop cross-species transmission of infectious diseases between humans and animals (Guan et al., 2003; WHO, 2020). Rodents nurture zoonoses that include hantavirus infection, plague, and leptospirosis. Therefore, the re-emergence of such zoonotic pathogens in humans is linked to the increase in urbanization in emergent nations and the prevalence of zoonotic infectious parasites or pathogens in rodent inhabitants (Himsworth et al., 2013; Guan et al., 2009; Evangelista & Coburn, 2010). Anthropogenic changes can directly link with urbanization bringing the bat population in direct contact with humans-livestock and changing the bionomics of diseases. Human activities increase exposure to wildlife species' urban habitats, further increasing the risk of spreading pathogen to livestock and humans (Pulliam et al., 2012; Plowright et al., 2011). If these conditions continue, the ecosystem cannot sustain urbanization in the long run, leading to the decline of the ecosystem quality with natural habitat loss.

DYNAMICS OF THE INTERFACE BETWEEN WILDLIFE, LIVESTOCK, AND HUMAN

Most of the emerging infectious pathogens circulate in communities comprised of hosts that are infected with various pathogens, and can infect multiple varieties of hosts (Hall et al., 2005; Telfer et al., 2010; Tomczuk et al., 2015). Several epidemiological studies have been done on zoonotic diseases, including Lyme disease (Keesing et al., 2016), Nipah virus (Pulliam et al., 2012; Daszak et al., 2006; Plowright et al., 2015), Cowpox virus (Begon et al., 2009) and Hendra virus (Pulliam et al., 2012; Daszak et al., 2006; Plowright et al., 2015).

Wildlife exists as maintenance or non-maintenance host within a community in a multihomed host-pathogen structure. The kinetics of a zoonosis pathogen involves two-aspect (i) transmission of the pathogen between the maintenance host or non-maintenance host and (ii) spillover transmission (Caron et al., 2015; Viana et al., 2014). The basic reproduction number (R_0) is a pathogen's transmission capability that reflects the host's tenacity to sustain the zoonotic pathogen by acting as a maintenance community. R_0 is associated with the degree of contact between an infectious and affected person. The mortality rate or recovery rate, as well as human spillover transmission from an individual, is regulated by the recurrence of infection in the maintenance or non-maintenance host. The mortality rate of an infected person and the force of infection depends on the degree of contact between contaminated individuals and humans. Therefore, contact is a major factor in disease emergence and reservoir kinetics (Davis et al., 2005; Begon et al., 2002; Lloyd-Smith et al., 2009).

Land-use changes are human-induced ecosystem changes that can function through several mechanisms inducing environmental stressors influencing the distribution and abundance of wildlife, the release of pathogens into the environment, creating novel pathogens to spread between species, shaping the dynamics of exposure of wildlife and vulnerability to pathogens that can further lead to spreading of human infection. Habitat fragmentation, degradation, and consequences lead to changes in ecological structure, affecting resource distribution. This has led to the movement of wildlife near human habitations, altered population structure, and change in food webs. A decrease in the natural resource can also lead to the introduction of biological toxicants and other pollutants to the environment, wildlife will serve the host to infectious agents, and higher susceptibility to infection. Illegal wildlife trade and improper management of habitats force the wild animals hosting pathogens to come nearby to humans. As the environmental condition changes that improve the pathogen or vector survival, humans get exposed to the pathogen or vector, leading to increased human susceptibility to infection. Increased crowding and movement of people, products, and services cause various infectious diseases, leading to epidemics and pandemics (Plowright et al., 2021). An increase in population and host ecological traits plays a key role in transmitting pathogens in wildlife and livestock species. The risk of disease emergence by zoonotic pathogen at each particular interface may depend on the influence of urbanization on coevolution and community ecology linkages between humans, wildlife, and livestock (Hayman et al., 2013).

To understand the kinetics of interspecies transmission at an urban juncture, abiotic factors must be considered as they influence the biotic slot where both host type and infectious agent exist. Abiotic niches allow host perseverance within a definite geographical radius, whereas the biotic niches are identified by the inhibitory effect of other species on survival (Estrada-Peña et al., 2014). Biotic factors are scale-dependent, and the epidemiological and ecological processes can also occur at different scales (Himsworth et al., 2013). Understanding the part of the geographical distribution of pathogen transmission has led to the emergence of landscape epidemiology (Meentemeyer et al., 2012).

In a hypothetical condition, the interface of wildlife-livestock-human can create conditions where a cascade of biotic and abiotic changes makes the situation suitable for emerging infectious diseases at different scales. At multiple scales, abiotic factors, i.e., climate, pollution, and habitat alteration within the communities, impede the endurance of new and extant species and hosts' fitness. Abiotic changes can directly influence antimicrobial resistance (Olson et al., 2013; Wellington et al., 2013), while physiological, ecological, and genetic similarities play essential parts in the interactions of host and infectious agents (Pilosof et al., 2014; Dobson, 2004). Competition or predation, a direct ecological interaction, will alter the population dynamics and the host reservoir's dispersal. Hence, such factors sway the interconnection rate within the source-island distance, source areas, and island areas (Begon et al., 2009; Poulin, 2004). Association of these factors with spillover to a host like wildlife or humans relies upon the strength of infection. When the changes are induced by human action, increased imbrication between wildlife, society, and livestock is possible though not enough to secure interspecies transmission. The spillover can be subdued if a human is the target host by certain variables like economic status, social structure, living conditions, health, and behaviour. From a genetic context of a pathogen, a leap of an infectious agent into a new host species requires adaptive or non-adaptive changes, including ecological fitting (Pepin et al., 2010). However, the trait of the pathogen and the host pool dictates the infection of new hosts by either an adaptive or non-adaptive process.

Certain pathogens like RNA viruses adapt fast and take advantage of the latest epidemiological events than other pathogens due to higher mutation rate (Woolhouse et al., 2005; Johnson et al., 2015), while for other pathogens, the reservoir's structural properties play a crucial role in facilitating ecological fitting (Johnson et al., 2015). Successful host jumps are also determined by the phylogenetic distance between reservoir hosts and humans. Some species belonging to some phylogenetic clades contain different cellular contents that make them liable for the invasion of pathogens (Johnson et al., 2015; Longdon et al., 2011).

A study showed that the mutation rate increases in a higher density of sensitive hosts, increasing the occurrence of viral host jumps. In contrast, interspecies transmission at higher levels can lead to the acceptance of more generalized strategies for the virulent pathogen (Ellis & Ramankutty, 2008). Anthropogenic activities lead to the segmentation of natural biomes, forming a mixture of different habitats. The fragments of this original biome are considered as patches that exist, unlike in the matrix of original habitats (Ellis & Ramankutty, 2008; Reisen, 2010; Wilcove et al., 1986). The interfaces between the original biome patches and the overall matrix are classified as ecotones at local scales (Despommier et al., 2006). Studies showed that expanding ecotonal areas can alter the niche of the pathogen by interactions with humans, wildlife, and domesticated animals, thus getting infected and increasing the possibility of transmission through contact (Reisen, 2010). Rodents undergoing ecological releases at forest interfaces can be drawn to farmlands and human settlements for resources and human settlements may provide suitable breeding habitats for vectors such as mosquitoes (Despommier et al., 2006; Brown et al., 2008). Certain documented evidence shows the interconnection between the emergence of diseases and ecotones for various zoonoses with wildlife reservoirs such as Lyme disease, *Escherichia coli* infection, yellow virus, rabies, Nipah virus encephalitis, cholera, hantavirus pulmonary syndrome, and influenza (Eisen et al., 2007; Hahn et al., 2014). Therefore, ecotones exemplify local scale interfaces where the zoonotic agents circulate and infect human, domestic, and wildlife animals (Despommier et al., 2006; Goldberg et al., 2008; Eisen et al., 2007; Hahn et al., 2014).

Resource provisioning is a crucial factor that significantly impacts urban surroundings concerning wildlife-livestock-human contact and interspecific wildlife contacts (Deplazes et al., 2004; Wright & Gompper, 2005). Keeping livestock informally in African cities is common, and the mixture of the livestock species is usually kept near humankind. In Asia, the Nipah virus and pathogenic avian H5N1 influenza virus are instances of zoonotic emergence events (Daszak et al., 2006; Sims et al., 2005). Bovine tuberculosis and hepatitis E, stable zoonoses, implies that livestock acts as a bridge host, thus linking humans and wildlife epidemiologically (Daszak et al., 2006; Sims et al., 2005; Meng et al., 2009).

Resource clustering occurs widely at landscape and local scales across urban environments because of either alteration in sanitation, feeding garden birds, livestock keeping practices, usage of agricultural by-products, or household food availability. *Pteropus* spp of bats, also known as flying foxes, has shown to increase in large urban colonies co-existing with horse and human populations in Eastern Australia due to decline in food resources in urban gardens (Bradley & Altizer, 2007; Jobbins et al., 2014; Lawson et al., 2014; Swirski et al. 2014; Becker et al., 2015). The increasing contact between humans and horses could upsurge the risk of Hendra virus emergence in humans and horses (Plowright et al., 2011). Studies have shown that certain urban environmental stressors like pesticide pollutants and heavy metals affect hosts' immunological function, which will have a long-term prevalence (Padgett & Glaser, 2003).

Jones et al. (2013) developed a conceptual structure for human, livestock, and wildlife interfaces in an emergent nation like Nairobi. He described that in metropolitan ecotonal interfaces and fragmented natural bio-diversities such as urban wetlands, forests, and agricultural edges, contact between wildlife, humans, and livestock, both domesticated and undomesticated, is increasing. In evolving urban landscapes such as backyard farming and casual refuse dumps, there is high connectivity between humans, wildlife, and livestock. However, in maintained urban landscapes such as sewage plants, there is little and controlled connection between wildlife, humans, and livestock. Wildlife habitats in recreational areas such as parks and gardens have less contact between humans, livestock, and wildlife, but there is an increase in contact between humans and domesticated wildlife (Jones et al., 2013).

Urban Factors Affecting the Epidemiology of Emerging Diseases

The most important factor affecting epidemiology is population density. The rise of migration of people to cities results in overcrowding, leading to informal urban settlements, shanty towns or slums. These slums having poor housing, poor sanitation facilities, and no fresh water, can pose a threat to the resident's /neighbour's health and could be a reproducing site for the surge of contagious diseases like Tuberculosis (TB), as seen in Dhaka, Bangladesh, or cholera as happened in Dar es Salaam, Tanzania (WHO, 2010; Banu et al., 2013; Penrose et al., 2010). Often the position of slums is farther from the city in a more perilous place where these populations experience a severe lack of economic and social opportunities. In 2012, 62% of the sub-Saharan African urban people lived in shantytowns (WHO, 2010). In 2009, 96% of the urban population in the Central African Republic lived in the slums (World Urban Forum, 2012). Places like Nairobi have about 60% of their population living in slums, with an infant mortality rate double that of other cities (WHO, 2019). A high prevalence of antimicrobial drug-resistant diarrhoeagenic, asymptomatic carriers of *Escherichia coli* has been found in Brazil's children of slums (Souza et al., 2009). The high prevalence of Tuberculosis in rural slum areas of Dhaka city, Bangladesh, is two-fold higher in contrast to the nationwide average and four-fold higher compared to the altogether urban score (Banu et al., 2013). Nevertheless, some countries showed different patterns, such as in Poland, the rate of Tuberculosis incidence is lower in rural compared to the urban population (Korzeniewska-Kosela, 2013). Factors like improved nutritional status, community health and hygiene, and socioeconomic status in the United States have shown to decline in Tuberculosis in the twentieth century (Burzynski & Schlunger, 2008).

The rise of urbanization is throwing challenges to proper city planning. Due to the rapid migration, the local municipal authorities may be unable to supply adequate sewage facilities, proper sanitation, safe housing, and freshwater, and lack of such amenities can give rise to diseases such as soil-transmitted helminthiases, trachoma, and schistosomiasis (Mara et al., 2010). Such infections cause mental and physical growth retardation and greatly influence public health and hygiene (Bethony et al., 2006). More than half of the world's populations have piped water connections to their homes. Over 2 billion individuals have access to better drinking water facilities, hygiene, and sanitation. Nevertheless, over 700 million people still face an acute shortage in access to better drinking water facilities, hygiene, and sanitation. Although open defecation decreased from 24 to 14% globally, almost a billion still go for open defecation. The population, which lacked all the sanitation and drinking water facilities, between 1990 and 2012 significantly increased from 215 to 756 million, a steep growth (WHO, 2014). Populations lacking basic facilities and sanitation are highly susceptible to soil-transmitted helminthiases (de Silva et al., 2003). However, good hygiene, sanitation, and cleanliness practices can prevent the proliferation of contagious diseases, as seen in Salvador, a Brazilian city, where sewage coverage in all households reduced diarrhoeal diseases by 22% (Barreto et al., 2007). Certain tropical diseases in developing countries have a faecal-oral transmission pathway, a severe threat to health. Such diseases are soil-transmitted helminthiases, trachoma, and schistosomiasis, and proper sanitation could certainly improve public health. Nonetheless, in many countries, the focus is on medication rather than improving cleanliness, sanitation, and hygiene practices. Therefore, safe and secure drinking water, cleanliness, proper hygiene, and sanitation must be considered in city planning (Mara et al., 2010).

Living in close contact with domestic livestock and poor hygienic conditions can be a risk factor for causing Chagas disease (i.e., vector proliferation) and parasitic infection caused by *Trypanosoma cruzi* (Ventura-Garcia et al., 2013). The Chagas disease poses serious health challenges affecting almost eight million people yearly in Latin America (Rassi et al., 2010). The most affected countries by this disease are Spain and Europe, where the cases range from 45,000 to 67,000 (Gascon et al., 2010).

The urban environment favours the rat population, facilitating close contact between humans and rats and possibly zoonotic transmission. Studies showed that rats carry pathogens like *Streptobacillus moniliformis, Angiostrogylus cantonensis, Leptospira spp.,* Seoul hantavirus, *Rickettsia typhi, Yersinia pestis,* and *Bartonella spp* (Himsworth et al., 2013). Close contact between humans and rats has been associated with juxtaposed open free public spaces (Wash, 2014). Leptospirosis was previously known as a rural disease, but increased urbanization has also led this disease to enter urban centres (Ko et al., 1999; Evangelista & Coburn, 2010). Seoul hantavirus causing haemorrhagic pyrexia with the nephritic syndrome in Chinese cities is due to increased urban expansion, increased rodent population, and increased human-rodent contact (Guan et al., 2009). Better knowledge and understanding of preventing the increased rat population can decline infectious zoonotic diseases (Guan et al., 2009).

Urbanization leading to a lack of hygiene and sanitation has turned traditional rural infectious diseases into urban diseases. According to WHO, 17 neglected tropical diseases have entered the urban environment (WHO, 2013). One such tropical neglected disease is lymphatic filariasis (LF), which has a significant impact on rural surroundings. But due to growing urbanization, lymphatic filariasis infections can now be found in urban areas, with 394.5 million people infected and 2 billion people at risk globally. Increasing lymphatic filariasis is due to improper hygiene and sanitation facilities (Erlanger et al., 2005). One parasite species, *Wuchereria bancrofti* and *Culex quinquefasciatus* mosquito, the sole vector, mainly develop in overcrowded areas and places where e drainage and sanitation facilities are poor. The

Culex spp. is found in Asia, East Africa, and parts of Central and South America (Simonsen & Mwakitalu, 2013). Another neglected tropical disease on WHO's list is Dengue. *Aedes aegypti* mosquito, the vector that transmits the diseases, lays eggs in urban environments. Lack of mosquito control, sanitation, poor drainage, and globalisation are several elements that have played a role in the upsurge of transmissible diseases (Gubler, 2011). The number of infections is rising worldwide and has significantly increased in tropical regions over the last 40 years. The latest projection per year is 390 million cases and a high rise in new infections in India (Bhatt et al., 2013).

Dengue has become a global problem; despite better knowledge of its transmission, it seems difficult to control it since vectors and people live in close contact in urban areas. All four serotypes of the dengue virus are found in Thailand. The outbreak of haemorrhagic pyrexia linked with Dengue is known to be originated in Bangkok and then spread to the rural provinces (Cummings et al., 2004). Certain aspects must be considered to influence the urban environment for vector-borne diseases to complete the transmission cycle. One vector-borne infection that has proliferated in metropolitan surroundings is West Nile Virus infection (WNV). The primary vector responsible for this infection is the *Culex pipens* mosquito, which lays eggs in artificial containers made by humans. For a successful transmission cycle, West Nile Virus also needs avian hosts such as *Turdus migratorius* (American robin) (Savage et al., 2008). An epidemic of 25% of WNV infection was reported in Dallas, Texas, in the United States in 2012 (Haley, 2012).

Another vector-borne disease is Leishmaniasis, caused by *Leishmania* protozoa. Leishmaniasis is transmitted by phlebotomine sand flies vector. This transmission happens when rural people migrate with their domesticated animals to urban surroundings, generating a favourable condition for transmitting vector-borne infectious diseases to the urban environment (Desjeux, 2001). Leishmaniasis affects almost 12 million people in 88 countries (Kedzierski, 2011). Widespread urbanization has led to increased transmission of infectious zoonotic diseases, which in turn increases public health problems (Ashford, 2000). Control programs, better knowledge, and better facilities must be implemented in growing cities, although it would be tough for slums and shantytowns to take such measures.

International travel generates a possible menace to large numbers of evolving contagious infections. World Tourism Organization (WTO) forecasted the latest increase in international tourism arrivals and predicted that by 2030 the international tourism arrivals would be about 1.8 billion (World Tourism Organisation, 2014). Overseas trade and exploration catalyze the manifestation of a global pandemic, as seen in recent events of SARS coronavirus in 2019. Previously, SARS also emerged in 2003 as a universal menace. It was thought that SARS originated from Bat SARS-like coronavirus and transmitted to human hosts during the hunting and trading of bats for food in wet markets in China (Li et al., 2005). SARS was initially reported in wet live animal markets for wildlife in China in Guandong. The initial cases involved the individuals handling and preparing exotic dishes with these animals. Studies seem to have concluded that the virus might have crossed over to the human host (WHO, 2013). SARS then spread out worldwide through international travel. It spread from urban cities to rural areas and led to a halt in travelling, which caused enormous commercial losses that hurt many nations worldwide (WHO, 2007). Therefore, to avoid such circumstances in the near future, it is essential to control and prevent the epidemic before it gets out of control.

Dengue is now a more frequent illness and affects the urban populations than malaria. Dengue is considered endemic in most subtropical and tropical regions, which are tourist destinations and travellers contribute to the spread (Wilder-Smith, 2012). *Staphylococcus aureus*, an antibiotic-resistant microbe spread worldwide, turned out to be associated with travel and touristry and greatly affects overall overseas health (Chroboczek et al., 2013; Tristan et al., 2007). Besides, research has linked *Enterobacteriaceae* to travellers, which greatly impacts international health (Ostholm-Balkhed et al., 2013; Ta¨ngde´n et al., 2010; Kennedy & Collignon, 2010). While it is crucial that preventive and control measures should be taken at the site of occurrence, it is also essential that proper knowledge of the particular disease is circulated among the local health workers and the global community.

CONCLUSION AND FUTURE CHALLENGES

In this chapter, the authors have considered the role of urbanization in zoonotic disease emergence while investigating the convolution of human-livestock-wildlife interfaces. We argue that interfaces should be regarded as the key component in altering the urban surroundings and the onset of diseases. Uncontrolled urbanization is rising globally, changing the cities' living conditions, resulting in the risk of emerging infectious diseases that will significantly affect human well-being globally. Close contact with wildlife could be an origin of infectious zoonosis diseases. Better housing, proper hygiene and sanitation, proper city planning, and better surveillance can help monitor and control the transmission of pathogens. Most studies use elementary concepts to study a small number of single pathogen and host species when exploring the transmission of diseases within multi-host systems. The disclosure of pathogen development surrounded by a convoluted system might get eclipsed if the emphasis is on single host species and a single pathogen. Therefore, new approaches need to be considered to understand the underlying mechanism of disease emergence. Ecologists have adapted new methods such as network theory and trait-based approaches. The development of such new techniques focuses on examining disease emergence's causation.

To study epidemiology within environmental networks, the application of network theory assures us to concentrate on the limitations of traditional approaches, which are utilized to apprehend the action of pathogens in communities belonging to hosts, such as experimental studies and mathematical modelling. Epidemiological networks can be applied to understand the factors that affect the transmission of pathogen and spread of the pathogen through the community since the structural networks within a multi-host community remain uncharted.

Population genetics or gene dispersal measures allocate genetic variation between different populations and explain the connectivity between humans, livestock, and wildlife. Phylogenetic approaches aim to study the evolutionary relationships among genetic pedigrees, and through pathogen sequence data, it can also be used to reform the epidemiological histories. Given the data such as host and date of the sequenced data, it can elucidate transmission pathways, populations of origin, and past events of host switching over a long period. Studies need to focus on population genetic and phylogenetic trait mapping approaches to study the epidemiological structures of multihost wildlife communities that can depict the epidemiological connectivity at an intercommunity, individual, and intracommunity scale since that area remains largely un-investigated. There are still some unresolved questions on the environment needed for the infectious agent to be injected into a novel host; what is pushing (the driven force) in creating these interfaces? Also, according to metropolitan land-tenure change, how does the pathogen pool in wildlife species vary? These areas, still needing some light, can contribute to developing better prevention control strategies.

Zoonotic diseases have become a challenge for the future. Between 1940 and 2004, more than 60% of 335 contagious infections have been zoonosis diseases. Deforestation is a factor that makes contact between humans and animals like bats or even primates, potential hosts for new viruses, closer to each other. The worldwide migration trend to cities causes urbanization to push the population into the previously unscathed ecosystems. The global trend of humanity migrating to towns is inflating, causing more intrusions to the natural ecosystems and encounters with newer pathogens that are causing harm to the urban health environments. In 2014, the world witnessed the Ebola virus, a zoonotic disease epidemic. Countries in Western Africa, i.e., Guinea, Liberia, and Sierra Leone, were the infestation hubs. The mortality rate was up to 90%, and the fatality rate of Western Africa was 70%. Closer contact of humans with fruit bats released the threat of Ebola Virus Disease, creating a global health crisis. Specific pathogens adapt to different conditions and continue to grow, creating a challenge for city governments and nations worldwide to prevent infectious disease transmission. Surveillance, proper knowledge of the public, resources and control programs are of primary importance for local government authorities and the global community to monitor and prevent the spread of diseases and for a quick response to public health threats.

REFERENCES

Alberti, M. (2005). The Effects of Urban Patterns on Ecosystem Function. *International Regional Science Review*, *28*(2), 168–192. doi:10.1177/0160017605275160

Ashford, R. W. (2000). The leishmaniases as emerging and reemerging zoonoses. *International Journal for Parasitology*, *30*(12-13), 1269–1281. doi:10.1016/S0020-7519(00)00136-3 PMID:11113254

Banu, S., Rahman, M. T., Uddin, M. K., Khatun, R., Ahmed, T., Rahman, M. M., Husain, M. A., & Leth, F. V. (2013). Epidemiology of Tuberculosis in an urban slum of Dhaka City, Bangladesh. *PLoS One*, *8*(10), 1–8. doi:10.1371/journal.pone.0077721 PMID:24204933

Barreto, M. L., Genser, B., Strina, A., Teixeira, M. G., Assis, A. M., Rego, R. F., Teles, C. A., Prado, M. S., Matos, S. M., Santos, D. N., dos Santos, L. A., & Cairncross, S. (2007). Effect of city-wide sanitation program on reduction in the rate of childhood diarrhea in northeast Brazil: Assessment by two cohort studies. *Lancet*, *370*(9599), 1622–1628. doi:10.1016/S0140-6736(07)61638-9 PMID:17993362

Becker, D. J., Streicker, D. G., & Altizer, S. (2015). Linking anthropogenic resources to wildlife-pathogen dynamics: Are view and meta-analysis. *Ecology Letters*, *18*(5), 483–495. doi:10.1111/ele.12428 PMID:25808224

Begon, M., Bennett, M., Bowers, R. G., French, N. P., Hazel, S. M., & Turner, J. (2002). A clarification of transmission terms in host-microparasite models: Numbers, densities, and areas. *Epidemiology and Infection*, *129*(1), 147–153. doi:10.1017/S0950268802007148 PMID:12211582

Begon, M., Telfer, S., Smith, M. J., Burthe, S., Paterson, S., & Lambin, X. (2009). Seasonal host dynamics drive the timing of recurrent epidemics in a wildlife population. *Proceedings of the Royal Society B*, *276*(1662), 1603-1610. 10.1098/rspb.2008.1732

Bethony, J., Brooker, S., Albonico, M., Geiger, S. M., Loukas, A., Diemert, D., & Hotez, P. J. (2006). Soil-transmitted helminth infections: Ascariasis, trichuriasis, and hookworm. *Lancet*, *367*(9521), 1521–1532. doi:10.1016/S0140-6736(06)68653-4 PMID:16679166

Bhatt, S., Gething, P. W., Brady, O. J., Messina, J. P., Farlow, A. W., Moyes, C. L., Drake, J. M., Brownstein, J. S., Hoen, A. G., Sankoh, O., Myers, M. F., George, D. B., Jaenisch, T., Wint, G. R., Simmons, C. P., Scott, T. W., Farrar, J. J., & Hay, S. I. (2013). The global distribution and burden of Dengue. *Nature*, *496*(7446), 504–507. doi:10.1038/nature12060 PMID:23563266

Boyce, M. R., Osman, R., & Katz, R. (2019). A research agenda for urban pandemic preparedness: a call to action. Washington: Center for Global Health Science & Security, Georgetown University.

Bradley, C. A., & Altizer, S. (2007). Urbanization and the ecology of wildlife diseases. *Trends in Ecology & Evolution*, *22*(2), 95–102. doi:10.1016/j.tree.2006.11.001 PMID:17113678

Brown, H. E., Duik-Wasser, M., Andreadis, T., & Fish, D. (2008). Remotely-sensed vegetation indices identify mosquito clusters of West Nile virus vectors in an urban landscape in the northeastern United States. *Vector Borne and Zoonotic Diseases (Larchmont, N.Y.)*, *8*(2), 197–206. doi:10.1089/vbz.2007.0154 PMID:18452400

Burzynski, J., & Schlunger, N. W. (2008). The epidemiology of Tuberculosis in the United States. *Seminars in Respiratory and Critical Care Medicine*, *29*(5), 492–498. doi:10.1055-0028-1085701 PMID:18810683

Caron, A., Cappelle, J., Cumming, G. S., Garine-Wichatitsky, M. D., & Gaidet, N. (2015). Bridge hosts, a missing link for disease ecology in multi-host systems. *Veterinary Research*, *46*(1), 1–11. doi:10.118613567-015-0217-9 PMID:26198845

Chroboczek, T., Boisset, S., Rasigade, J. P., Meugnier, H., Akpaka, P. E., Nicholson, A., Nicolas, M., Olive, C., Bes, M., Vandenesch, F., Laurent, F., Etienne, J., & Tristan, A. (2013). Major West Indies MRSA clones in human beings: Do they travel with their hosts? *Journal of Travel Medicine*, *20*(5), 283–288. doi:10.1111/jtm.12047 PMID:23992570

Cohen, B. (2015). Urbanization, City growth, and the New United Nations development agenda. *Cornerstone*, *3*(2), 4–7.

Cummings, D. A., Irizarry, R. A., Huang, N. E., Endy, T. P., Nisalak, A., Ungchusak, K., & Burke, D. S. (2004). Traveling waves in the occurrence of Dengue hemorrhagic fever in Thailand. *Nature*, *427*(6972), 344–347. doi:10.1038/nature02225 PMID:14737166

Daszak, P., Plowright, R. K., Epstein, J. H., Pulliam, J., Rahman, S. A., Field, H. E., Jamaluddin, A., Sharifah, S. H., Smith, C. S., Olival, K. J., Luby, S., Halpin, K., Hyatt, A. D., & Cunningham, A. A. (2006). Disease Ecology: Community Structure and Pathogen Dynamics. In S. Collinge & C. Ray (Eds.), *The emergence of Nipah and Hendra virus: pathogen dynamics across a wildlife-livestock-human continuum* (pp. 186–201). Oxford University Press.

Davis, S., Calvet, E., & Leirs, H. (2005). Fluctuating rodent populations and risk to humans from rodent-borne zoonoses. *Vector Borne and Zoonotic Diseases (Larchmont, N.Y.)*, *5*(4), 305–314. doi:10.1089/vbz.2005.5.305 PMID:16417426

de Silva, N. R., Brooker, S., Hotez, P. J., Montresor, A., Engels, D., & Savioli, L. (2003). Soil-transmitted helminth infections: Updating the global picture. *Trends in Parasitology*, *19*(12), 547–551. doi:10.1016/j.pt.2003.10.002 PMID:14642761

Deplazes, P., Hegglin, D., Gloor, S., & Romig, T. (2004). Wilderness in the city: The urbanization of Echinococcus multilocularis. *Trends in Parasitology*, *20*(2), 77–84. doi:10.1016/j.pt.2003.11.011 PMID:14747021

Desjeux, P. (2001). The increase risk factors for Leishmaniasis worldwide. *Transactions of the Royal Society of Tropical Medicine and Hygiene*, *95*(3), 239–243. doi:10.1016/S0035-9203(01)90223-8 PMID:11490989

Despommier, D., Ellis, B. R., & Wilcox, B. A. (2006). The role of ecotones in emerging infectious diseases. *EcoHealth*, *3*(4), 281–289. doi:10.100710393-006-0063-3

Dobson, A. (2004). Population dynamics of pathogens with multiple host species. *American Naturalist*, *164*(S5), 64–78. doi:10.1086/424681 PMID:15540143

Eisen, R. J., Glass, G. E., Eisen, L., Cheek, J., Enscore, R. E., Ettestad, P., & Gage, K. L. (2007). A spatial model of shared risk for plague and hantavirus pulmonary syndrome in the southwestern United States. *The American Journal of Tropical Medicine and Hygiene*, *77*(6), 999–1004. doi:10.4269/ajtmh.2007.77.999 PMID:18165511

Ellis, E. C., & Ramankutty, N. (2008). Putting people on the map: Anthropogenic biomes of the world. *Frontiers in Ecology and the Environment*, *6*(8), 439–447. doi:10.1890/070062

Erlanger, T. E., Keiser, J., Caldas De Castro, M., Bos, R., Singer, B. H., Tanner, M., & Utzinger, J. (2005). Effect of water resource development and management on lymphatic filariasis, and estimates of population at risk. *The American Journal of Tropical Medicine and Hygiene*, *73*(3), 523–533. doi:10.4269/ajtmh.2005.73.523 PMID:16172476

Estrada-Peña, A., Ostfeld, R. S., Peterson, A. T., Poulin, R., & Fuente, J. D. L. (2014). Effects of environmental change on zoonotic disease risk: An ecological primer. *Trends in Parasitology*, *30*(4), 205–214. doi:10.1016/j.pt.2014.02.003 PMID:24636356

Evangelista, K. V., & Coburn, J. (2010). Leptospira as an emerging pathogen: A review of its biology, pathogenesis, and host immune responses. *Future Microbiology*, *5*(9), 1413–1425. doi:10.2217/fmb.10.102 PMID:20860485

Gandon, S. (2004). Evolution of multi-host parasites. *Evolution; International Journal of Organic Evolution*, *58*(3), 455–469. PMID:15119430

Gascon, J., Bern, C., & Pizano, M. J. (2010). Chagas disease in Spain, the United States, and other non-endemic countries. *Acta Tropica*, *115*(1-2), 22–27. doi:10.1016/j.actatropica.2009.07.019 PMID:19646412

Goldberg, T. L., Gillespie, T. R., Rwego, I. B., Estoff, E. L., & Chapman, C. A. (2008). Forest fragmentation as the cause of bacterial transmission among non-human primates, humans, and livestock, Uganda. *Emerging Infectious Diseases*, *14*(9), 1375–1382. doi:10.3201/eid1409.071196 PMID:18760003

Guan, P., Huang, D., He, M., Shen, T., Guo, J., & Zhou, B. (2009). Investigating the effects of climatic variables and reservoir on the incidence of hemorrhagic fever with renal syndrome in Huludao City, China: a17-year data analysis based on structural equation model. *BMC Infectious Diseases*, *9*(109), 109. doi:10.1186/1471-2334-9-109 PMID:19583875

Guan, Y., Zheng, B. J., He, Y. Q., Liu, X. L., Zhuang, Z. X., Cheung, C. L., Luo, S. W., Li, P. H., Zhang, L. J., Guan, Y. J., Butt, K. M., Wong, K. L., Chan, K. W., Lim, W., Shortridge, K. F., Yuen, K. Y., Peiris, J. S., & Poon, L. L. (2003). Isolation and characterization of viruses related to the SARS coronavirus from animals in southern China. *Science*, *302*(5643), 276–278. doi:10.1126cience.1087139 PMID:12958366

Gubler, D. J. (2011). Dengue, urbanization, and globalization: The unholy trinity of the 21st century. *Tropical Medicine and Health*, *39*(4), 3–11. doi:10.2149/tmh.2011-S05 PMID:22500131

Hahn, M. B., Gurley, E. S., Epstein, J. H., Islam, M. S., Patz, J. A., Daszak, P., & Luby, S. P. (2014). The role of landscape composition and configuration on Pteropus giganteus roosting ecology and Nipah virus spillover risk in Bangladesh. *The American Journal of Tropical Medicine and Hygiene*, *90*(2), 247–255. doi:10.4269/ajtmh.13-0256 PMID:24323516

Haley, R. W. (2012). Controlling urban epidemics of West Nile virus infection. *Journal of the American Medical Association*, *308*(13), 1325–1326. doi:10.1001/2012.jama.11930 PMID:22922679

Hall, S. R., Duffy, M. A., & Caceres, C. E. (2005). Selective predation and productivity jointly drive complex behavior in host-parasite systems. *American Naturalist*, *165*(1), 70–81. doi:10.1086/426601 PMID:15744671

Halliday, J. E. B., Knobel, D. L., & Allan, K. J., de C. Bronsvoort, B. M., Handel, I., Agwanda, B., Cutler, S., Olack, B., Ahmed, A., Hartskeerl, R. A., Njenga, M. K., Cleaveland, S. & Breiman, R. F. (2013). Urban leptospirosis in Africa: Across-sectional survey of Leptospira infection in rodents in the Kibera urban settlement, Nairobi, Kenya. *The American Journal of Tropical Medicine and Hygiene*, *89*(6), 1095–1102. doi:10.4269/ajtmh.13-0415 PMID:24080637

Hassell, J. M., Begon, M., Ward, M. J., & Fèvre, E. M. (2017). Urbanization and Disease Emergence: Dynamics at the Wildlife-Livestock-Human Interface. *Trends in Ecology & Evolution*, *32*(1), 55–67. doi:10.1016/j.tree.2016.09.012 PMID:28029378

Hayman, D. T. S., Bowen, R. A., Cryan, P. M., McCracken, G. F., O'Shea, T. J., Peel, A. J., Gilbert, A., Webb, C. T., & Wood, J. L. N. (2013). Ecology of zoonotic infectious diseases in bats: Current knowledge and future directions. *Zoonoses and Public Health*, *60*(1), 2–21. doi:10.1111/zph.12000 PMID:22958281

Himsworth, C. G., Parsons, K. L., Jardine, C., & Patrick, D. M. (2013). Rats, cities, people, and pathogens: A systemic review and narrative synthesis of literature regarding the ecology of rat-associated zoonoses in urban centers. *Vector Borne and Zoonotic Diseases (Larchmont, N.Y.)*, *13*(6), 349–359. doi:10.1089/vbz.2012.1195 PMID:23590323

Jobbins, S. E., Sanderson, C. E., & Alexander, K. A. (2014). Leptospira interrogans at the human-wildlife interface in northern Botswana: A newly identified public health threat. *Zoonoses and Public Health*, *61*(2), 113–123. doi:10.1111/zph.12052 PMID:23672285

Johnson, C. K., Hitchens, P. L., Evans, T. S., Goldstein, T., Thomas, K., Clements, A., Joly, D. O., Wolfe, N. D., Daszak, P., Karesh, W. B., & Mazet, J. K. (2015). Spillover and pandemic properties of zoonotic viruses with high host plasticity. *Scientific Reports*, *5*, 1–8. PMID:26445169

Jones, B. A., Grace, D., Kock, R., Alonso, S., Rushton, J., Said, M. Y., McKeever, D., Mutua, F., Young, J., McDermott, J., & Pfeiffer, D. U. (2013). Zoonosis emergence is linked to agricultural intensification and environmental change. *Proceedings of the National Academy of Sciences of the United States of America*, *110*(21), 8399–8404. doi:10.1073/pnas.1208059110 PMID:23671097

Jones, K. E., Patel, N. J., Levy, M. A., Storeygard, A., Balk, D., Gittleman, J. L., & Daszak, P. (2008). Global trends in emerging infectious diseases. *Nature*, *451*(7181), 990–993. doi:10.1038/nature06536 PMID:18288193

Karesh, W. B., Dobson, A., Lloyd-Smith, J. O., Lubroth, J., Dixon, M. A., Bennett, M., Aldrich, S., Harrington, T., Formenty, P., Loh, E. H., Machalaba, C. C., Thomas, M. J., & Heymann, D. L. (2012). Ecology of zoonoses: Natural and unnatural histories. *Lancet*, *380*(9857), 1936–1945. doi:10.1016/S0140-6736(12)61678-X PMID:23200502

Kedzierski, L. (2011). Leishmaniasis. *Human Vaccines*, *7*(11), 1204–1214. doi:10.4161/hv.7.11.17752 PMID:22048116

Keesing, F., Holt, R. D., & Ostfeld, R. S. (2006). Effects of species diversity on disease risk. *Ecology Letters*, *9*(4), 485–498. doi:10.1111/j.1461-0248.2006.00885.x PMID:16623733

Kennedy, K., & Collignon, P. (2010). Colonisation with Escherichia coli resistant to “critically important” antibiotics: A high risk for international travellers. *European Journal of Clinical Microbiology & Infectious Diseases*, *29*(12), 1501–1506. doi:10.100710096-010-1031-y PMID:20835879

Ko, A. I., Galvao Reis, M., Ribeiro Dourado, C. M., Johnson, W. D. Jr, & Riley, L. W. (1999). Urban epidemic of severe leptospirosis in Brazil. Salvador Leptospirosis Study Group. *Lancet*, *354*(9181), 820–825. doi:10.1016/S0140-6736(99)80012-9 PMID:10485724

Korzeniewska-Kosela, M. (2013). Tuberculosis in Poland in 2011. *Przegląd Epidemiologiczny*, *67*(2), 277–281, 375–378. PMID:24040733

Kyte, R. (2014). *Climate Change is a Challenge for Sustainable Development*. The World Bank.

Lawson, B., de Pinna, E., Horton, R. A., Macgregor, S. K., John, S. K., Chantrey, J., Duff, J. P., Kirkwood, J. K., Simpson, V. R., Robinson, R. A., Wain, J., & Cunningham, A. A. (2014). Epidemiological evidence that garden birds are a source of human salmonellosis in England and Wales. *PLoS One*, *9*(2), e88968. doi:10.1371/journal.pone.0088968 PMID:24586464

Li, N., Ulucak, R., & Danish. (2022). Turning points for environmental sustainability: The potential role of income inequality, human capital, and globalization. *Environmental Science and Pollution Research International*, *29*(27), 40878–40892. doi:10.100711356-021-18223-6 PMID:35083689

Li, W., Shi, Z., Yu, M., Ren, W., Smith, C., Epstein, J. H., Wang, H., Crameri, G., Hu, Z., Zhang, H., Zhang, J., McEachern, J., Field, H., Daszak, P., Eaton, B. T., Zhang, S., & Wang, L. F. (2005). Bats are natural reservoirs of SARS-like coronaviruses. *Science*, *310*(5748), 676–679. doi:10.1126cience.1118391 PMID:16195424

Lloyd-Smith, J. O., George, D., Pepin, K. M., Pitzer, V. E., Pulliam, J. R. C., Dobson, A. P., Hudson, P. J., & Grenfell, B. T. (2009). Epidemic dynamics at the human-animal interface. *Science*, *326*(5958), 1362–1367. doi:10.1126cience.1177345 PMID:19965751

Longdon, B., Hadfield, J. D., Webster, C. L., Obbard, D. J., & Jiggins, F. M. (2011). Host phylogeny determines viral persistence and replication in novel hosts. *PLoS Pathogens*, *7*(9), e1002260. Advance online publication. doi:10.1371/journal.ppat.1002260 PMID:21966271

Mara, D., Lane, J., Scott, B., & Trouba, D. (2010). Sanitation and health. *PLoS Medicine*, *7*(11), 1–7. doi:10.1371/journal.pmed.1000363 PMID:21125018

Meentemeyer, R. K., Haas, S. E., & Vaclavik, T. (2012). Landscape epidemiology of emerging infectious diseases in natural and human-altered eco- systems. *Annual Review of Phytopathology*, *50*(1), 379–402. doi:10.1146/annurev-phyto-081211-172938 PMID:22681449

Meng, X. J., Lindsay, D. S., & Sriranganathan, N. (2009). Wild boars as sources of infectious diseases in livestock and humans. *Philosophical Transactions of the Royal Society of London. Series B, Biological Sciences*, *364*(1530), 2697–2707. doi:10.1098/rstb.2009.0086 PMID:19687039

Olson, D. H., Aanensen, D. M., Ronnenberg, K. L., Powell, C. I., Walker, S. F., Bielby, J., Garner, T. W. J., Weaver, G., Group, B. M., & Fisher, M. C. (2013). Mapping the Global Emergence of *Batrachochytrium dendrobatidis*, the Amphibian Chytrid Fungus. *PLoS One*, *8*(2), 1–13. doi:10.1371/journal.pone.0056802 PMID:23463502

Ostholm-Balkhed, A., Ta¨rnberg, M., Nilsson, M., Nilsson, L. E., Hanberger, H., & Ha¨llgren, A. (2013). Travel-associated fecal colonization with ESBL-producing Enterobacteriaceae: Incidence and risk factors. *The Journal of Antimicrobial Chemotherapy*, *68*(9), 2144–2153. doi:10.1093/jac/dkt167 PMID:23674762

Padgett, D. A., & Glaser, R. (2003). How stress influences the immune response. *Trends in Immunology*, *24*(8), 444–448. doi:10.1016/S1471-4906(03)00173-X PMID:12909458

Penrose, K., de Castro, M. C., Werema, J., & Ryan, E. T. (2010). Informal urban settlements and cholera risk in Dar es Salaam, Tanzania. *PLoS Neglected Tropical Diseases*, *4*(3), 1–11. doi:10.1371/journal.pntd.0000631 PMID:20300569

Pepin, K. M., Lass, S., Pulliam, J. R. C., Read, A. F., & Llyod-Smith, J. O. (2010). Identifying genetic markers of adaptation for surveillance of viral host jumps. *Nature Reviews. Microbiology*, *8*(11), 802–813. doi:10.1038/nrmicro2440 PMID:20938453

Phillips, D. R. (1993). Urbanization and human health. *Parasitology*, *106*(S1), 93–107. doi:10.1017/S0031182000086145 PMID:8488075

Pickett, S. T. A., Cadenasso, M. L., Grove, J. M., Boone, C. G., Groffman, P. M., Irwin, E., Kaushal, S. S., Marshall, V., McGrath, B. P., Nilon, C. H., Pouyat, R. V., Szlavecz, K., Troy, A., & Warren, P. (2011). Urban ecological systems: Scientific foundations and a decade of progress. *Journal of Environmental Management*, *92*(3), 331–362. doi:10.1016/j.jenvman.2010.08.022 PMID:20965643

Pilosof, S., Fortuna, M. A., Cosson, J. F., Galan, M., Kittipong, C., Ribas, A., Segal, E., Krasnov, B. R., Morand, S., & Bascompte, J. (2014). The host-parasite network structure is associated with community-level immunogenetic diversity. *Nature Communications*, *5*(1), 1–9. doi:10.1038/ncomms6172 PMID:25312328

Plowright, R. K., Eby, P., Hudson, P. J., Smith, I. L., Westcott, D., Bryden, W. L, Middleton, D., Reid, P. A., McFarlane, R. A., Martin, G., Tabor, G. M., Skerratt, L. F., Anderson, D. L., Crameri, G., Quammen, D., Jordan, D., Freeman, P., Wang, L. F., Epstein, J. H., Marsh, G. A., Kung, N. Y., & McCallum, H. (2015). Ecological dynamics of emerging bat virus spillover. *Proceedings. Biological Sciences*, *282*(1798).

Plowright, R. K., Foley, P., Field, H. E., Dobson, A. P., Foley, J. E., Eby, P., & Daszak, P. (2011). Urban habituation, ecological connectivity, and epidemic dampening: the emergence of Hendra virus from flying foxes (Pteropus spp.). *Proceedings of the Royal Society B: Biological Sciences*, *278*(1725), 3703-3712.

Plowright, R. K., Reaser, J. K., Locke, H., Woodley, S. J., Patz, J. A., Becker, D. J., Oppler, G., Hudson, P. J., & Tabor, G. M. (2021). Land use-induced spillover: A call to action to safeguard environmental, animal, and human health. *The Lancet. Planetary Health*, *5*(4), e237–e245. doi:10.1016/S2542-5196(21)00031-0 PMID:33684341

Poulin, R. (2004). Macroecological patterns of species richness in parasite assemblages. *Basic and Applied Ecology*, *5*(5), 423–434. doi:10.1016/j.baae.2004.08.003

Pulliam, J. R. C., Epstein, J. H., Dushoff, J., Rahman, S. A., Bunning, M., Jamaluddin, A. A., Hyatt, A. D., Field, H. E., Dobson, A. P., & Daszak, P. (2012). Agricultural intensification, priming for persistence and the emergence of Nipah virus: A lethal bat-borne zoonosis. *Journal of the Royal Society, Interface*, *9*(66), 89–101. doi:10.1098/rsif.2011.0223 PMID:21632614

Rassi, A. Jr, Rassi, A., & Marin-Neto, J. A. (2010). Chagas disease. *Lancet*, *375*(9723), 1388–1402. doi:10.1016/S0140-6736(10)60061-X PMID:20399979

Reisen, W. K. (2010). Landscape epidemiology of vector-borne diseases. *Annual Review of Entomology*, *55*(1), 461–483. doi:10.1146/annurev-ento-112408-085419 PMID:19737082

Reperant, L. A. (2010). Applying the theory of island biogeography to emerging pathogens: Toward predicting the sources of future emerging zoonotic and vector-borne diseases. *Vector Borne and Zoonotic Diseases (Larchmont, N.Y.)*, *10*(2), 105–110. doi:10.1089/vbz.2008.0208 PMID:19589061

Rosselló, J., Gallego, M. S., & Awan, W. (2017). Infectious disease risk and international tourism demand. *Health Policy and Planning*, *32*(4), 538–548. doi:10.1093/heapol/czw177 PMID:28104695

Savage, H. M., Anderson, M., Gordon, E., McMillen, L., Colton, L., Delorey, M., Sutherland, G., Aspen, S., Charnetzky, D., Burkhalter, K., & Godsey, M. (2008). Host-seeking heights, host-seeking activity patterns, and West Nile virus infection rates for members of the Culex pipiens complex at different habitat types within the hybrid zone, Shelby County, TN, 2002 (Diptera: Culicidae). *Journal of Medical Entomology*, *45*(2), 276–288. doi:10.1093/jmedent/45.2.276 PMID:18402144

Simonsen, P. E., & Mwakitalu, M. E. (2013). Urban lymphatic filariasis. *Parasitology Research*, *112*(1), 35–44. doi:10.100700436-012-3226-x PMID:23239094

Sims, L. D., Domenech, J., Benigno, C., Kahn, S., Kamata, A., Lubroth, J., Martin, V., & Roeder, P. (2005). Origin and evolution of highly pathogenic H5N1 avian influenza in Asia. *The Veterinary Record*, *157*(6), 159–164. doi:10.1136/vr.157.6.159 PMID:16085721

Souza, T. B., Morais, M. B., Tahan, S., Melli, L. C., Rodrigues, M. S., & Scaletsky, I. C. (2009). High prevalence of antimicrobial drug-resistant diarrheagenic Escherichia coli in asymptomatic children living in an urban slum. *The Journal of Infection, 59*(4), 247–251. doi:10.1016/j.jinf.2009.08.007 PMID:19706305

Swirski, A. L., Pearl, D. L., Williams, M. L., Homan, H. J., Linz, G. M., Cernicchiaro, N., & LeJeune, J. T. (2014). Spatial epidemiology of Escherichia coli O157:H7 in dairy cattle about night roosts of Sturnus vulgaris (European starling) in Ohio, USA (2007-2009). *Zoonoses and Public Health, 61*(6), 427–435. doi:10.1111/zph.12092 PMID:24279810

Ta¨ngde'n, T., Cars, O., Melhus, A., & Lo¨wdin, E. (2010). Foreign travel is a major risk factor for colonization with Escherichia coli producing CTX-M-type extended-spectrum beta-lactamases: A prospective study with Swedish volunteers. *Antimicrobial Agents and Chemotherapy, 54*(9), 3564–3568. doi:10.1128/AAC.00220-10 PMID:20547788

Telfer, S., Lambin, X., Birtles, R., Beldomenico, P., Burthe, S., Paterson, S., & Begon, M. (2010). Species interactions in a parasite community drive infection risk in a wildlife population. *Science, 330*(6001), 243–246. doi:10.1126cience.1190333 PMID:20929776

Tomczuk, K., Grzybek, M., Szczepaniak, K., Studzinska, M., Demkowska-Kutrzepa, M., Roczen-Karczmarz, M., & Klockiewicz, M. (2015). Analysis of intrinsic and extrinsic factors influencing the dynamics of bovine Eimeria spp. from central–eastern Poland. *Veterinary Parasitology, 214*(1-2), 22–28. doi:10.1016/j.vetpar.2015.09.027 PMID:26455571

Tristan, A., Bes, M., Meugnier, H., Lina, G., Bozdogan, B., Courvalin, P., Reverdy, M. E., Enright, M. C., Vandenesch, F., & Etienne, J. (2007). Global distribution of Panton-Valentine leucocidin - positive methicillin-resistant Staphylococcus aureus, 2006. *Emerging Infectious Diseases, 13*(4), 594–600. doi:10.3201/eid1304.061316 PMID:17553275

United Nations, Department of Economic and Social Affairs. (2014). World urbanization prospects: 2014 revision (ST/ESA/SER.A/366). Population Division.

Ventura-Garcia, L., Roura, M., Pell, C., Posada, E., Gascón, J., Aldasoro, E., Munoz, J., & Pool, R. (2013). Socio-cultural aspects of Chagas disease: A systematic review of qualitative research. *PLoS Neglected Tropical Diseases, 7*(9), 1–8. doi:10.1371/journal.pntd.0002410 PMID:24069473

Viana, M., Mancy, R., Biek, R., Cleaveland, S., Cross, P. C., Lloyd-Smith, J. O., & Haydon, D. T. (2014). Assembling evidence for identifying reservoirs of infection. *Trends in Ecology & Evolution, 29*(5), 270–279. doi:10.1016/j.tree.2014.03.002 PMID:24726345

Walsh, M. G. (2014). Rat sightings in New York City are associated with neighborhood sociodemographics, housing characteristics, and proximity to open public spaces. *PeerJ, 2*, e533. Advance online publication. doi:10.7717/peerj.533 PMID:25237595

Wang, S., Zhang, W., Wang, H., Wang, J., & Jiang, M. J. (2021). How Does Income Inequality Influence Environmental Regulation in the Context of Corruption? A Panel Threshold Analysis Based on Chinese Provincial Data. *International Journal of Environmental Research and Public Health, 18*(15), 8050. doi:10.3390/ijerph18158050 PMID:34360341

Wellington, E. M. H., Boxall, A. B., Cross, P., Feil, E. J., Gaze, W. H., Hawkey, P. M., Johnson-Rollings, A. S., Jones, D. L., Lee, N. M., Otten, W., Thomas, C. M., & Williams, A. P. (2013). The role of the natural environment in the emergence of antibiotic resistance in gram-negative bacteria. *The Lancet. Infectious Diseases, 13*(2), 155–165. doi:10.1016/S1473-3099(12)70317-1 PMID:23347633

Wilcove, D. S., McLellan, C. H., & Dobson, A. P. (1986). Habitat fragmentation in the temperate zone. In Conservation Biology: The Science of Scarcity and Diversity (pp. 237-256). Sinauer Associates, Inc. Publishers.

Wilder-Smith, A. (2012). Dengue infections in travellers. *Paediatrics and International Child Health, 32*(sup1, s1), 28–32. doi:10.1179/2046904712Z.00000000050 PMID:22668447

Woolhouse, M. E. J., & Gowtage-Sequeria, S. (2005). Host range and emerging and reemerging pathogens. *Emerging Infectious Diseases, 11*(12), 1842–1847. doi:10.3201/eid1112.050997 PMID:16485468

Woolhouse, M. E. J., Haydon, D. T., & Antia, R. (2005). Emerging pathogens: The epidemiology and evolution of species jumps. *Trends in Ecology & Evolution, 20*(5), 238–244. doi:10.1016/j.tree.2005.02.009 PMID:16701375

World Health Organization. (2007). *The world health report 2007- a safer future: global public health security in the 21st century*. World Health Organization.

World Health Organization. (2008). *Our cities, our health, our future - acting on social determinants for health equity in urban settings*. World Health Organization.

World Health Organization. (2010). *Hidden cities: unmasking and overcoming health inequities in urban settings*. World Health Organization.

World Health Organization. (2013). *Sustaining the drive to overcome the global impact of neglected tropical diseases _ second WHO report on neglected diseases*. World Health Organization.

World Health Organization. (2014). *Progress on drinking water and sanitation - Joint Monitoring Programme update 2014*. World Health Organization.

World Health Organization. (2019). *Urban settings as a social determinant of health*. World Health Organization.

World Health Organization. (2020). *Novel Coronavirus (2019-nCoV): situation report, 11*. World Health Organization.

World Tourism Organization. (2014). UNWTO Tourism Highlights, 2014 Edition (eISBN 9789284416226). UNTWO Publications.

World Urban Forum. (Ed.). (2012). *State of the world's cities 2012/2013*. United Nations Human Settlements Programme (UN-HABITAT).

Wright, A. N., & Gompper, M. E. (2005). Altered parasite assemblages in raccoons in response to manipulated resource availability. *Oecologia, 144*(1), 148–156. doi:10.100700442-005-0018-3 PMID:15891856

KEY TERMS AND DEFINITIONS

Basic Reproduction Number (R_0): The predicted figure of occurrence directly caused by an infection in a vulnerable inhabitant throughout their infectious period. It measures the transmission capability of a disease.

Bridge Host: The ability of a non-maintenance host community of people to transfer a foreign infectious agent from the maintenance host to the target group of people. To absorb this role, (i) the bridge host must be capable of excretion, infection of a pathogen, or capable of transmitting a pathogen mechanically. (ii) close encounter or share habitation with maintenance host or target group of people.

Interface: A common boundary of a community of species between two bodies (wildlife – livestock – human) across which a pathogen can be transmitted between the biological communities.

Maintenance Host: Population or species that make up a disease pool and within which a pathogen continues to stay without any restoration from another host.

Network Theory: From an epidemiological perspective, network theory consists of some nodes representing the interaction between individuals, which provides a structure for envisioning possible mechanisms of pathogen transmission within a community of people.

Phylogenetics: The study of development and evolutionary relationships between a particular species of organisms. These associations are signified as evolution or phylogenetic tree consisting of branching points also known as nodes and edges.

Population Genetics: It is a study of genetic variation and composition within and between populations including modifications in allele frequencies over generations.

Spillover: the process of transmission of a foreign infectious agent from a host population or its reservoir into other species or hosts thus resulting in an outbreak.

Target Host: A host population within a reservoir of diseases which is the spotlight for controlling the disease.

Chapter 4
Urbanization and the Mosquito Population:
Threats to Human Health

Debraj Biswal
https://orcid.org/0000-0002-7317-6774
Government General Degree College at Mangalkote, India

ABSTRACT

Mosquitoes have always been a source of threat to human health because of their ability to transmit deadly diseases. Despite several efforts to curb their population, they have managed to survive and spread havoc. Urbanization has emerged to be one of the major contributors to their reproductive success followed by climate change. Urbanization has stimulated opportunistic breeding behaviour in mosquitoes by generating a vast array of temporary breeding habitats. The effect has been compounded by climate change allowing the mosquitoes to breed throughout the year and spread to regions which were previously inhospitable to them. Further, a study of mosquito breeding behaviour shows that gravid females follow several visual and olfactory cues to select their breeding habitats to ensure the well-being of their young ones. Thus, study of mosquito breeding habitats in urban areas helps to identify their markers to be used in mosquito larval source management. Similar problems can be largely overcome in the future by taking a more inclusive approach during town planning.

DOI: 10.4018/978-1-7998-9414-8.ch004

INTRODUCTION

Mosquitoes have always been an important topic of research because of their ability to transmit deadly diseases that cause millions of human deaths annually. The female members of the species spread the diseases. The physiology of the female mosquitoes is such that they require animal blood for the maturation of their eggs. This triggers their blood-feeding drive as a result of which they visit their animal hosts regularly (Kolimenakis et al., 2021). Humans are their easy target since they remain confined to their habitats. Moreover, the tendency of humans to modify their surroundings indirectly encourages the mosquito population to multiply. Literature reveals that the mosquitoes have evolved with the burgeoning human population and have travelled worldwide, taking the help of the airways and waterways mainly used for trading in the earlier days. Thus, humans have unknowingly helped the mosquito population thrive, flourish, evolve, and disperse across continental barriers (Biswal, 2018; Wang et al., 2020; Kolimenakis et al., 2021).

Apart from the medical issues related to mosquitoes, they are a source of agony for their bites and painful human experiences, both indoors and outdoors, especially in the tropical and sub-tropical countries. With time, they have evolved into domestic insects encountered in every household (Augustina et al., 2021). Studies on mosquito-biting behavior show that they prefer to hide in the dark corners of human homes or livestock sheds and bite at their appropriate periods. It needs to be mentioned that mosquitoes have broadly been classified into day-biting and night-biting mosquitoes based on their activity timings (Karlekar and Andrew, 2016). Research on their biting behavior has revealed that it is related to the transmission of parasites or has driven a host-parasite association.

The last few years have witnessed human population growth at alarming rates and the simultaneous bloom of urbanization. Recent studies show that the growth of towns and cities has taken place neglecting several ecological factors, including the potential breeding habitats of mosquitoes (Rose et al., 2020). The survival and expansion of the mosquito population have intrigued the scientific community to delve deeper into the lifecycle and the breeding habitat ecology of mosquitoes (Zahouli et al., 2017). Though the primary aim of such studies was to identify the vulnerable points in their lifecycle that could be manipulated or targeted, they reveal the tremendous abilities of the mosquitoes to adjust and adapt to their changing surroundings that have, in a way, ensured their breeding success (Zahouli et al., 2017). This has resulted in the development of unhealthy cities with dire consequences for the human population; the problem has been compounded by climate change, which is also a by-product of population growth (Rose et al., 2020). The current chapter discusses the aspects that have prevented the mosquitoes from being obliterated.

It also discusses what can be done in the future to develop inclusive vector control strategies and protect the human population from life-threatening diseases.

MOSQUITO BREEDING SITES IN URBAN AREAS

One of the primary requirements for mosquito breeding is water where it lays eggs and the larvae metamorphose into adults. Therefore, any site and object that can hold water for some time encourage the mosquitoes to breed. Their breeding sites range from natural water bodies like ponds to artificial structures such as reservoirs (Ranasinghe and Amarasinghe, 2020). They can breed in temporary structures with water besides permanent ones (Augustina et al., 2021). Studies show that they select breeding sites close to human habitations to ensure their blood meal. For this reason, a greater population of mosquito larvae is encountered in paddy fields close to urban areas than in those that are far away (Augustina et al., 2021).

Interestingly, if the urban habitats are closely studied one can easily identify a range of permanent and/or temporary structures that can accumulate water and stimulate the mosquito population. The drainage systems provide the largest breeding habitats for mosquitoes (Augustina et al., 2021). The drains, especially the stagnant ones, hold water and nutrients for the developing larvae. Various other structures, overhead tanks, reservoirs, concrete designs in buildings allowing for water accumulation, pots, cracks and crevices in the walls, dustbins, discarded objects and containers, cans, bottles, tyres, glass vessels, coconut shells, peels of fruits or vegetables or anything that can hold water may act as mosquito breeding sites (Karlekar and Andrew, 2016; Wilke et al., 2019). Earlier, container breeding was found mostly in the case of *Aedes* sp. But, over the last few years, others like *Culex* sp. and *Anopheles* sp. have also been reported to breed in discarded containers (Karlekar and Andrew, 2016). Many researchers have termed this as 'opportunistic breeding behavior' given the fact that it is not the usual mode of breeding by mosquitoes other than *Aedes* sp. and this particular breeding behavior has considerably augmented their reproductive success (Wilke et al., 2019).

Studies over the last two decades on the breeding habitats and the breeding trends of mosquitoes show that time has added newer sites to the list helping the mosquitoes to diversify their breeding habitats (Zahouli et al., 2017; Owolabi and Bagbe, 2019; Ranasinghe and Amarasinghe, 2020; Kolimenakis et al., 2021). From the perspective of a mosquito, this offers two major advantages – 1) to breed in the oddest location, and 2) to escape the conventional mosquito control strategies adopted by humans from time to time. Sometimes, the breeding habitats are so unique that they become almost impossible to be reached for vector management. In other cases, ecological issues prevent the application of certain compounds that target the mosquito population at grass-root levels. The residual effects of several insecticides, even though marketed as eco-friendly, have not been completely evaluated restricting their mass acceptance (Conti et al., 2014). Additionally, the development of biological resistance against certain insecticides has jeopardized many effective mosquito control agents (Araújo et al., 2015). All of these have acted in favour of mosquitoes and enhanced their survival strategies.

The urban areas have emerged as breeding havens for mosquitoes mainly for three reasons, 1) the presence of opportunistic breeding habitats in addition to the conventional ones, 2) the presence of abundant hosts due to the dense human population, and 3) routes of escape from conventional vector management practices due to unique breeding locations. In addition to these, the ability of the *Culex* sp. larvae to thrive in the turbid waters with the help of their siphons has given the mosquito species a competitive edge over others (Amini et al., 2020). Since the mosquitoes need water for breeding, the temporary breeding habitats like the pits by the streets or even the tyre marks on mud become fully functional during the rainy season when they are filled with water (Owolabi and Bagbe, 2019). The same applies to the discarded containers or the trash bins that become temporary vessels holding water (Wilke et al., 2019). Studies conducted in the arid and semi-arid regions of the world have put forward another important observation. They found that leaky water pipes distributing fresh drinking water to different parts of a city or a town are important sources of water for mosquito breeding (Alkhayat et al., 2020) even during the dry seasons of the year. This helps the nuisance of the mosquitoes to perpetuate in a year-round manner (Rose et al., 2020). Global climate change has worsened these situations and that is being discussed in this chapter.

ECOLOGY OF THE MOSQUITO BREEDING HABITATS

Over the last few years, there has been a growing interest in deciphering the ecology of breeding in mosquitoes. Initially, the studies were conducted to find if one or more of the ecological factors, that govern their breeding habitats, could be manipulated to control the mosquito population as a part of environment-friendly vector control strategies. As information kept pouring in, it developed into an independent field of research. Earlier experiments conducted in this field wanted to know about larval feeding habits (Biswal, 2017). The gut contents analysis of the mosquito larvae found that they are filter feeders feeding mainly on algae / cyanobacteria (e.g. *Spirulina major, Anabaena affinis,* etc.) and even sediments (Merritt et al., 1992). This prompted the ecologists to believe that other organisms with similar feeding habits might act as competitors to the mosquito larvae. Eventually, the association of mosquito larvae with micro invertebrates (e.g. *Metacyclops minutus*, *Brachionus forticula*,

Diaphanosoma brachyurum, *Sida crystallina*, *Pandorina morum*, *Paramecium caudatum*) in their habitat proved the presence of such competition (Ranasinghe and Amarasinghe, 2020). Scientists tried to exploit this information to their advantage and attempted to control the mosquito population by introducing indigestible algae (e.g. *Kirchneriella* sp., *Scenedesmus quadricauda*) into the water bodies that could surpass the growth of the digestible ones (e.g. *Pediastrum* sp., *Spirogyra* sp., *Ankistrodesmus* sp.). In some other experiments, they tried to make the aquatic body unsuitable for the survival of mosquito larvae by introducing zooplanktons (e.g. *Daphnia* sp., *Mesocyclops longisetus*) that could outcompete the mosquito larvae in grazing the easily digestible algae (e.g. *Spirogyra* sp.) (Biswal, 2017). Though the results seemed encouraging in the laboratory set-ups, its applicability on large scales became doubtful.

A more common approach to utilizing the mosquito breeding habitat ecology to control them biologically has been the exploitation of the prey-predator relationships (Ranasinghe and Amarasinghe, 2020). It is a well-known fact that mosquito larvae are preyed upon by several aquatic creatures. Among them, the fishes (e.g. *Aphanius dispar*, *Oreochromis* sp., *Aplocheilus panchax*) are better known while the dragonfly (e.g. *Crocothemis erythraea*) and damselfly (e.g. *Ischnura evansi*) nymphs are a bit lesser-known (Roberts, 2014; Poerwanto et al., 2020). Introducing these predatory organisms in aquatic bodies has given good results, even on large scales, but this can be applied only to the large water bodies. The smaller or the temporary ones cannot support fish and, therefore, the mosquitoes are spared from their predators in such ephemeral pools. Large-scale application of dragonfly nymphs as mosquito biocontrol agents never received popularity probably because they are preyed upon by several aquatic creatures as well as birds (Ranasinghe and Amarasinghe, 2020).

Another ecological association, concerning eco-friendly mosquito control, that has been much investigated is the host-parasite relationship. In the natural environment, the mosquito larvae are known to be attacked by several parasitic organisms such as protozoans (e.g. *Chilodinella* sp., *Vorticella microstoma*), nematodes (e.g. *Romanomermis iyengari*), bacteria (e.g. *Bacillus cereus*) and fungi (e.g. *Metarhizium robertsii*) (Ragavendran and Natarajan, 2018; Abagli et al., 2019; Ranasinghe and Amarasinghe, 2020; Gabrieli et al., 2021). Researchers studied these organisms and tried to apply them to control the population of mosquito larvae (Ranasinghe and Amarasinghe, 2020). There were several advantages of using them, of which two important ones were, 1) they could be applied to small or even temporary breeding habitats, and 2) they are eco-friendly, although the eco-friendly aspect of many of these agents has sometimes been doubted giving room for further research.

Among the organisms examined, bacteria have garnered much research attention because of their multiple roles (Gabrieli et al., 2021). They have been also explored for their symbiotic associations with the mosquito larvae exerting beneficial functions in their life. These bacteria present in the breeding habitat water enter the mosquito body through the larval gut surviving the digestive enzymes (Duguma et al., 2015). However, many researchers believe that the consistent presence of such bacteria in different larval instars is not because of the bacteria's ability to resist the digestive enzymes of the larvae but because of continual ingestion from the surroundings. Many other scientists believe that some bacteria take refuge in the larval Malpighian tubules during the molting of instars (Dada et al., 2014; Duguma et al., 2015). Though the bacterial population fluctuates in different instars, some of them manage to survive within their guts even up to their adult stages. Thus, some bacteria present in the breeding habitat attain the status of permanent gut microbiota of mosquitoes. A section of this bacterial flora has also been recorded to undergo transstadial and transovarial transmission in mosquitoes (Dada et al., 2014; Duguma et al., 2015). This reflects the importance of bacterial flora as symbionts in these vectors. Apart from being a source of nutrition, the bacterial flora regulates the gut health often preventing the establishment of pathogenic bacteria. The bacterial population is also required for the development of vector competence (Souza et al., 2019). More recent studies are giving further insights into the role of the bacterial population in breeding, including oviposition by gravid females (Girard et al., 2021). They are likely to have several other developmental and immunological roles aiding their host's survival. The application of anti-bacterial compounds targeted against the mosquito gut bacteria has also been explored extensively to control the vector population. The basic principle behind this vector control strategy is disruption of the larval physiology to halt their development, thereby, blocking the completion of their lifecycle (Gabrieli et al., 2021). This portion is being separately dealt with under the breeding behavior of mosquitoes.

Beyond exploring these ecological relationships namely, competition (Ranasinghe and Amarasinghe, 2020), predation (Roberts, 2014), parasitism (Abagli et al., 2019), and symbiosis (Duguma et al., 2015), the physicochemical parameters of these natural breeding habitats (Hery et al., 2021) have also been studied. The information obtained, thereby, helps to understand the effect of these factors on the presence and/or absence of mosquito larvae in the susceptible mosquito breeding habitats. Records show that mosquitoes prefer a more or less neutral pH for their larval development although conflicting reports are also come across (Amini et al., 2020; Hery et al., 2021). Salinity has been found closely related to the abundance of mosquito larvae (of freshwater species) since high chloride concentration or high salinity is associated with low density of mosquito larvae (Panigrahi et al., 2014; Amini et al., 2020). The role of light on mosquito larvae has been much investigated although the results are contradictory. Some state that light has a positive influence on mosquito larvae (Owolabi and Bagbe, 2019), while others found no such definite relationships (Gimnig et al., 2001). A few reports claim that light helps in the maintenance of optimum temperature conditions of the breeding habitats required by the larvae to survive, and according to some light helps determine the abundance of larval food in the water body by stimulating the growth of photosynthetic organisms like bacteria and/or algae (Alkhayat et al., 2020). Here, it needs to be mentioned that the more diverse the feeding habit of the mosquito larvae the greater their ability to adjust to different habitats. This hypothesis has been supported by many studies, which state that mosquito larvae have a wide range of feeding preferences (Merritt et al., 1992). Probably, this is the reason why the presence of light has not been reported to have any strong association with the presence of mosquito larvae in the breeding habitat.

The study of the ecology of mosquito breeding habitats would have been incomplete had the influence of the associated aquatic plants not been deciphered. The presence of plants like *Cynodon dactylon*, *Acrostichum aurium*, *Nypa frusticans*, *Derris heterophylla,* and *Acanthus ilicifolius* have been reported from mosquito breeding habitats in Indonesia (Poerwanto et al., 2020) and elsewhere. Studies show that the presence of aquatic plants does have a positive influence on the survival of the eggs and the immature stages of mosquitoes (Poerwanto et al., 2020; Ranasinghe and Amarasinghe, 2020). Their presence can provide a sort of refuge for eggs and larvae from predators (Ranasinghe and Amarasinghe, 2020). The bacteria and microalgae growing on the plant surfaces, especially the leaves and the stem, are known to be grazed upon by the mosquito larvae (Merritt et al., 1992), thus, the plants become an indirect source of nutrition. Directly, the decomposing plant parts can also be a source of nutrition for the mosquito larvae (Rejmánková et al., 2013). The aquatic macrophytes can act as guards protecting the mosquito eggs and/or larvae from insecticides, thereby, reducing their efficiency. Furthermore, they can also affect the invertebrate, algal, and bacterial population by secreting certain chemicals (e.g. phenols and alkaloids) (Rejmánková et al., 2013). Apart from these effects, many studies report the direct impacts of aquatic macrophytes on mosquito breeding behavior. For example, Turnipseed et al. (2018) observed the presence of certain macrophytes like water lettuce (*Pistia stratioles*) and water hyacinth (*Eichhornia crassipes*) stimulated oviposition in *Culex quinquefasciatus* and *Aedes aegypti* by releasing certain volatile compounds. They further found that the effect of volatile substances emitted by the aquatic macrophytes varied with the mosquito species. For instance, gravid females of *Anopheles quadrimaculatus* are not attracted to the volatiles from aquatic plants (Turnipseed et al., 2018). Eventually, they suggested that the presence of aquatic macrophytes might serve as markers of particular mosquito species, or the extracted plant volatiles could serve as traps for the gravid females (Turnipseed et al., 2018). Subsequent studies showed that the abundance of mosquito larvae in an aquatic body was inversely related to the diversity of aquatic macrophytes (Lumpkin et al., 2020).

Different species of aquatic plants have their impacts in two ways - firstly, they increase the diversity of competitors for mosquito larvae, and secondly, they increase the abundance of ambush predators (e.g. the dragonflies and damselflies belonging to the families Aeshnidae and Coenagrionidae respectively). The abundance and diversity of competitors increase because the increased plant surface areas provide greater scope for foraging and plenty of food. The resting ambush predators, on the other hand, get ample hiding opportunities to hunt their prey that mainly include mosquito larvae (Lumpkin et al., 2020). These ecological relationships are important to understand because clearing of the vegetation in the water bodies may indirectly make the aquatic habitats favorable for mosquitoes. Thus, uprooting and/or mowing of aquatic plants should be carefully considered in and around the potential mosquito breeding habitats during the rainy season, more so in the urban areas. However, the presence of aquatic plants does not seem to be obligatory for mosquitoes to breed, especially in urban habitats (Lumpkin et al., 2020), but they exploit the benefits whenever they get an opportunity. This reflects the flexibility of mosquitoes in choosing their breeding habitats and is highly responsible for their survival in urban landscapes where they choose human-made structures over natural ones for breeding (Rose et al., 2020).

There are several other important observations regarding the feeding ecology of different species of mosquito larvae. Usually, the mosquitoes prefer their young ones to be raised with the conspecifics although the occurrence of multiple species of mosquito larvae at the same site is not rare (Amini et al., 2020). Under such circumstances, it is observed that differences in the feeding behavior of larvae belonging to different species reduce the interspecific competition to a large extent. Some species graze at the surface of the water while others graze from either the water column or even the bottom (Merritt et al., 1992). It has been suggested that these differences depend on the minute structural dissimilarities of the mouthparts among the larvae of different mosquito species. However, the structures are such that they offer a degree of flexibility in their feeding modes. Accordingly, the larvae exhibit more than one feeding mode like filter-feeding (from column), bottom browsing, surface grazing, suspension-feeding, scavenging, etc. Their behavior, however, depend on the habitat, the available food, and the co-occurrence of larvae of different species (Merritt et al., 1992). For example, *Anopheles* sp. larvae feed in the air-water interface while those of *Culex* sp. feed in the water column. The larvae of *Mansonia* sp. usually feed near the roots of aquatic plants and are often referred to as detritus feeders. These feeding microhabitats have been encountered in several breeding habitats like in natural lakes or artificial containers (Merritt et al., 1992).

In the case of containers, the larvae divide the food source among themselves by grazing either from the water surface (air-water interface) or the surface of the container walls. For example, larvae of *Culex* sp. are known to feed in the water column while those of *Aedes* sp. and *Ochlerotatus* sp. forage on the container surfaces and the bottom when these three species co-occur (Yee et al., 2004). Under special conditions, mosquito larvae are also known to change their feeding habits because of certain chemicals released by the predators. These chemicals are called kairomones. Their presence has been reported to increase surface-feeding behavior in *Culex* sp. larvae because it involves less movement and lowers the chance of being attacked by predators compared to active bottom foraging (Roberts, 2014). Moreover, the mosquito larvae can sense the alarm pheromones being released from their injured conspecifics and change their feeding behavior (Roberts, 2014). These are some common protective strategies against predators.

BREEDING BEHAVIOR OF MOSQUITOES

Researchers began to take interest in the breeding behavior of mosquitoes when they found that many susceptible aquatic habitats having all the features to qualify as mosquito-breeding habitats were devoid of their larvae. This indicated the presence of specific factors that influenced the egg-laying behavior of mosquitoes. Subsequent research in this field gave further insights into how mosquito breeding behavior guaranteed their reproductive success.

The most important observation to date has been the presence of visual and olfactory cues that influence the egg-laying behavior of mosquitoes. In many cases, especially the females of those mosquito species that are active by the day are stimulated by visual cues like colors to lay eggs (Pavlovich and Rockett, 2000; Chakraborty and Chatterjee, 2015; Girard et al., 2021). However, definite and strong correlations between colors and egg-laying behavior in mosquitoes have not been reported though! Chemical and/or sensory cues have given more discernible results (Girard et al., 2021). It has been reported earlier that insects, in general, use their antennae to test their surroundings before egg-laying and lay eggs only when they are assured of the well-being of their young ones. In case they sense something negative, which they sometimes do, they select a different site for ovipositing (Girard et al., 2021). Experiments in the field have detected the roles of certain chemicals (e.g. semiochemicals) being emitted from the breeding habitats to attract the mosquitoes for egg-laying (Girard et al., 2021). Investigations on the source of such chemicals showed that bacteria present in the breeding waters produce some volatile compounds that may act as oviposition attractants. Most experiments conducted on this have studied the effect of one bacterium at a time, on the egg-laying behavior of mosquitoes, though some experiments have used a mixture of two or more bacteria for the same. They have always been found to have a positive influence on the mosquito ovipositing behavior when compared with plain distilled water (Panigrahi et al., 2014; Chakraborty and Chatterjee, 2015). The experiments conducted on this line have been known as oviposition assay experiments and the calculation of the oviposition assay index (OAI) helps to determine the oviposition attractancy of the experimental breeding water (Chakraborty and Chatterjee, 2015). Some common bacteria recovered from mosquito breeding waters with oviposition stimulating effects include *Lactobacillus* sp., *Enterobacter* sp., *Bacillus cereus*, *Pseudomonas* sp., *Shigella* sp., and *Escherichia coli* (Pavlovich and Rockett, 2000; Girard et al., 2021). Hence, researchers have tested eco-friendly vector control by targeting these bacteria (Nilsson et al., 2018). Then, some experiments have tried adding chemicals (repellents) to breeding waters to test whether they can deter the females from laying eggs. The application of neem oil and its various formulations have been quite popular in these trials (Kala et al., 2019). Several other plant-derived compounds including nanoparticles have found a place in these experiments as well. Using these chemo-attractants, which have often been termed as 'pheromones' by many workers, to trap gravid females in small tubes followed by killing them has also been employed as an eco-friendly vector control strategy. Various other terms are also used to describe these techniques, of which ovitraps and eave tubes are two examples (Achee et al., 2019).

Attraction toward certain bacteria present in the water can be considered as an adaptive strategy of mosquitoes to enhance their survival since some bacteria are their symbionts. The presence of symbiotic bacteria in different parts of the mosquito body (other than their guts) has not been studied much except for the salivary glands, ovaries, and Malpighian tubules (Sharma et al., 2014; Tchioffo et al., 2016; Berhanu et al., 2019). Some common bacteria isolated from mosquitoes include *Pseudomonas* sp., *Streptococcus* sp., *Methylobacterium* sp., *Chlorobi* sp., *Acinetobacter* sp., *Lactobacillus* sp., *Rubrobacter* sp. and *Chloriflexi* sp. and certain Cyanobacterial species (Sharma et al., 2014; Berhanu et al., 2019). Most of such species reportedly enhance the vector competence of the mosquitoes and have been studied mainly for targeting in vector control strategies. The composition of bacterial colonies in mosquito bodies has been found to differ from their collection sites except for some that seem to be universally present in almost all breeding waters (Tchioffo et al., 2016). This shows that the mosquitoes acquire a section of bacteria, which finally shape up the microbial flora in their body, from their surroundings. Stage-specific temporal fluctuations in the composition of these bacterial colonies have also been observed (Duguma et al., 2015). These evidences point toward the importance of bacteria and their acquisition from the surroundings in the development of mosquitoes (Nilsson et al., 2018). Probably, they can evaluate the bacterial composition of their breeding habitat waters by sensing the volatile compounds being emitted from the microbial community. Following this, the gravid females select the site to lay eggs only if they can smell enough beneficial bacteria that can support the proper development of their offspring (Girard et al., 2021). This is further reinforced by the fact that the presence of certain algae (e.g. *Tolypothrix* sp., *Cladophora holsaltica*) and bacteria (Comamonadaceae, Burkholderiaceae, Xanthomonadaceae) is associated with the absence/low density of mosquito larvae because either they are harmful or they do not support the mosquito larval population in any way (Marten, 2007; Nilsson et al., 2018). This fine-tuning seems to be a major contributor to the microbial conditioning of the mosquito body.

Delving deeper into the breeding behavior, many other important facts have come to light. It has been observed that mostly a breeding habitat shows the presence of larvae belonging to a particular mosquito species (Panigrahi et al., 2014). Mixed-species larval communities have also been observed, especially in container breeding habitats frequently encountered in urban landscapes (Nilsson et al., 2018). That has also been observed in ephemeral habitats and larger water bodies or permanent structures (Ranasinghe and Amarasinghe, 2020). However, experimental verifications of the breeding behavior of mosquitoes under laboratory conditions have revealed that they usually choose to lay eggs in water containing eggs and/or larvae of their species and they have a tendency to avoid waters containing eggs and/or larvae of different mosquito species (Panigrahi et al., 2014). Successive experiments have proved that apart from the bacteria in the breeding waters, the eggs and/or larvae of the mosquitoes themselves produce pheromones, which act as olfactory cues helping the gravid females to differentiate between the occurrences of the same or different species of mosquito larvae in the aquatic bodies. Subsequently, the mosquitoes choose the water bodies having eggs or larvae of similar species. Probably, they have the luxury to do so when a large number of breeding habitats are available, or when the aquatic surface is large (Panigrahi et al., 2014).

Comparative oviposition assay experiments between plain distilled water without any eggs or larvae (of same or different mosquito species) versus containers having eggs or larvae (of same or different mosquito species) showed that gravid females usually avoid the aquatic habitats that have not been pre-occupied by other mosquitoes even if they belong to different species (Panigrahi et al., 2014). Based on these observations, it can be said that mosquitoes look for living creatures in the water bodies and are stimulated by the presence of life. This finding gave further insights into the habitat ecology of mosquitoes and partly explained the reason for the absence of mosquito larvae from some sites. Probably, the presence of eggs or the immature stages of other mosquitoes assures the safety of an aquatic body as far as their survivability is concerned (Panigrahi et al., 2014). The absence of any stages of any mosquito species would mean that something is wrong with the site for which it has not been chosen by others to lay eggs. However, the constraints of the surface area of the aquatic habitats were found to be equally applicable in this case too. When the surface areas of the available water bodies are small, the mosquitoes preferred to lay eggs in containers devoid of eggs or developmental stages of any other mosquito. This could be interpreted as an adaptive strategy to minimize competition by avoiding overcrowding in small breeding habitats (Panigrahi et al., 2014).

Therefore, it can be said that the mosquitoes are far more intelligent than the scientific community had imagined them to be. While laying eggs, they take every precaution to ensure the perpetuation of their species and that largely works in their favor. The combination of factors (as described above) has become the inherent yardsticks of measurement of the suitability of seemingly fine breeding habitat by a gravid female. Considering all these cues, the females, before releasing their eggs, have taken them a long way to achieving the kind of breeding success they enjoy. The precautionary measures taken by female mosquitoes eliminate certain habitats to qualify as 'mosquito breeding habitats'. These necessitate further study of their breeding habitat ecology and identifying suitable markers to be used for efficient vector control programs. In the last few years, the characterization of the bacterial communities of the natural breeding habitat of mosquitoes has emerged as one such marker (Day, 2016; Nilsson et al., 2018).

OLFACTORY BEHAVIOR OF MOSQUITOES

A thorough understanding of the olfactory behavior of mosquitoes is important for devising appropriate control strategies. There are two important aspects to this. The molecular part regulates the signal transduction pathways of behavioral control in response to odorant substances and the part that deals with the learning behavior in mosquitoes by associating the odorant substances with advantages and drawbacks. Olfactory behavioral responses have long been used for mosquito control. For instance, 1-octen-3-ol, carbon dioxide (CO_2), and toxic sugar baits are some of the common attractants used for controlling adult mosquitoes (Müller et al., 2010; Qualls et al., 2015). Picaridin and DEET are common repellents used for the same purpose (Bohbot et al., 2011). However, the underlying mechanisms elucidated only lately have provided some important insights into their modes of feeding, host preference, breeding, and egg-laying behavior in mosquitoes (Liu et al., 2010).

In *Anopheles gambiae,* two major olfactory signal transduction pathways have been unearthed. They include the family of *An. gambiae* odorant receptors (AgORs) and *An. gambiae* ionotropic receptors (AgIRs) (Liu et al., 2010). These receptors are selectively expressed in olfactory receptor neurons (ORNs) that bind to odorant molecules. The members of the AgORs, however, are expressed in a lifecycle stage-specific manner (Xia et al., 2008). According to reports, the AgORs are activated by compounds present in human sweat, which influence the host-seeking, egg-laying, and other behavioral responses essential for maintaining the lifecycles of the anophelines. They include carboxylic acids, lactic acid, and ammonia (Cork and Park, 1996; Bernier et al., 2000). Their activation does not necessarily mean that the mosquitoes are drawn toward those odorants; substances like N, N-diethyl-meta-toluamide (DEET) evoke repellent behavior (Liu et al., 2010). According to Ditzen et al. (2008), different subsets of the odorant receptors (ORs) are differentially activated by the attractants and the repellents. Apart from *An. gambiae*, this has also been observed in *Aedes aegypti* and *Culex* sp. mosquitoes (Stanczyk et al., 2010).

The signaling pathways of AgIRs, on the other hand, are independent of those of the AgORs which respond to several odorant cues (other than those present in human sweat) that determine various aspects of mosquito behavior (Liu et al., 2010). Experiments show that the AgIRs are primarily stimulated by amines that fail to excite the AgOR signaling pathways (Liu et al., 2010). The AgIRs also mediate the behavioral attraction and repulsion in mosquitoes by different mechanisms, which may contribute to the behavioral and physiological responses to attractants and/or repellents. Thus, more than one signal transduction pathways regulate the olfactory responses in mosquitoes. This is an emerging field and needs further studies for an elaborate understanding of the molecular bases of olfaction in mosquitoes, the details obtained from which can be used for mosquito control strategies by manipulating their olfactory behavior (Liu et al., 2010).

Studies on mosquito olfaction observe that their olfactory behavior can be modified over time through associative and non-associative learning, which increases their fitness in the long run (Lutz et al., 2017). The species-specific behavioral responses of mosquito larvae to predators have been described as a product of evolution, which includes learning, by the larvae to recognize and avoid their predators (Lutz et al., 2017). Even the adults can identify their larval predators and avoid laying eggs in those aquatic habitats (Silberbush et al., 2010).

Their olfactory stimuli, to a large extent, regulate the feeding behavior of larval mosquitoes. For example, the detritivorous mosquito larvae are innately attracted to substances liberated by the decomposition of organic substances (e.g. 2-methyl phenol, indole, and other cresol derivatives) (Xia et al., 2008; Liu et al., 2010). Similarly, the volatile compounds released by microbes also stimulate sensory cues in mosquito larvae influencing their feeding behavior (Lutz et al., 2017). In adult mosquitoes, the semiochemicals released from the microbes in the aquatic habitats influence their egg-laying behavior (Audrain et al., 2015). Olfactory cues that help the adult mosquitoes identify their food sources are also the effects of associative learning (Lutz et al., 2017). This can be best exemplified by the species-specific preferences of male and female mosquitoes to plants for carbohydrates. Interestingly, it has been reported that the mosquitoes learn to associate floral and extra floral scents with a sugar meal that determines their behavioral responses towards those sources (Lutz et al., 2017).

Learning behavior in female mosquitoes of the blood-feeding species can determine their host preference and aid them in their host-seeking activities (Lutz et al., 2017). According to Zwiebel and Takken (2004) mosquitoes can differentiate between human and animal hosts (i.e. anthropophilic and zoophilic species) by learning through association with human and animal odors. The mosquito species and their blood probing behaviors are probably stimulated by certain chemical compounds present in animal skin secretions (Zwiebel and Takken, 2004).

Evolutionary studies can explain the reasons for species-specific differences observed in mosquito olfactory behavior based on phylogenetics (Afify and Potter, 2020). They showed that divergence between the two mosquito sub-families (Anophelinae and Culicinae) occurred approximately 145 to 200 million years ago allowing independent evolution of their olfactory systems (Moreno et al., 2010). As such, they respond differently to different olfactory stimulants (Afify and Potter, 2020). Although they may share some homologous receptors, their chemosensory receptors might have evolved so the receptor of one species can detect a chemical that the receptor of a different species cannot (Afify and Potter, 2020).

INFLUENCE OF CLIMATE CHANGE ON MOSQUITO BREEDING

Mosquitoes prefer warm and humid climates that dominate the world's tropical and sub-tropical regions (Nosrat et al., 2021). They are deterred by cold climates, which kept them away from the colder temperate regions. However, the global rise of temperatures in the last few years and the alterations in the precipitation patterns changed this scenario to a large extent (Nosrat et al., 2021). The temperate regions have now become slightly warmer than they used to be, making them habitable for the mosquitoes (Nosrat et al., 2021). The global climate change that the world is currently facing has been triggered by anthropogenic activities like deforestation, industrialization, intensification of agriculture, and other operations that add greenhouse gases (GHGs) to the atmosphere (Wang et al., 2020). Despite its damaging effects on the environment, the mosquitoes have found it beneficial (Kolimenakis et al., 2021). Nevertheless, report of mosquito-borne diseases from those parts of the world where they did not exist earlier incites fear and anguish.

Changes in precipitation patterns, as well as change in quantity of rainfall, have been recorded across the globe in the last few years. In many parts of the world where the span of the wet season has increased, functional temporary breeding habitats have also increased. Additionally, irregular rainfall patterns have made water available for mosquito breeding even in drier seasons of the year. Climate change has thus facilitated year-round breeding in mosquitoes that make their appearances and outbreak of epidemics unpredictable (Giesen et al., 2020). The information on the seasonality of mosquitoes helps to design the vector management programs and their time of implementation. Usually, the mosquito breeding season, the wet season, is chosen to target the immature stages to control the population of future adults. However, global climate change has led to irregularities in the seasonal patterns of mosquitoes, which has a lot to do with rendering the vector management programs ineffective.

The effect of climate change on mosquito breeding behavior is not completely understood (Filho et al., 2019). According to some reports, increased rainfall events may result in increased incidences of floods and/or availability of standing water that may encourage the mosquitoes to lay eggs (Wu et al., 2016). However, flash floods arising from excessive rainfall may cause the mosquito eggs and/or larvae to get washed away or die (Asad and Carpenter, 2018). This could exert a negative impact on their population. Soaring temperatures in arid environments may cause the mosquito eggs or young ones to die of desiccation. However, unpredictable rainfall events in the drier parts may positively influence the vector population by generating diverse breeding habitats (Filho et al., 2019). Literature shows that warmer temperatures augment the hatching rates of the mosquito eggs and fasten the rates of larval development thereby reducing the breeding time of mosquitoes (Deichstetter, 2017). This has a stimulatory effect on mosquito reproduction. Warmer temperatures have also made the winters bearable for mosquitoes and in some parts, they can breed even in the cold season in case of unexpected rainfalls. The dispersal of the mosquitoes from the tropical to the temperate regions has been facilitated by developments in transport and communication, tourism, trade, etc. but their survivability in those regions has been possible because of the global rise in annual temperatures (Biswal, 2018). Thus, it may be said that global climate change has facilitated their global dispersal. However, the seriousness of contracting mosquito-borne diseases does not lie in their worldwide dispersal alone. Interestingly, it has been observed that increased ambient temperatures have increased the metabolic rates of mosquitoes and the digestibility of blood by females (Deichstetter, 2017). This has been found to accelerate mosquitoes' blood probing behavior, which increases the chances of spreading diseases (Filho et al., 2019). The early emergence of adults gives them more time to breed while increased blood meals allow them to produce more broods (Biswal, 2018). Thus, the overall effect is increased abundance of mosquito vectors. Additionally, the lifecycle of the pathogens and the parasites shortens in warmer climates, which eventually reduces the duration of their extrinsic incubation period and increases their chances of infecting humans (Deichstetter, 2017). The role of urbanization in contributing to global climate change is not an unknown fact (Wang et al., 2020). Considered in that way, urbanization does promote vector survival and their behavioral modifications that augment the incidences of epidemics in the human population. The effect of urbanization on the spatiotemporal dynamics of the mosquito population has been elaborated in the following section.

SPATIOTEMPORAL TRENDS IN MOSQUITO POPULATION AND MOSQUITO-BORNE DISEASES ACROSS THE WORLD

Several studies have correlated the distribution of mosquito vectors with environmental and demographic factors. Correlations between urbanization and climate change effects show that both these factors act synergistically to increase the fitness of vectors in new habitats (Gubler, 2011). Romeo-Aznar et al. (2018) showed that the spatial distribution of the mosquito vectors does not depend on urbanization alone. The density of the human population also determines the vector abundance by increasing the source of food and breeding habitats. Additionally, the socioeconomic conditions of human populations are also a determining factor for the distribution of mosquitoes. It has been observed that poor socioeconomic conditions are usually related to irregular freshwater supply forcing the practice of water storage in containers (Vikram et al., 2015). The vessels, in turn, act as breeding grounds for mosquitoes. In the affluent locality, the use of containers (like flower vases and plant pots) for decoration, the drinking water supply pipes, the drainage system, and the associated structures can act as breeding sites. However, in the second case, the number of probable mosquito breeding habitats does not increase beyond a point after the high-rise buildings are constructed and occupied (Romeo-Aznar et al., 2018). Therefore, diverse socioeconomic and environmental conditions in urban areas lead to heterogeneous assemblages of mosquito vector populations. These factors cause differential spatial distribution of mosquitoes within a city and among different cities (Romeo-Aznar et al., 2018). The close association of the mosquito lifecycle and physiology to seasonal factors is responsible for their temporal variations and, thereby, the occurrence of epidemics (Romeo-Aznar et al., 2018).

Liu-Helmersson et al. (2019) showed that the abundance of *Ae. aegypti* was high in regions with a dominant tropical and subtropical climate. They recorded the greatest abundance from South America, South Asia, and mid-Africa with seasonal variations. Their reports also depicted conspicuous clustering of mosquito vectors around the populous urban areas. Unfortunately, the seasonal variations of the global patterns of mosquito distribution have been scarcely documented (Liu-Helmersson et al., 2019). The major drawback of the studies trying to register the global patterns of spatio-temporal vector abundance is that they fail to consider the natural weather and climatic variations while recording the vector population dynamics (Semenza et al., 2016; Santos and Meneses, 2017). It becomes very difficult to understand and/or predict the variation trends of vector populations worldwide (Liu-Helmersson et al.,

2019). According to Liu-Helmersson et al. (2019), the vector abundance is highest in the summer and autumn seasons and lowest in the winter and spring. They have further reported that the highest global increase of *Ae. aegypti* (by as much as 8.2%) has been recorded since 1990 which coincides with the rise of global temperature. It has been proposed that the global temperature is predicted to rise further in the coming years with the increase in the abundance of *Ae. aegypti* by about 20% to 30% (Liu-Helmersson et al., 2019).

Even though urbanization determines the trend of epidemic diseases like dengue, the effect of climate change can modify the outcomes depending on the altitudes of those urban regions (Bhatt et al., 2013). However, they have not been documented enough and according to the reports of Colón-González et al. (2021), the length of the transmission season of malaria will increase by more than one month in the tropical highlands of Africa, the Americas, and the Eastern Mediterranean region. Similarly, the length of the transmission season of dengue will increase by approximately four months in the lowlands of the Western Pacific and Eastern Mediterranean region (Colón-González et al., 2021). Colón-González et al. (2021) showed that the epidemic belt of malaria would move northwards towards Central Europe, North America, and northern Asia. Likewise, the epidemic belt of dengue has also been projected to move northwards in the northern USA and central northern Europe (Colón-González et al., 2021). Studies show if the increase in the mean global temperature could be managed below 2°C then the world population can be protected, to some extent, from these public health issues (Colón-González et al., 2018).

Thus, in general, the spatiotemporal distribution of mosquito population and mosquito-borne diseases depends mainly on climatic factors and socioeconomic development. Urbanization, an important phase in socioeconomic development, forms the foci of mosquito clustering. The patchiness of mosquito distribution in urban areas is influenced by the variations in the microclimate and the socioeconomic conditions of the inhabitants. Furthermore, the altitude of the urban areas also governs the spatiotemporal patterns of mosquito distribution by its modifying effect on the weather and climate conditions.

LARVAL SOURCE MANAGEMENT

The entire concept of larval source management (LSM) is largely derived from the apparent vulnerability of the mosquito larvae making them easy targets of vector management strategies compared to the winged adults. Moreover, since the larvae remain confined to their breeding habitat, i.e., water, they can be easily trapped and eliminated. Their vulnerability has made them the target of choice (Koenraadt, 2021). However, in recent years, the research has partly shifted from directly controlling the larval populations by eliminating them to modifying their habitats or their sources as a more effective vector control technique (Antonio-Nkondjio et al., 2018; Koenraadt, 2021). Experiments have been targeted to modify their habitats to cut their suitability for ovipositing. It may either be by adding ovipositing deterrents, by killing the oviposition attractant microbes in habitat breeding water or by altering the salinity of the aquatic bodies (Day, 2016).

The use of chemicals in their probable breeding habitats has already been reported to raise environmental concerns (as elaborated in the following section) and cause the development of biological resistance in mosquitoes. This has urged the scientific community to develop eco-friendly alternatives (Achee et al., 2019). Plant derivatives like citronella oil (Yadav et al., 2014), nirgundi oil (Ahuja et al., 2015), marigold petals extract (Rana et al., 2017), Tulsi oil (Kumar et al., 2017), castor oil (Wamaket et al., 2018), neem oil (Ayinde et al., 2020), etc. have shown positive results but the assessment of the effects of their residues on nature and the non-target organisms have given contradictory results restricting their commercial application on a wide-scale (Conti et al., 2014). Moreover, their stability in natural conditions and the complications involved in their preparation have been cited as other constraints (Kala et al., 2019).

Eradication, modification, and treatment of larval sources must first be identified. The urbanized locales offer several unconventional and unique breeding habitats, most of which are hidden allowing them to survive for long. This is the most challenging part of larval source management. The second limitation is the cost involved because the development of breeding habitat markers and applying them in field conditions will become a financial burden for the resource-deprived countries, because of which their mass acceptance will decline. Thus, cheaper and simpler techniques are required for this purpose. Once identified, the treatment of the breeding habitats, especially the larger ones, becomes an ecological issue. Despite the advancements in scientific technologies, this field requires further research to make handy formulations that can be used by common people whenever they come across any susceptible mosquito larval habitats in their surroundings. The development of plant-based nanoparticles and/or encapsulated plant oils has a promising future in this field (Ghayempour and Montazer, 2016).

ENVIRONMENTAL IMPACTS OF MOSQUITOCIDES

Organophosphates (e.g. naled, and malathion) and pyrethroids (e.g. resmethrin, d-phenothrin, and permethrin) are the most commonly employed insecticides for the control of adult mosquitoes (Davis et al., 2007). Sometimes pyrethroids require additives like piperonyl butoxide (PBO) to increase their effectiveness (Davis et al.,

2007). The adulticides are usually applied as ultra-low volume (ULV) sprays with droplet sizes varying from 8μ to 30μ. Since a majority of them are broad-spectrum insecticides, they may exert negative impacts on non-target arthropods if they happen to enter the spray zone. Of these, the arthropods whose period of activity matches the blood-feeding regime of female mosquito vectors are most likely to be affected (Davis et al., 2007). The insecticides can easily be absorbed by the cuticle of the insects and exert their toxic effects within a short period following exposure. Additionally, the toxic effects of these insecticides depend on their ability to persist in the environment. The longer they persist, the greater the risks of their uptake by non-target organisms. For instance, pyrethroids have been reported to exert low toxicity on birds and mammals, but have prolonged persistence in the environment with severe toxic effects on aquatic animals (Davis et al., 2007). Organophosphates, on the other hand, are short-lived with low toxicity on aquatic animals but exert high toxic effects on terrestrial species (Rochlin et al., 2022).

Research shows that the effect of the insecticide sprays is much wider than expected. The winds carry them to the nearby vegetation, fruits, soil, and even the water bodies. They can also be carried by rainwater (as surface runoffs) to adjacent aquatic bodies contaminating them. Leaching the insecticides into the soil below can pollute the groundwater (Ponting et al., 2021). In this way, the aquatic animals (both invertebrates and vertebrates) can be affected (Davis et al., 2007). The major routes of entry to aquatic birds are through their food sources and accidental ingestion while preening. Similar observations have been made in mammals where they can acquire the chemicals from the nearby vegetation while feeding or grooming. The aquatic creatures like *Daphnia* sp. and fishes, however, collect the contaminants directly from their surrounding water medium (Davis et al., 2007).

MOSQUITO CONTROL: A DILEMMA

Literature on mosquito research is overloaded with articles on the role of mosquitoes as vectors stressing the importance of their eradication. However, a few scattered publications talk about the ecosystem services the mosquitoes provide directly and indirectly. For example, male mosquitoes are not involved in disease transmission and feed on nectars of flowers or plant-sap. In this process, they pollinate several flowering plants. Research shows that even mosquitoes are species-specific regarding their choice of plant source for nutrition (Lutz et al., 2017). For example, the males of *Anopheles gambiae* mosquitoes prefer certain plant species such as *Thevetia neriifolia*, *Senna siamea*, *Mangifera indica*, *Cassia sieberiana,* and *Dolonix regia* (Gouagna et al., 2010; Lutz et al., 2017). The paucity of information regarding the plant species pollinated by mosquitoes has restricted their recognition as pollinators like honeybees. An interesting fact about this observation is the role of mosquitoes as alternative pollinators which gains importance under the current scenario of the rapidly declining honeybee (or other pollinators) population due to several factors, including urbanization and climate change (Cunningham et al., 2022). Here, it is essential to mention that even the female mosquitoes of the anthropophilic species require carbohydrate sources in the form of flower nectar, honeydew, plant-sap and juice of ripe fruits to maintain their normal metabolic activities other than the maturation of oocytes, for which they require blood meal (Lutz et al., 2017). Thus, the male and female mosquitoes of the blood-feeding species may also contribute to biodiversity conservation. This may be more important in urban ecosystems, where the population of pollinators is far worse than in rural ecosystems (Cunningham et al., 2022). Given the mosquitoes' tremendous ability to adapt to the changing environment, in the future, their role as pollinators, would gain significance. However, the lack of sufficient data makes the topic debatable.

Besides, they support several other animals being a part of the food web. They are fed upon by several species of lizards, salamanders, frogs, spiders, and insectivorous birds (Fang, 2010). In aquatic ecosystems, their larvae form a source of nutrition for many aquatic organisms including fishes (e.g. *Gambusia affinis*) (Fang, 2010). Though similar studies are not available for urban ecosystems, their ecological importance cannot be negated altogether. In fact, the loss of their natural predators (and thence the biodiversity erosion) associated with urbanization may also be an important factor contributing to the prevalence of mosquitoes in urban areas, an issue that needs further investigations.

In general, the disease-spreading activity of mosquito vectors is regarded as an ecosystem disservice. According to Gutierrez-Arellano and Mulligan (2018), the concept of ecosystem disservice is new and rapidly gaining popularity because researchers believe that both the positive (ecosystem service) and negative (ecosystem disservice) effects of the components of the ecosystem drive the balanced ecological functioning and maintenance of biodiversity. According to Shackleton et al. (2016), a better understanding of the well-being of humans in relation to their surroundings requires deeper information on both ecosystem services and disservices of the ecosystem constituents. Since the mosquitoes directly exert negative impacts on human health, they are a source of a disservice, and ecosystem alterations like changes in land use and land cover can augment such ecosystem disservices (Gutierrez-Arellano and Mulligan, 2018). Thus, urbanization (which involves modifications of land use and land cover) provides a positive impetus to the mosquito population (due to reasons described earlier) and thereby the ecosystem disservices they provide.

FUTURE PERSPECTIVES

From the preceding section, it may appear that mosquitoes are necessary for ecosystem functioning. However, the disease-spreading ability of the mosquitoes is undoubtedly dangerous causing the loss of several human lives, annually, on a global scale. Therefore, the eradication of their population has emerged as the primary target of humans. In this context, the failure of conventional vector control strategies may be attributed to various reasons some of which are directly and indirectly related to the human population growth and the concomitant rise in urbanization. This can be prevented to a large extent by taking an inclusive approach during town-planning. The ecological and epidemiological aspects should be seriously considered by the architects while designing buildings. This would require cooperation from all levels – policy makers, architects, ecologists, and epidemiologists. The list should also include community participation by educating the inhabitants of a town or a city about the opportunistic breeding habitats of mosquitoes. Furthermore, supplying them with adequate eco-friendly insecticides to be applied in the probable breeding habitats could also accelerate the vector control programs. Keeping the climate change in mind, this operation can be continued throughout the year to alleviate the vector population.

Geo-information technologies or the geographic information system (GIS), developed tremendously in the last few years, are being used widely in vector biology research (Jemal and Al-Thukair, 2018). The efficacy of this technique in locating the most susceptible mosquito breeding habitats in a town or a city has already been established. Drone technologies with high-resolution cameras can be used to plot the sites of those breeding habitats that are to be made available to the personnel involved in vector eradication programs. However, the major drawback of these technologies is financial that needs to be worked out since the technology has the potential to become a very efficient aid in larval source management strategies in the near future.

CONCLUSION

The ecology of natural breeding habitats of mosquitoes is governed by general ecological principles like competition, predation, parasitism, and symbiosis that have been exploited from time to time to devise vector control strategies. Interestingly, the mosquitoes are tolerant to a wide range of ecological factors allowing them to adjust and adapt themselves to the changes in their surroundings quite easily. Moreover, the gravid females have mechanisms to test the water in the breeding habitat before laying eggs to ensure the survival of their progeny. These have been the keys to their reproductive success even in the time of human population explosion, urbanization, and climate change. While population rise has provided them with more hosts, urbanization has encouraged their opportunistic breeding behaviors and climate change has allowed them to expand their population across continental borders. Larval source management can be very effective in controlling the vector population in urban areas but that would require the characterization and identification of these habitats. The utilization of unconventional and unique breeding habitats by mosquitoes in urban settings is the greatest obstacle to vector management. This necessitates extensive research and fieldwork to gather information and use it for sustainable town planning in the future.

REFERENCES

Abagli, A. Z., Alavo, T. B. C., Perez-Pacheco, R., & Platzer, E. G. (2019). Efficacy of the mermithid nematode, *Romanomermis iyengari*, for the biocontrol of *Anopheles gambiae*, the major malaria vector in Sub-Saharan Africa. *Parasites & Vectors, 12*(1), 253. doi:10.118613071-019-3508-6 PMID:31118105

Achee, N. L., Grieco, J. P., Vatandoost, H., Seixas, G., Pinto, J., Ching-Ng, L., Martins, A. J., Juntarajumnong, W., Corbel, V., Gouagna, C., David, J. P., Logan, J. G., Orsborne, J., Marois, E., Devine, G. J., & Vontas, J. (2019). Alternative strategies for mosquito-borne arbovirus control. *PLoS Neglected Tropical Diseases, 13*(1), e0006822. doi:10.1371/journal.pntd.0006822 PMID:30605475

Afify, A., & Potter, C. J. (2020). Insect repellents mediate species-specific olfactory behaviours in mosquitoes. *Malaria Journal, 19*(1), 127. doi:10.118612936-020-03206-8 PMID:32228701

Ahuja, S. C., Ahuja, S., & Ahuja, U. (2015). Nirgundi (*Vitex negundo*) – Nature's Gift to Mankind. *Asian Agri-History, 19*(1), 5–32.

Alkhayat, F. A., Ahmad, A. H., Rahim, J., Dieng, H., Ismail, B. A., Imran, M., Sheikh, U. A. A., Shahzad, M. S., Abid, A. D., & Munawar, K. (2020). Characterization of mosquito larval habitats in Qatar. *Saudi Journal of Biological Sciences, 27*(9), 2358–2365. doi:10.1016/j.sjbs.2020.07.006 PMID:32884417

Amini, M., Hanafi-Bojd, A. A., Aghapour, A. A., & Chavshin, A. R. (2020). Larval habitats and species diversity of mosquitoes (Diptera: Culicidae) in West Azerbaijan Province, Northwestern Iran. *BMC Ecology, 20*(1), 60. doi:10.118612898-020-00328-0 PMID:33213441

Antonio-Nkondjio, C., Sandjo, N. N., Awono-Ambene, P., & Wondji, C. S. (2018). Implementing a larviciding efficacy or effectiveness control intervention against malaria vectors: Key parameters for success. *Parasites & Vectors*, *11*(1), 57. doi:10.118613071-018-2627-9 PMID:29368633

Araújo, H. R., Carvalho, D. O., Ioshino, R. S., Costa-da-Silva, A. L., & Capurro, M. L. (2015). *Aedes aegypti* control strategies in Brazil: Incorporation of new technologies to overcome the persistence of dengue epidemics. *Insects*, *6*(2), 576–594. doi:10.3390/insects6020576 PMID:26463204

Asad, H., & Carpenter, D. O. (2018). Effects of climate change on the spread of zika virus: A public health threat. *Reviews on Environmental Health*, *33*(1), 31–42. doi:10.1515/reveh-2017-0042 PMID:29500926

Audrain, B., Farag, M. A., Ryu, C. M., & Ghigo, J. M. (2015). Role of bacterial volatile compounds in bacterial biology. *FEMS Microbiology Reviews*, *39*(2), 222–233. doi:10.1093/femsre/fuu013 PMID:25725014

Augustina, I., Jabal, A. R., Permana, G. I., & Ratnasari, A. (2021). Distribution and ecology of mosquito larvae in Pahandut subdistrict, Palangkaraya city. *Journal of Physics: Conference Series*, *1918*(5), 052018. doi:10.1088/1742-6596/1918/5/052018

Ayinde, A. A., Morakinyo, O. M., & Sridhar, M. K. C. (2020). Repellency and larvicidal activities of *Azadirachta indica* seed oil on *Anopheles gambiae* in Nigeria. *Heliyon*, *6*(5), e03920. doi:10.1016/j.heliyon.2020.e03920 PMID:32420486

Berhanu, A., Abera, A., Nega, D., Mekasha, S., Fentaw, S., Assefa, A., Gebrewolde, G., Wuletaw, Y., Assefa, A., Dugassa, S., Tekie, H., & Tasew, G. (2019). Isolation and identification of microflora from the midgut and salivary glands of *Anopheles* species in malaria endemic areas of Ethiopia. *BMC Microbiology*, *19*(1), 85. doi:10.118612866-019-1456-0 PMID:31035931

Bernier, U. R., Kline, D. L., Barnard, D. R., Schreck, C. E., & Yost, R. A. (2000). Analysis of human skin emanations by gas chromatography/mass spectrometry. 2. Identification of volatile compounds that are candidate attractants for the yellow fever mosquito (*Aedes aegypti*). *Analytical Chemistry*, *72*(4), 747–756. doi:10.1021/ac990963k PMID:10701259

Bhatt, S., Gething, P. W., Brady, O. J., Messina, J. P., Farlow, A. W., Moyes, C. L., Drake, J. M., Brownstein, J. S., Hoen, A. G., Sankoh, O., Myers, M. F., George, D. B., Jaenisch, T., Wint, G. R., Simmons, C. P., Scott, T. W., Farrar, J. J., & Hay, S. I. (2013). The global distribution and burden of dengue. *Nature*, *496*(7446), 504–507. doi:10.1038/nature12060 PMID:23563266

Biswal, D. (2017). Mosquito bio-control by indigestible green algae: Securing humans in a non-traditional way. In M. Gupta (Ed.), Non-traditional security: Problems and Prospects (pp. 184-193). Avenel Press.

Biswal, D. (2018). Impact of climate change on parasitic diseases: Do we need to bother? In M. Bose, S. A. Bandyopadhyay, & S. Poddar (Eds.), *Contemporary Health Issues and Environmental Impact* (pp. 32–40). Lincoln University College. doi:10.31674/books.2018.chiei.ch05

Bohbot, J. D., Fu, L., Le, T. C., Chauhan, K. R., Cantrell, C. L., & Dickens, J. C. (2011). Multiple activities of insect repellents on odorant receptors in mosquitoes. *Medical and Veterinary Entomology*, *25*(4), 436–444. doi:10.1111/j.1365-2915.2011.00949.x PMID:21395633

Breidenbaugh, M. S., & de Szalay, F. A. (2010). Effects of aerial applications of naled on nontarget insects at Parris Island, South Carolina. *Environmental Entomology*, *39*(2), 591–599. doi:10.1603/EN09087 PMID:20388292

Chakraborty, A., & Chatterjee, S. (2015). Studies on the fitness components and comparative oviposition preferences of *Aedes albopictus* in various larval microhabitats of Burdwan, West Bengal. *International Journal of Mosquito Research*, *2*(3), 156–160.

Colón-González, F. J., Harris, I., Osborn, T. J., Steiner São Bernardo, C., Peres, C. A., Hunter, P. R., & Lake, I. R. (2018). Limiting global-mean temperature increase to 1.5-2 °C could reduce the incidence and spatial spread of dengue fever in Latin America. *Proceedings of the National Academy of Sciences of the United States of America*, *115*(24), 6243–6248. doi:10.1073/pnas.1718945115 PMID:29844166

Colón-González, F. J., Sewe, M. O., Tompkins, A. M., Sjödin, H., Casallas, A., Rocklöv, J., Caminade, C., & Lowe, R. (2021). Projecting the risk of mosquito-borne diseases in a warmer and more populated world: A multi-model, multi-scenario intercomparison modelling study. *The Lancet. Planetary Health*, *5*(7), e404–e414. doi:10.1016/S2542-5196(21)00132-7 PMID:34245711

Conti, B., Flamini, G., Cioni, P. L., Ceccarini, L., Macchia, M., & Benelli, G. (2014). Mosquitocidal essential oils: Are they safe against non-target aquatic organisms? *Parasitology Research*, *113*(1), 251–259. doi:10.100700436-013-3651-5 PMID:24146210

Cork, A., & Park, K. C. (1996). Identification of electrophysiologically-active compounds for the malaria mosquito, *Anopheles gambiae*, in human sweat extracts. *Medical and Veterinary Entomology*, *10*(3), 269–276. doi:10.1111/j.1365-2915.1996.tb00742.x PMID:8887339

Cunningham, M. M., Tran, L., McKee, C. G., Polo, R. O., Newman, T., Lansing, L., Griffiths, J. S., Bilodeau, G. J., Rott, M., & Guarna, M. M. (2022). Honey bees as biomonitors of environmental contaminants, pathogens, and climate change. *Ecological Indicators*, *134*, 108457. doi:10.1016/j.ecolind.2021.108457

Dada, N., Jumas-Bilak, E., Manguin, S., Seidu, R., Stenström, T.-A., & Overgaard, H. J. (2014). Comparative assessment of the bacterial communities associated with *Aedes aegypti* larvae and water from domestic water storage containers. *Parasites & Vectors*, *7*(1), 391. doi:10.1186/1756-3305-7-391 PMID:25151134

Davis, R. S., Peterson, R. K., & Macedo, P. A. (2007). An ecological risk assessment for insecticides used in adult mosquito management. *Integrated Environmental Assessment and Management*, *3*(3), 373–382. doi:10.1002/ieam.5630030308 PMID:17695110

Day, J. F. (2016). Mosquito oviposition behavior and vector control. *Insects*, *7*(4), 65. doi:10.3390/insects7040065 PMID:27869724

Deichstetter, P. (2017). The effect of climate change on mosquito-borne diseases. *The American Biology Teacher*, *79*(3), 169–173. doi:10.1525/abt.2017.79.3.169

Ditzen, M., Pellegrino, M., & Vosshall, L. B. (2008). Insect odorant receptors are molecular targets of the insect repellent DEET. *Science*, *319*(5871), 1838–1842. doi:10.1126cience.1153121 PMID:18339904

Duguma, D., Hall, M. W., Rugman-Jones, P., Stouthamer, R., Terenius, O., Neufeld, J. D., & Walton, W. E. (2015). Developmental succession of the microbiome of *Culex* mosquitoes. *BMC Microbiology*, *15*(1), 140. doi:10.118612866-015-0475-8 PMID:26205080

Fang, J. (2010). Ecology: A world without mosquitoes. *Nature*, *466*(7305), 432–434. doi:10.1038/466432a PMID:20651669

Filho, W. L., Scheday, S., Boenecke, J., Gogoi, A., Maharaj, A., & Korovou, S. (2019). Climate Change, Health and Mosquito-Borne Diseases: Trends and Implications to the Pacific Region. *International Journal of Environmental Research and Public Health*, *16*(24), 5114. doi:10.3390/ijerph16245114 PMID:31847373

Gabrieli, P., Caccia, S., Varotto-Boccazzi, I., Arnoldi, I., Barbieri, G., Comandatore, F., & Epis, S. (2021). Mosquito Trilogy: Microbiota, Immunity and Pathogens, and Their Implications for the Control of Disease Transmission. *Frontiers in Microbiology*, *12*, 630438. doi:10.3389/fmicb.2021.630438 PMID:33889137

Ghayempour, S., & Montazer, M. (2016). Micro/nanoencapsulation of essential oils and fragrances: Focus on perfumed, antimicrobial, mosquito-repellent and medical textiles. *Journal of Microencapsulation*, *33*(6), 1–14. doi:10.1080/02652048.2016.1216187 PMID:27701985

Giesen, C., Roche, J., Redondo-Bravo, L., Ruiz-Huerta, C., Gomez-Barroso, D., Benito, A., & Herrador, Z. (2020). The impact of climate change on mosquito-borne diseases in Africa. *Pathogens and Global Health*, *114*(6), 287–301. doi:10.1080/20477724.2020.1783865 PMID:32584659

Gimnig, J. E., Ombok, M., Kamau, L., & Hawley, W. A. (2001). Characteristics of larval anopheline (Diptera: Culicidae) habitats in Western Kenya. *Journal of Medical Entomology*, *38*(2), 282–288. doi:10.1603/0022-2585-38.2.282 PMID:11296836

Girard, M., Martin, E., Vallon, L., Raquin, V., Bellet, C., Rozier, Y., Desouhant, E., Hay, A.-E., Luis, P., Moro, C. V., & Minard, G. (2021). Microorganisms associated with mosquito oviposition sites: Implications for habitat selection and insect life histories. *Microorganisms*, *9*(8), 1589. doi:10.3390/microorganisms9081589 PMID:34442667

Gouagna, L. C., Poueme, R. S., Dabiré, K. R., Ouédraogo, J. B., Fontenille, D., & Simard, F. (2010). Patterns of sugar feeding and host plant preferences in adult males of *An. gambiae* (Diptera: Culicidae). *Journal of Vector Ecology*, *35*(2), 267–276. doi:10.1111/j.1948-7134.2010.00082.x PMID:21175931

Gubler, D. J. (2011). Dengue, Urbanization and Globalization: The Unholy Trinity of the 21(st) Century. *Tropical Medicine and Health*, *39*(4, Suppl.), 3–11. doi:10.2149/tmh.2011-S05 PMID:22500131

Gutierrez-Arellano, C., & Mulligan, M. (2018). A review of regulation ecosystem services and disservices from faunal populations and potential impacts of agriculturalisation on their provision, globally. *Nature Conservation*, *30*, 1–39. doi:10.3897/natureconservation.30.26989

Hery, L., Guidez, A., Durand, A. A., Delannay, C., Normandeau-Guimond, J., Reynaud, Y., Issaly, J., Goindin, D., Legrave, G., Gustave, J., Raffestin, S., Breurec, S., Constant, P., Dusfour, I., Guertin, C., & Vega-Rúa, A. (2021). Natural variation in physicochemical profiles and bacterial communities associated with *Aedes aegypti* breeding sites and larvae on Guadeloupe and French Guiana. *Microbial Ecology*, *81*(1), 93–109. doi:10.100700248-020-01544-3 PMID:32621210

Jemal, Y., & Al-Thukair, A. A. (2018). Combining GIS application and climatic factors for mosquito control in Eastern Province. *Saudi Arabia. Saudi Journal of Biological Sciences*, *25*(8), 1593–1602. doi:10.1016/j.sjbs.2016.04.001 PMID:30591775

Kala, S., Naik, S. N., Patanjali, P. K., & Sogan, N. (2019). Neem oil water dispersible tablet as effective larvicide, ovicide and oviposition deterrent against *Anopheles culicifacies*. *South African Journal of Botany*, *123*, 387–392. doi:10.1016/j.sajb.2019.03.033

Karlekar, S., & Andrew, R. (2016). Behavioral resilience of *Culex quinquefasciatus* Say 1823 at Nagpur district of Maharashtra. *International Journal of Mosquito Research*, *3*(5), 25–30.

Koenraadt, C. J. M. (2021). Larval source management for malaria control: Prospects for new technologies and community involvement. In C. J. M. Koenraadt, J. Spitzen, & W. Takken (Eds.), *Innovative strategies for vector control: Progress in the global vector control response, Ecology and Control of Vector-borne Diseases* (Vol. 6, pp. 155–167). Wageningen Academic Publishers. doi:10.3920/978-90-8686-895-7_8

Kolimenakis, A., Heinz, S., Wilson, M. L., Winkler, V., Yakob, L., Michaelakis, A., Papachristos, D., Richardson, C., & Horstick, O. (2021). The role of urbanisation in the spread of *Aedes* mosquitoes and the diseases they transmit—A systematic review. *PLoS Neglected Tropical Diseases*, *15*(9), e0009631. doi:10.1371/journal.pntd.0009631 PMID:34499653

Kumar, S., Warikoo, R., Mishra, M., Samal, R. R., Shrankhla, Panmei, K., Dagar, V. S., & Sharma, A. (2017). Impact of *Ocimum basilicum* leaf essential oil on the survival and behaviour of an Indian strain of dengue vector, *Aedes aegypti* (L.). *Vector Biology Journal*, *2*(2), 2. doi:10.4172/2473-4810.1000122

Kwan, J. A., Novak, M. G., Hyles, T. S., & Niemela, M. K. (2009). Mortality of nontarget arthropods from an aerial application of pyrethrins. *Journal of the American Mosquito Control Association*, *25*(2), 218–220. doi:10.2987/08-5858.1 PMID:19653509

Liu, C., Pitts, R. J., Bohbot, J. D., Jones, P. L., Wang, G., & Zwiebel, L. J. (2010). Distinct olfactory signaling mechanisms in the malaria vector mosquito *Anopheles gambiae*. *PLoS Biology*, *8*(8), e1000467. doi:10.1371/journal.pbio.1000467 PMID:20824161

Liu-Helmersson, J., Brännström, Å., Sewe, M. O., Semenza, J. C., & Rocklöv, J. (2019). Estimating past, present, and future trends in the global distribution and abundance of the arbovirus vector *Aedes aegypti* under climate change scenarios. *Frontiers in Public Health*, *7*, 148. doi:10.3389/fpubh.2019.00148 PMID:31249824

Lumpkin, W. P., Stirek, K. R., & Dyer, L. A. (2020). Macrophyte diversity and complexity reduce larval mosquito abundance. *Journal of Medical Entomology*, *57*(4), 1041–1048. doi:10.1093/jme/tjaa012 PMID:32006421

Lutz, E. K., Lahondère, C., Vinauger, C., & Riffell, J. A. (2017). Olfactory learning and chemical ecology of olfaction in disease vector mosquitoes: A life history perspective. *Current Opinion in Insect Science*, *20*, 75–83. doi:10.1016/j.cois.2017.03.002 PMID:28602240

Marten, G. G. (2007). Larvicidal algae. *Journal of the American Mosquito Control Association*, *23*(sp2), 177–183. doi:10.2987/8756-971X(2007)23[177:LA]2.0.CO;2 PMID:17855939

Merritt, R. W., Dadd, R. H., & Walker, E. D. (1992). Feeding behavior, natural food, and nutritional relationships of larval mosquitoes. *Annual Review of Entomology*, *37*(1), 349–376. doi:10.1146/annurev.en.37.010192.002025 PMID:1347208

Moreno, M., Marinotti, O., Krzywinski, J., Tadei, W. P., James, A. A., Achee, N. L., & Conn, J. E. (2010). Complete mtDNA genomes of *Anopheles darlingi* and an approach to anopheline divergence time. *Malaria Journal*, *9*(1), 127. doi:10.1186/1475-2875-9-127 PMID:20470395

Müller, G. C., Beier, J. C., Traore, S. F., Toure, M. B., Traore, M. M., Bah, S., Doumbia, S., & Schlein, Y. (2010). Successful field trial of attractive toxic sugar bait (ATSB) plant-spraying methods against malaria vectors in the *Anopheles gambiae* complex in Mali, West Africa. *Malaria Journal*, *9*(1), 210. doi:10.1186/1475-2875-9-210 PMID:20663142

Nilsson, L. K. J., Sharma, A., Bhatnagar, R. K., Bertilsson, S., & Terenius, O. (2018). Presence of *Aedes* and *Anopheles* mosquito larvae is correlated to bacteria found in domestic water-storage containers. *FEMS Microbiology Ecology*, *94*(6), fiy058. doi:10.1093/femsec/fiy058 PMID:29617987

Nosrat, C., Altamirano, J., Anyamba, A., Caldwell, J. M., Damoah, R., Mutuku, F., Ndenga, B., & LaBeaud, A. D. (2021). Impact of recent climate extremes on mosquito-borne disease transmission in Kenya. *PLoS Neglected Tropical Diseases*, *15*(3), e0009182. doi:10.1371/journal.pntd.0009182 PMID:33735293

Owolabi, D. O., & Bagbe, A. S. (2019). Assessment of physico-chemical and ecological variables in selected natural breeding sites of mosquitoes in Ibadan, Oyo State, Nigeria. Archives of Pharmacy and Pharmacology Research, 1(5).

Panigrahi, S. K., Barik, T. K., Mohanty, S., & Tripathy, N. K. (2014). Laboratory evaluation of oviposition behavior of field collected *Aedes* mosquitoes. *Journal of Insects*, *207489*, 1–8. doi:10.1155/2014/207489

Pavlovich, S. G., & Rockett, C. L. (2000). Color, bacteria, and mosquito eggs as ovipositional atiractants for *Aedes aegypti* and *Aedes albopictus* (Diptera: Culicidae). *Great Lakes Entomologist*, *33*(2), 7, 141–153.

Poerwanto, S. H., Ahmad, S., Giyantolin, & Windyaraini, D. H. (2020). Characterization of mosquitoes breeding site (Diptera: Culicidae) in Bogowonto Lagoon Ecosystem, Kulonprogo Districts, special region of Yogyakarta, Indonesia. AIP Conference Proceedings, 2260.

Ponting, J., Kelly, T. J., Verhoef, A., Watts, M. J., & Sizmur, T. (2021). The impact of increased flooding occurrence on the mobility of potentially toxic elements in floodplain soil - A review. *The Science of the Total Environment*, *754*, 142040. doi:10.1016/j.scitotenv.2020.142040 PMID:32916489

Qualls, W. A., Müller, G. C., Traore, S. F., Traore, M. M., Arheart, K. L., Doumbia, S., Schlein, Y., Kravchenko, V. D., Xue, R. D., & Beier, J. C. (2015). Indoor use of attractive toxic sugar bait (ATSB) to effectively control malaria vectors in Mali, West Africa. *Malaria Journal*, *14*(1), 301. doi:10.118612936-015-0819-8 PMID:26242186

Ragavendran, C., & Natarajan, D. (2018). Larval toxicity of entomopathogenic bacteria *Bacillus cereus* against *Aedes aegypti, Anopheles stephensi*, and *Culex quinquefasciatus* mosquitoes. In B. K. Tyagi & D. Dhanasekaran (Eds.), *Microbial Control of Vector-Borne Diseases* (pp. 113–135). CRC Press. doi:10.1201/b22203-7

Rana, M., Singh, S. S. J., & Yadav, S. (2017). Effect of microencapsulated plant extracts on mosquito repellency. *Journal of Applied and Natural Science*, *9*(4), 2127–2131. doi:10.31018/jans.v9i4.1498

Ranasinghe, H. A. K., & Amarasinghe, L. D. (2020). Naturally occurring microbiota associated with mosquito breeding habitats and their effects on mosquito larvae. *BioMed Research International*, *4065315*, 1–11. doi:10.1155/2020/4065315 PMID:33381553

Rejmánková, E., Grieco, J., Achee, N., & Roberts, D. R. (2013). Ecology of larval habitats. In S. Manguin (Ed.), *Anopheles mosquitoes: New insights into malaria vectors* (pp. 397–446). IntechOpen. doi:10.5772/55229

Roberts, D. (2014). Mosquito larvae change their feeding behavior in response to kairomones from some predators. *Journal of Medical Entomology*, *51*(2), 368–374. doi:10.1603/ME13129 PMID:24724285

Rochlin, I., White, G., Reissen, N., Martheswaran, T., & Faraji, A. (2022). Effects of aerial adulticiding for mosquito management on nontarget insects: A Bayesian and community ecology approach. *Ecosphere*, *13*(1), e3896. doi:10.1002/ecs2.3896

Romeo-Aznar, V., Paul, R., Telle, O., & Pascual, M. (2018). Mosquito-borne transmission in urban landscapes: the missing link between vector abundance and human density. *Proceedings. Biological Sciences, 285*(1884), 20180826.

Rose, N. H., Sylla, M., Badolo, A., Lutomiah, J., Ayala, D., Aribodor, O. B., Ibe, N., Akorli, J., Otoo, S., Mutebi, J.-P., Kriete, A. L., Ewing, E. G., Sang, R., Gloria-Soria, A., Powell, J. R., Baker, R. E., White, B. J., Crawford, J. E., & McBride, C. S. (2020). Climate and urbanization drive mosquito preference for humans. *Current Biology, 30*(18), 3570–3579. doi:10.1016/j.cub.2020.06.092 PMID:32707056

Santos, J., & Meneses, B. M. (2017). An integrated approach for the assessment of the *Aedes aegypti* and *Aedes albopictus* global spatial distribution, and determination of the zones susceptible to the development of Zika virus. *Acta Tropica, 168*, 80–90. doi:10.1016/j.actatropica.2017.01.015 PMID:28111132

Semenza, J. C., Lindgren, E., Balkanyi, L., Espinosa, L., Almqvist, M. S., Penttinen, P., & Rocklöv, J. (2016). Determinants and drivers of infectious disease threat events in Europe. *Emerging Infectious Diseases, 22*(4), 581–589. doi:10.3201/eid2204.151073 PMID:26982104

Shackleton, C. M., Ruwanza, S., Sinasson Sanni, G. K., Bennett, S., De Lacy, P., Modipa, R., Mtati, N., Sachikonye, M., & Thondhlana, G. (2016). Unpacking pandora's box: Understanding and categorising ecosystem disservices for environmental management and human wellbeing. *Ecosystems (New York, N.Y.), 19*(4), 587–600. doi:10.100710021-015-9952-z

Sharma, P., Sharma, S., Maurya, R. K., De, T. D., Thomas, T., Lata, S., Singh, N., Pandey, K. C., Valecha, N., & Dixit, R. (2014). Salivary glands harbor more diverse microbial communities than gut in *Anopheles culicifacies*. *Parasites & Vectors, 7*(1), 235. doi:10.1186/1756-3305-7-235 PMID:24886293

Silberbush, A., Markman, S., Lewinsohn, E., Bar, E., Cohen, J. E., & Blaustein, L. (2010). Predator-released hydrocarbons repel oviposition by a mosquito. *Ecology Letters, 13*(9), 1129–1138. doi:10.1111/j.1461-0248.2010.01501.x PMID:20618841

Souza, R. S., Virginio, F., Riback, T. I. S., Suesdek, L., Barufi, J. B., & Genta, F. A. (2019). Microorganism-based larval diets affect mosquito development, size and nutritional reserves in the yellow fever mosquito *Aedes aegypti* (Diptera: Culicidae). *Frontiers in Physiology, 10*, 152. doi:10.3389/fphys.2019.00152 PMID:31024326

Stanczyk, N. M., Brookfield, J. F., Ignell, R., Logan, J. G., & Field, L. M. (2010). Behavioral insensitivity to DEET in Aedes aegypti is a genetically determined trait residing in changes in sensillum function. *Proceedings of the National Academy of Sciences of the United States of America*, *107*(19), 8575–8580. doi:10.1073/pnas.1001313107 PMID:20439757

Tchioffo, M. T., Boissière, A., Abate, L., Nsango, S. E., Bayibéki, A. N., Awono-Ambéné, P. H., Christen, R., Gimonneau, G., & Morlais, I. (2016). Dynamics of Bacterial Community Composition in the malaria mosquito's epithelia. *Frontiers in Microbiology*, *6*, 1500. doi:10.3389/fmicb.2015.01500 PMID:26779155

Turnipseed, R. K., Moran, P. J., & Allan, S. A. (2018). Behavioral responses of gravid *Culex quinquefasciatus, Aedes aegypti*, and *Anopheles quadrimaculatus* mosquitoes to aquatic macrophyte volatiles. *Journal of Vector Ecology*, *43*(2), 252–260. doi:10.1111/jvec.12309 PMID:30408300

Vikram, K., Nagpal, B. N., Pande, V., Srivastava, A., & Gupta, S. K., Anushrita, Singh, V. P., Singh, H., Saxena, R., Tuli, N. R., Yadav, N. K., Paul, R., Valecha, N., & Telle, O. (2015). Comparison of *Ae. aegypti* breeding in localities of different socio-economic groups of Delhi, India. *International Journal of Mosquito Research*, *2*(2), 83–88.

Wamaket, N., Dieng, H., Komalamisra, N., Apiwathnasorn, C., Morales, R. E., Thanomsub, B. W., Srisawat, R., & Attrapadung, S. (2018). Larvicidal and adulticidal activities of castor oil against the dengue vector, *Aedes aegypti*. *Tropical Biomedicine*, *35*(3), 610–618. PMID:33601748

Wang, Y., Yim, S., Yang, Y., & Morin, C. W. (2020). The effect of urbanization and climate change on the mosquito population in the Pearl River Delta region of China. *International Journal of Biometeorology*, *64*(3), 501–512. doi:10.100700484-019-01837-4 PMID:31811391

Wilke, A. B. B., Chase, C., Vasquez, C., Carvajal, A., Medina, J., Petrie, W. D., & Beier, J. C. (2019). Urbanization creates diverse aquatic habitats for immature mosquitoes in urban areas. *Scientific Reports*, *9*(1), 15335. doi:10.103841598-019-51787-5 PMID:31653914

Wu, X., Lu, Y., Zhou, S., Chen, L., & Xu, B. (2016). Impact of climate change on human infectious diseases: Empirical evidence and human adaptation. *Environment International*, *86*, 14–23. doi:10.1016/j.envint.2015.09.007 PMID:26479830

Xia, Y., Wang, G., Buscariollo, D., Pitts, R. J., Wenger, H., & Zwiebel, L. J. (2008). The molecular and cellular basis of olfactory-driven behavior in *Anopheles gambiae* larvae. *Proceedings of the National Academy of Sciences of the United States of America, 105*(17), 6433–6438. doi:10.1073/pnas.0801007105 PMID:18427108

Yadav, N. P., Rai, V. K., Mishra, N., Sinha, P., Bawankule, D. U., Pal, A., Tripathi, A. K., & Chanotiya, C. S. (2014). A novel approach for development and characterization of effective mosquito repellent cream formulation containing citronella oil. *BioMed Research International, 786084*, 1–11. doi:10.1155/2014/786084 PMID:25379509

Yee, D. A., Kesavaraju, B., & Juliano, S. A. (2004). Larval feeding behavior of three co-occurring species of container mosquitoes. *Journal of Vector Ecology, 29*(2), 315–322. PMID:15707290

Zahouli, J. B. Z., Koudou, B. G., Muèller, P., Malone, D., Tano, Y., & Utzinger, J. (2017). Urbanization is a main driver for the larval ecology of *Aedes mosquitoes* in arbovirus-endemic settings in south-eastern Côte d'Ivoire. *PLoS Neglected Tropical Diseases, 11*(7), e0005751. doi:10.1371/journal.pntd.0005751 PMID:28704434

Zwiebel, L. J., & Takken, W. (2004). Olfactory regulation of mosquito-host interactions. *Insect Biochemistry and Molecular Biology, 34*(7), 645–652. doi:10.1016/j.ibmb.2004.03.017 PMID:15242705

Chapter 5
Men, Mosquito, and Malaria:
The Circle of Life in Plasmodium

Arunava Mukherjee
https://orcid.org/0000-0002-0649-3219
Ramakrishna Mission Vivekananda Centenary College, Rahara, India

Pradipta Kumar Ghosh
Ecosystem and Ecology Laboratory, Ramakrishna Mission Vivekananda Centenary College, Kolkata, India

Debojyoty Ghosh
Sundarban Tiger Reserve, India

Banani Mandal
Jogesh Chandra Chaudhuri College, India

ABSTRACT

The chapter is dedicated to malaria, plasmodium, their ecology, evolution, life cycle, and host interaction. Authors divided the chapter into different subheadings to easily elucidate several aspects of the parasite and disease, which also include disease model evolution of zoonotic malaria of past and present, rise and fall of malaria drugs due to resistant plasmodium strains, and also situations of urban malaria. This may be useful for the reader as the authors also included many scientific anecdotes regarding the history of malaria, plasmodium, and human society.

DOI: 10.4018/978-1-7998-9414-8.ch005

INTRODUCTION

Noble laureate and Buddhist spiritual leader, the Dalai Lama (14th), has once famously said, "If you think you are too small to make a difference, try sleeping with a mosquito." It is quite evident that from the dawn of the Paleolithic age to the modern Anthropocene timescale, a tiny dipteran insect had piled up more human kill than any other large predator, including humans itself, on this little blue planet. This remarkable yet infamous achievement by mosquitos could not be possible unless with the presence of a pathogen or/and parasite. Of them, Plasmodium is a top-tier executive always at work. With a tiny mobile insect vector and a globally distributed enormous host population of naked apes, the business of Malaria is still booming even after the epic chloroquine and Artemisinin era. To this date, humans as a society knowingly and accidentally learned a few things about their oldest foe, malaria, which helped them to survive in the long run. A brief glimpse of this knowledge and occasional insights are placed here, ranging from the origin of malaria, its global consequences, counteracting between host and pathogens, and the new-old problems of malaria with some foreshadowing of nature.

GENERAL OUTLINE OF *PLASMODIUM SPECIES* AND THEIR LIFE CYCLE INSIDE HOSTS

Plasmodium is a prominent genus of protozoan endoparasite under the large phylum Apicomplexa that they share with other parasitic honchos such as Babesia, Cryptosporidium, Toxoplasma, etc. Plasmodium asexually initiates merogonic replication inside the host's erythrocytes and after digesting host haemoglobin produces haemozoin pigment as a byproduct. Due to the presence of apical complex and apicoplast (product of ancient secondary endosymbiosis), they are grouped into the phylum Apicomplexa (McFadden & Yeh, 2017). This obligate endoparasitic apicomplexan affects a vast range of terrestrial vertebrates as their definitive host and dipteran invertebrates as an intermediate host to accomplish their complex life cycle. Despite having more than 170 extant species and four subgenera, only 4 to 5 species are known to infect human beings with their severe pathogenicity, namely *Plasmodium vivax*, *P. ovale*, *P. malariae*, *P. knowlesi*, and *P. falciparum* and have distinct virulence and fever cycles. Infection is solely transferred by the vector, female *Anopheles sp.* mosquitoes, during their obligatory feeding of warm blood before oviposition. 10–100 sporozoites are injected with insect saliva into the host bloodstream during a blood meal. Sporozoites invade the host liver and migrate through macrophages (Kupffer stellate cells) and hepatic parenchymal cells within a short period. After 6-16 days, the mature sporozoites in the liver lead to 10 000-30

000 merozoites forming through erythrocytic schizogony. In the case of *P. vivax* and *P. ovale,* dormant hypnozoites also develop parallel to schizogony.

Based on the species of *Plasmodium*, erythrocytic schizogony lasts for 48-72 h in *P. vivax*, *P. ovale*, and *P. malariae*, but *P. falciparum* lacks this strict cycle. After several weeks, some of the merozoites differentiate into gametocytes, and consequently, both can be found in the host blood simultaneously. In vertebrates, cell types such as hepatocytes, erythrocytes and endothelial cells are heavily exploited during schizogony (Lucius *et al.,* 2017).

Following the ingestion by a mosquito, the gametes are developed by quick exflagellation from the gametocytes developing eight plasma protrusions to fertilize the macrogametes. The mobile zygote (ookinete) lodges itself between epithelial cells and the basement membrane of the mosquito gut. Sporogony starts when oocyst is formed from ookinete, resulting in sporozoites by way of meiotic division within weeks. The sporozoites then migrate to the vector mosquito's salivary gland through the haemocoel, from where they slip into host blood, and the cycle goes on.

ORIGIN OF THE PLASMODIUM, FROM PARASITES OF GREAT APES, TO OUT OF AFRICA BOTTLENECK LINEAGE

Malaria has been affecting Humans for more than five hundred thousand years. It was around a similar time scale when primitive humans discovered the fire-making processes. It has been rediscovered that rain forests of central Africa were the cradle of *Plasmodium* spp. even before the settlement of the hunter-agriculturist tribal community there. It is quite a logical speculation that the ancestral form of the Plasmodium evolved an acute parasitic nature due to the existence of potential host groups for parasitism around half a billion years ago in that very place. Molecular pieces of evidence strongly suggest that this ancestor of Plasmodium was a chloroplast-containing, sexually reproducing free-living protozoan that adapted to survive in the mid-gut of a group of aquatic invertebrates, presumably larvae of aquatic dipterans (appeared around 150-200 mya.). Gamete formation and fertilisation of *Plasmodium* spp. takes place inside the midgut of anopheline mosquito females, which evokes a presumption that the ancestors of plasmodium coexisted with the local anopheline mosquitos in the same water puddles and started hitchhiking with adult mosquitos. The mosquito naturally pierced an inattentive host for a blood meal, thus plunging the parasite into the crimson ocean of blood involuntarily; this repeated process possibly created a subset of parasites that became able to sustain the niche shift and utilised the new niche fully to get accommodation in an intermediate host. It has been speculated that old-world primates and tropical birds were the primary hosts of plasmodium even before the arrival of humans in the lush jungles of Africa. Still, parasites that resemble *P. falciparum* can be regularly found in chimps and Gorillas, such as malarial parasites under the Laverania subgenus, which regularly infect ape species. Interestingly, different *Laverania sp*. even with its great genetic diversity restricts itself to specific hosts despite sharing the same geographical range and niche (D.E. Loy *et al.*, 2017). It is possible that the widespread migration and inherited immunities of host humans led to the population bottleneck of *P. falciparum*.

HOW DO HUMAN HOSTS EVADE PLASMODIUM ATTACKS?

Despite all odds against the definitive host, it somehow managed to evade parasitic attack or at least dodge some ill effects by specific genotypic and phenotypic modifications. These modifications often cause a mild form of inherited diseases but provide blanket protection from the deadly malarial strains that can be lethal. It was first proposed by JBS Haldane in his famous 'Malaria hypothesis' (Haldane, 1948). The 'Malaria hypothesis' states that in homozygous forms these inherited diseases are mostly fatal (where the disease gene is inherited from both parents), while in the heterozygous state (where only one parent contributes the gene for the diseases trait) these inherited traits are largely beneficial to combat malaria in endemic regions. Here are some brief examples of such anti-malarial immunities.

Mediterranean Thalassemia

Thalassemia, often regarded as a vertically inherited haemoglobin deficient blood disorder, is a particular class of anaemia due to abnormalities in the encoded genes for haemoglobin (Cavalli-Sforza *et al.*, 1994). They involve the effective loss of either the alpha (HBA & HBA2) or beta (mutated HBB gene on Chromosome 11) chains of the molecule (Weatherall & Clegg, 2001). In Europe, specifically around the Mediterranean coastline, higher frequencies of thalassemia are found, most of them thalassemia major (Mediterranean anaemia). Haldane (1948) first observed this trend and hypothesised that populations in malaria-endemic regions like Mediterranean seashores also have a high rate of mutated haemoglobin disease. Beta thalassemia, also in heterozygous form, creates a natural immunity against plasmodium sporozoites and eventually mitigates malaria. Later, eminent geneticist Anthony C. Alison confirmed this hypothesis in 1954. Thalassemia occurs at high frequencies in most of Africa, the Indian subcontinent and Southeast Asia, which were the global limits of endemic malaria even 2,000 years ago.

Sickle Cell Trait

The sickle cell haemoglobin gene in heterozygous form (HbS) can provide more than 90% protection against malaria. Beet (1946) first noted the protective trait of Sickle Cell Anaemia (SCA) against falciparum malaria in Zambia, where a significantly reduced trend of malarial infection was found among SCA trait carriers than those who lacked the inherited genetic disorder altogether. After clinically and epidemiologically exploring SCA traits Alison and many others found that a single distinct point mutation in the haemoglobin beta chain, where the glutamate (Glu) is replaced by valine (Val) at the 6th location, changes into HbS, which, in homozygote condition, gives rise to lethal sickle cell anaemia (Alison, 1965) that cuts lifespan even before reaching puberty. The cost to HbS homozygote regarding reproductive expressions would have been considered a loss of fitness of nearly 100%. Such a fatal mutation can only sustain in the long run if the heterozygotes, carriers of one HbS gene and one normal HbA gene, have a more tremendous advantage to counterbalance the loss of the SCA homozygotes, which in turn provides good protection against the death risk from infection with *P. falciparum* malaria. In West Africa, the sickle cell gene children with heterozygosity are at much-reduced death risk from *P. falciparum* malaria, as against children with homozygosity for the normal gene. HbS is the prime example of the malaria hypothesis where specific genetic polymorphisms affecting red blood corpuscles (RBCs) are significantly selected due to their efficacy against malarial parasites.

G6PD Deficiency or Favism

A mutated gene for glucose-6-phosphate dehydrogenase (G6PD) can result in G6PD deficiencies. It is a congenital X-linked recessive metabolic error, thus mainly affecting male individuals. G6PD deficiencies, also known as favism, are a special type of haemolytic crisis triggered rapidly by consuming oxidant-rich foods such as fava bean (also from naphthalene, aspirin, antimalarial drugs, intravenous Vit-C doses and stressful viral infections, etc.). Fava beans (Vicia faba L.) contain a considerable amount of Vicine and Convicine that get metabolised into divicine and isouramil. Unfortunately, in G6PD deficient individuals, these compounds cause haemolysis by disrupting the RBC wall, often leading to jaundice, Haemoglobinurias and acute kidney failure in some severe cases (Giles *et al.,* 1967). During the Korean War (1950-1953), US soldiers of North African and Mediterranean descent who were largely G6PD deficient, suffered from haemolytic anaemia. They had taken preventive doses of anti-malarial primaquine before. In combination with both heterozygous and hemizygous populations, G6PD deficiency is associated with a level of congenital protection against severe *P. falciparum* malaria in about 50%

of instances. Hence, elevated records of G6PD deficiency around the globe can account for the effect of selection by malaria (Tishkoff *et al.,* 2001). That is why two major common variants of G6PD (G6PD-A & G6PD Mediterranean) are also present among African ethnicity, and the latter one is prevalent among the natives of the Mediterranean area and the Indian Subcontinent. However, the selection for G6PD deficiency seemed to have preceded the emergence of the falciparum event and coincided with the time when vivax malaria was a major selective force in the African continent (Alison, 1965).

Ovalocytosis

Melanesia is the second-highest malaria-prone area in terms of intensity after Africa, yet many tribal ethnic groups thrived there from time immemorial. Among the Melanesian aboriginals, there occurs a special polymorphism of 'band 3' structural proteins of the erythrocyte membrane. Unlike the loose binding of normal band 3 protein to Ankyrin, the mutated bond remains stronger than the normal type, resulting in greater robustness of the cell exterior membrane, which is the feature of South-East Asian ovalocytosis (SAO). The SAO differs greatly from the other hereditary elliptocytosis and is not interlinked with the scaffolding of the cytoskeleton, but results from modification in gene encoding, ultimately leading to a mutation of the band 3 protein. The condition was named after the ovoid shape of the RBCs in the heterozygote individuals. It also offers a form of malaria immunity as hereditary haemolytic anaemia. Due to mutation in the protein structure of the RBC, high ATP consumptions, less sensitivity to cellular spicule formation and osmotic pressures, increase in rigidity, ion exchange depletion and reduction of antigenic expressions can be observed, leading to significant resistance against the survival of *P. falciparum*. This physiological barrier acts as an effective process of natural selection, especially in the endemic malaria region characterised by high SAO (35%) cases. Band 3 gene deletion is lethal *in utero* regarding homozygote individuals, making the reproductive cost almost 100% by increasing the chance of ovalocytosis mutation. SAO occurs in New Guinea's malaria-infested lowlands at higher frequencies but nearly vanishes in the malaria-free cooler highlands. These trait carriers are at reduced risk of infection with *P. falciparum* and are mostly immune to vivax malaria (Genton *et al.,* 1995). Children with SAO mutation are also quite resistant to cerebral malaria (Carme *et al.,* 1993).

Duffy Negative RBC

Many indigenous populations of Central and West Africa are immune to *P. vivax* malaria, and that is one of the most striking examples of differential disease susceptibility among human populations. Thus, unlike in Asia, Europe and South America, vivax malaria is extremely rare in most of sub-Saharan Africa. These group of peoples with Duffy antigen negative RBC genotype FyFy are generally immune to P. vivax merozoite infections (Miller *et al.,* 1976). The Duffy antigen, encoded by the *FY* gene, is an active chemokine receptor expressed in various cell types, including erythrocytes in most human populations (except indigenous tribes of sub-Saharan Africa) having a promoter SNP that disrupts the binding site for the erythroid transcription factor GATA-1, which thwarts *P. vivax* from invading erythrocytes becoming the main factor for the remarkable absence of *P. vivax* in parts of Africa where other malaria parasites flourish (Miller *et al.*, 1976). This remarkable genetic discovery led to the observation of *P. vivax* Duffy-binding protein, a spiky molecule on the surface of the parasite that is critical for human erythrocyte invasion by *P. vivax* (Chitnis and Miller, 1994). The African Duffy-negative allele denoted as "FY*O," has the highest fixation index or F_{ST} (measure of variation of population due to its genotypic structures) observed in humans and has other features strongly suggestive of more recent positive selection (Hamblin and Di Rienzo, 2000). The apparent strength of biased selection at the *FY* locus is somewhat baffling for malariologists since *P. vivax* infection is mostly not lethal. It has even been suggested that prior infection from *P. vivax* may protect against the fatal consequences of its lethal cousin *P. falciparum.*

HOW DOES PLASMODIUM EVADE THE HOST'S IMMUNE SYSTEM?

Antigenic Variation of *Plasmodium falciparum*

Immune response against *P. falciparum* blood-stage infection is mainly determined by the immune system's recognition of the clonally variant surface molecule called PfEMP1 (*P. falciparum* Erythrocyte Membrane Protein 1) (Leech *et al.,* 1984). Genome sequencing revealed that *var* genes (~60) in the haploid genome of *P. falciparum* are crucial in the evasion process in the immunity system (Gardner *et al.,* 2002). Genome sequencing of another widely distributed human malaria species, *P. vivax* focused on the existence of a sizeable sub-telomeric gene family that is clonally variant at the surface of IEs (del Portillo *et al.,* 2001). Later, *P. falciparum* and *P. reichenowi* only were found to possess the genes encoding the virulence factor (encoded by the *var* multigene family). Different PfEMP1 adhesion ligands with varying functions can give rise to phenotypic variations to modify the antigenic makeup, resulting in significant recognition concealed by the host body's immune responses. A key factor in pathogenesis stems from the functional diversity and immune escape achieved by combining the surface of the IE molecule in *P. falciparum* (Miller *et al.,* 2002). The type of adherence trait expressed by IEs may determine disease severity and virulence. These variant adhesion molecules are critical players in the parasite's survival inside IEs, making them important targets for various intervention strategies. The causative factors for evolving the intracellular malaria parasites to expose immunodominant antigens at the surface of the IE are still unknown. It is generally accepted that mature blood-stage forms of *P. falciparum* increase the erythrocyte rigidity (Dondorp *et al.,* 2000), but apparently to a much lesser degree than other malaria species to avoid the culling of senescent RBC in the spleen. Using transgenic parasites, it was found that PfEMP1 is combined with the silencing of host body interferon-gamma during early immune response against plasmodium (D'Ombrain *et al.,* 2007), which may establish that a reduced level of interferon-gamma ensures the survival of parasite inside the host body.

Neutrophil Paralysis in *Plasmodium vivax* Malaria

The pathogenic activity of *P. vivax* malaria rises from the host-derived inflammatory mediators. Any infection by a foreign body activates neutrophils, the most abundant leucocyte in the peripheral blood system, with phagocytic activity and produces cell-destroying superoxides. However, despite that neutrophils remain poor sources of cytokines during malaria. An interaction between severe malaria and secondary bacterial infections (Bronzan *et al.,* 2007) suggests that the failure of neutrophils to migrate to the infection site may represent a mechanism that enhances the susceptibility of malaria patients to secondary bacterial infections. It is now also evident that Plasmodium can be recognised by the cells readily by TLR2, TLR4 and TLR9 (Coban *et al.,* 2005) and induce downstream regulation of CXCR2 via positively activating GRK2. Importantly, GRK2 also blocks CD88 (Langkabel *et al.,* 1999). Thus, decreased expression of CXCR2 and CD88 on active neutrophils from malaria patients may be a consequence of a boosted expression of GRK2. A primary mechanism for desensitization of activated GPCR is their phosphorylation by GRKs (Krupnick & Benovic, 1998). It is also described that transcription of GRK2 and GRK5 is regulated via LPS-mediated activation, leading to reduced expression of chemokine receptors and neutrophil chemotaxis, thus leading to neutrophil paralysis (Fan & Malik, 2003; Leoratti *et al.,* 2012). Enhanced GRK2 and GRK5 expressions were found linked with patients with sepsis, also associated with reduced migration of neutrophils and higher susceptibility to secondary microbial infection. A few studies have also revealed the link between *P. vivax* causing malaria with life-threatening manifestations, such as respiratory distress, severe thrombocytopenia and anaemia (Bassat & Alonso, 2011) that can cause even secondary infections ending untimely in septic shock (Franklin *et al.,* 2009) due to elevated secondary infections. Thus, extremely potent TLR agonist bacteria can easily cause significant damage to a malaria patient.

PAST, PRESENT AND FUTURE OF MALARIA AS A ZOONOTIC DISEASE

Living With Malaria: African Great Ape's Perspective

Many scientists correlate the aggressive virulence of *P. falciparum*, with its recent timeline in humans, having jumped to us probably due to a long-term spillover. Hence, the prime questions which continue to arise are, what was the source animal, and when did the leap occur? Studies have found the closest relative of *P. falciparum* is *P. reichenowi*, a malarial parasite regularly infecting chimpanzees, bonobos and gorillas of Central and West Africa. *P. reichenowi*, much older than *P. falciparum,* possesses greater genetic variety. In addition, all known variants of *P. falciparum* seem to be twigs within *P. reichenowi* family tree. Rich *et al,* (2009) proposed that falciparum has branched from *P. reichenowi* after spilling over from chimpanzees into human beings. The spillover probably occurred just once historically, as early as three mya or as lately as 10,000 years ago. The shifted strain of *P. reichenowi*, finding itself inside a novel host, mastered to survive and propagate successfully. It crowded the bloodstream of the first human hosts and then hitchhiked itself another mosquito ride under the same old tropical sun. From that accidental exposure, it travelled onward to other human settlements as they foraged in the forest and constantly revamped by mutations and adaptation to deviate from *P. reichenowi* branch to become regular human-infecting *P. falciparum*.

An alternative view was proposed by Krief *et al,* (2010) after locating another host of *P. falciparum*, a host in which that parasite appears to have evolved before spilling into humans, the bonobo (*Pan paniscus).* Screening blood samples from 42 bonobos from the Democratic Republic of the Congo (DRC), they found four individuals with blood parasites that are genetically indistinguishable from *P. falciparum.* The speculated outcome of the finding is that falciparum malaria was originally transmitted from bonobos into humans, probably sometime in the last 1.3 million years. The bonobos testing positive for *P. falciparum* had shown little signs of illness with lower parasitic loads in their blood and seemed consistent with an ancient host-parasite interaction. This finding led us to a hypothesis and an unspoken forewarning that if bonobos can carry a form of *P. falciparum* that is too similar to what humans have, those parasites may still be passing both sides between bonobos and us. Hence, falciparum malaria may be considered a true zoonotic disease. Humans in and around the forests of DRC might be infected on a regular basis with

P. falciparum from bonobos and vice versa, thus connecting man, mosquito and bonobos in a triad of *P. falciparum* infection. Although unspoken yet written on the wall of ecology and epidemiology, if cross-species or genus transmission occurs in this way, the great dream of malaria eradication becomes even less achievable.

***P. knowlesi* and the man monkey conflict:** A spillover of near future *Plasmodium knowlesi*, previously reported from Borneo and Peninsular Malaysia, can infect long-tailed macaque (*Macaca fascicularis*) and pig-tailed macaque (*Macaca nemestrina*). This lesser-known parasite is the fifth candidate responsible for human malaria (Collins, 2012). The parasite was first isolated from long-tailed macaque by HGM Campbell at Calcutta School of Tropical Medicine in Calcutta (Kolkata), India in 1931 (Napier and Campbell, 1932). Later, Robert Knowles and Biraj Mohan Das Gupta, injected the parasite into several other kinds of monkeys and followed the progress of infection and serial passages. The strain remained devastating to rhesus macaques, with high fevers and parasite loads in blood, killing or weakening the animals quickly. Knowles and Gupta also administered it to three human volunteers and found similar result with a distinctive one-day long intermittent high fever cycle. Later in 1932, it was named *Plasmodium knowlesi* by Sinton and Mulligan in honor of Robert Knowles (Coatney, *et al.*, 1971).

Around 1999, medical researchers Balbir Singh and Janet Cox-Singh reached Sarawak, the Malaysian side of Borneo, where they found seemingly unnatural cases of malaria, infected by *P. malariae* coming from the Ibn tribal community from Kapit to Sarawak. The skewed statistics of reported malaria with remarkable high incidence from Kapit were critical which is unusual for *P. malariae* infection. Instead of children, the victims of *P. malariae* majorly constitute healthy adults. Scientists soon unexpectedly found out using PCR data that *P. knowlesi* is present in the samples instead of *P. malariae*. Further investigations clarified that each patient was bitten by mosquitos that had bitten local macaques, a cross-species malarial infection in a nutshell. Later studies show that Ibn peoples, who practice agriculture, regularly stand guard to ward off crop-raiding macaques for several nights in harvest seasons (Quammen, 2013). During this course, they are regularly bitten by jungle mosquitoes like *Anopheles latens,* the vector of *P. knowlesi*. *P. knowlesi* was also found in blood samples from long-tailed macaques, confirming them as a regular reservoir. Three kinds of Asian macaques were reservoir hosts for the parasite, the long-tailed and pig-tailed macaque and the banded leaf monkey in Sarawak. Other monkeys, still unidentified, might also be potential candidates. While the macaques are arboreal, *Anopheles latens* has expanded its choice to biting humans, and that can be linked with ongoing forest fragmentation and poaching in SE Asia. It is logical to reason that with massive deforestation and wildlife depleting activities, humans indirectly influenced *P. knowlesi* to host switching. Thus, it should be considered a warning against future spillover irrespective of its nature as a transitional or overlapping stage.

DISEASE MODEL OF MALARIA: IN THE LIGHT OF ROSS MODEL AND OTHERS

During his tenure at Indian Medical Service in the 1890s, Sir Ronald Ross discovered and demonstrated the role of mosquitoes in the life cycle of malarial parasites. Ross also designed a mathematical model, widely known as the Ross model, to understand the malaria transmission process better (Ross, 1915). In epidemiological studies, the basic objective is to understand the transmission of infectious parasites within the host population. The individuals of the same host population can be grouped into diverse categories depending upon the nature of infection and the density of parasites within the host. Kermack and McKendrik (1922) classified the host population into different categories as, susceptible to infection (S), infected individuals unable to transfer the infection to other individuals of the population during the dormant period termed as Exposed (E), infectious individuals being able to transmit the parasite (I) and the individuals who have recovered from the infection (R). This S-E-I-R model is often called as epidemiological compartment model of Infectious diseases. Observing the nature of diseases and by using components like Susceptible (S), Exposed (E), Infected (I), or Recovered (R) at least 8 types of models can be generated such as SI, SIS, SEI, SEIS, SIR, SIRS, SEIR and SEIRS. The mathematical model described and designed by Sir Ronald Ross follows the SIS pattern. *Plasmodium* needs both human and mosquito hosts to complete their life cycle. Ross further classified human and vector mosquito populations into susceptible (Sh & Sm) and infected (Ih & Im) populations. While in the human population this model follows an SIS pattern, i.e., the infected class often being susceptible again (relapsing), within the vector mosquito population the model follows only the SI pattern, thanks to their short lifespan (Mandal *et al*, 2011).

In many endemic areas with malaria, an intimate relationship is observed between malaria and poverty. Modelling in these areas indicates the need for actual sustainable management systems for the ambient environment with good health care facilities and nutrient subsidies which indirectly create prophylaxis against malaria.

NATURAL CALAMITY AND OUTBREAK OF MALARIA: A CASE STUDY FROM ANDAMAN AND NICOBAR

Andaman and Nicobar were among the worst suffered areas of the 2004 Tsunami. Other than losing lives and economic losses, it caused excessive salinization of croplands and freshwater sources as long-term after-effects. After the tsunami, there were outbreaks of malaria in a few habitable islands among local inhabitants (Manimunda *et al.*, 2011). Though considered historically endemic in Andaman and

Nicobar Islands for over a century and in the penal colonies of Andaman, malaria was reported as the cause of high mortality and morbidity from 1858 to 1945. During this period, the British government constructed many engineering establishments along the shorelines that might have amplified the breeding of the vector mosquito, *Anopheles sundaicus* which prefers brackish-water bodies for laying eggs. In the post-tsunami months of 2004, government health officials observed that a surge of malaria cases occurs in several spots even in much interior of the islands. The areas being endemic to vivax and falciparum were a concern to health workers and ordinary people. As incursion of seawater during the tsunami filled several stagnant water bodies, paddy fields and puddles, fit for breeding *A. sundaicus* caused a surge in malaria (Tyagi *et al.,* 2013). The proximity of human settlements to these pools and lack of an alternative host, like cattle or poultry, may lead to a higher degree of man-vector contact raising the threat of malaria outbreak in those areas. It was also found that larvae of *Anopheles subpictus* (found in 71% of the favourable habitats sampled) and *Anopheles vagus*, coexist in saline pools with *A. sundaicus* larvae (Sunish *et al.,* 2015). Chronological data analysis from malaria-infected zones by the Directorate of Health Services showed that Slide Positivity Rate (SPR) for malaria in the Andaman district fluctuating between 0.2 and 0.4% in different months of 2003-2004, rocketed into a 5-6 fold increase in the post-tsunami months (Krishnamoorthy *et al.,* 2005).. This trend was also evident in the Nicobar district where the SPR was 10.59% during the post-tsunami months compared to 0.75% during 2004. In recent years, in Nicobar, it is also been documented that *A. sundaicus* also act as a vector for *Plasmodium knowlesi*. In the Nicobar, Long-tailed macaques, *Macaca fascicularis umbrosa*, are the only nonhuman primates (Vidhya *et al.,* 2019). These macaques are also an intermediate host of *P. knowlesi*. Areas of the Nicobar Islands are known for higher counts of malaria cases and for macaques that often come close to human settlements seem to be a perfect spillover point of zoonosis in not so distant future as anthropogenic interventions are rising in those small islands. During 2016-2018, researchers found traces of *P. knowlesi* in *A. sundaicus* using nested PCR, which is an alarming issue, though lack of concrete data made it open-ended and there are still lacunae regarding this transmission. Still, the ongoing anthropogenic pressure in those fragile islands may elevate the speed of the destruction of nature, which in turn can elevate the chances of a zoonotic disease outbreak (Sekhsaria, 2021).

DOMINANT MALARIA VECTORS IN INDIA

According to the WHO (WHO malaria report, 2021), India alone accounts for nearly 82% of malaria cases in the South-East Asian region. In 2011, 1.3 million confirmed malaria cases and 753 deaths were reported from India only, with approximated cases and death toll being 10-20 folds more (Dhingra *et al.,* 2010). However, for a developing country with endemic malaria, India fared well with a notable decline of malaria cases by 17.6% in 2019 than in 2018. Also, the application of the 'High Burden to High Impact (HBHI)' initiative to halt Malaria continued to make impressive gains in India with WHO and the RBM Partnership, resulting in an 18% reduction in the cases and a 20% reduction in death over the last two years with the aid of rapid rate of diagnostics, Long Lasting Insecticidal Nets (LLINs), etc. Nevertheless, the chances of the spread of chloroquine resistance and the emergence of multi-drug resistant strains remain a challenge. Even though the intensity is low to moderate regarding transmission, the disease is a serious health issue in some below poverty line marginalized populations (in Jharkhand, Odisha, Madhya Pradesh, etc.) throughout India.

Like other major insect groups, mosquito diversity is richer in the equator due to the tropical climate with numerous and diverse breeding resources (Nagpal & Sharma, 1995). Only six taxa out of 58 anophelines are major malaria vectors with wide regional spatial distribution (Dev & Sharma, 2013). *Anopheles culicifacies s.l.* is India's rural and countryside malaria vector with a larger range and generates about 65% of cases each year. *An. fluviatilis s.l.* is a stream breeder on the foothills contributing to 15% of malaria cases. *An. minimus* also breeds in the foothill streams but the northeastern regions. *An. dirus s.l.* is a forest dweller of northeastern states, while *An. sundaicus* is native to Andaman and Nicobar Islands and is a brackish water breeder. *An. stephensi* is the most well-known vector species of mosquito, which contributes largely to urban malaria (Subbarao, 1998). All other sibling species of the *An. culicifacies* complex are predominantly nocturnal zoophilic and endophagic. It can be controlled indoors by spraying pyrethroids (Singh *et al.,* 2012).

An. stephensi is most abundant during months of monsoon rainfall (June–August). In urban areas, being endophilic and endophagic, it breeds in domestic containers, building construction sites, overhead tanks, roadside boxes and discarded car tires (Sharma *et al.,* 1993). It is the major contributor to malaria outbreaks in urban areas related to construction projects. It can be controlled using legislative approved integrated control measures, reducing potential breeding pools, using chemical and biological larvicides and aerosol spraying to control adult mosquitoes.

Urban Malaria: A Hidden Problem in Rising India Regarding Construction Sectors

Though it was recorded that even today rural areas are reported as malaria's primary preferable habitat, the urban counterpart is also not too far. In spite of an agro- rural backed economy, India has steadily increased economic growth creating a change in the structure and mobility of rural Indian society. This also fueled the growth of migrant workers who had crossed state borders in search of work with better wages, especially in the rapidly growing cities. From some of the most underprivileged and malaria-exhausted parts of the country, migrant workers reach urban worksites (Kakkilaya, 2021). This seems to be also promising to malaria pathogens as it escapes their rural counterpart's period of long isolation and smaller host base. Now the pathogens of the countryside could freely blend in with those of the metropolis. As cities were expanding fast, both vertically and horizontally, they engulf multiple rural, forested areas and suburb sprawls. Displacement caused by these rapid speed of habitat fragmentation has made the vector mosquito communities ecological refugees that got fast acclimatized to the metropolis environment, where the construction industry plays a pivotal role in the sustainability of mosquitoes. Across the cities, anopheline mosquitoes found a new undisturbed brooding niche for their eggs in the water tanks' stagnant fresh waters (Sharma, 2004). The trouble becomes critical as due to the abundance of cheap laborers with meager wages for hire, developers mostly chose reinforced concrete where the mixture of cement needed to be "cured" by being submerged in water for several days. As the number of construction sites multiplied, so did the number of safe stagnant pools of curing waters, which act as a breeding sanctuary for mosquito larvae. Thus, the building developers would raise not only luxury apartment complexes but also large armies of mosquitoes as an unnoticed byproduct that leads to dire consequences. As malarial mosquitoes prodigally reproduce in stagnant curing waters at construction sites and thrive on the worker's blood in the makeshift slums nearby, it becomes a perfect epidemiological triad (Shivalli *et al.*, 2016). Also in the monsoons, the holes and crevices in half-built construction sites, are filled with rainwater followed by sporadic malaria outbreaks that spread through the city like wildfire (Kakkilaya, 2021). Extrapolating the data on migrant construction workers' health it was found that other than occupational hazards, the other major illness was singular, the high fevers of tropical malaria (Balkrikshna *et al.*, 2011).

Despite these discoveries, malaria is still persisted mostly with a tag of a rural problem. In many rural areas, health officials very actively investigate malaria cases to diagnose, treat and tally for their official statistics. However, such comparable efficient mechanism is mostly non-existent in urban areas and malaria cases remain largely undocumented and uncounted. Urban people with malaria who are treated by private doctors or are under self-medication (with over-the-counter medicines) were simply invisible in the official statistics. Because of this biased surveillance, it was easy to maintain the illusion that malaria is a problem restricted to the countryside. This illusion, in turn, served an unspoken but powerful socio-political purpose. It illogically portrays the disease as a problem of rural backwardness, which could be swept away by rapid economic development, an idea triumphantly pursued and patronized by corporate and political leaders. In fact, malaria may have been just the opposite, a problem of unplanned and unbridled urbanization. It was speculated earlier by experts that between 2018 and 2024, another construction boom, would arise with sixty million new homes under construction to fulfill the government's push for more affordable housing (Dhingra *et al.,* 2010). Thus India, the world's second biggest consumer of cement (> 60% of total used in residential real estate) will become a superpower in urban malarial distribution due to mosquito friendly construction procedures. In continuation of climate change with increased temperature the curing of the concrete for a longer time nurture more growth and reproduction of mosquitoes unknowingly (Sreedhar, 2012). Thus, mere human negligence can lead to possible epidemics in bustling urban landscapes, which may go berserk in long term.

Malaria and How It Shaped the History of Humankind

It often seems to be unreal yet quite astonishing to consider how infectious diseases have shaped human history. Unfortunate outbreaks of diseases not only piled up corpses but also tinkered with human society, culture, demography and general history. Malaria is also a notable candidate that somehow alters and tinkers the human history from global to local perspectives and also has sculpted our history, in more delicate yet equally long-lasting ways.

The Bantu Monopoly in Africa

The ethnic population of the Bantu linguistic group, ranging from western Africa and throughout the interior of southern Africa was among the foremost people, whose relative immunity to *P. falciparum* likely overpowered potential rivals of other ethnic groups. Linguistic evidences reveal that more than 300 different languages have stemmed from a proto-Bantu language that was used widely in southern Africa till 3000- 4000 years ago. However, there was, at one time, a larger diversity of linguistic groups on the continent than in recent times. Remnants of these non-Bantu-derived languages are still practiced by nomadic hunter-gatherer populations that live in more marginal lands.

Several theories were drawn on how Bantu groups achieve this kind of rapid leadership, some accounts for their regular better farmed diets, some rely more on their war tactics, but none fully makes a definite reason for their unbeaten long-term dominance. It is now known that the early Bantu people had far greater long-term tolerance to falciparum malaria than the other non-Bantu hunter-gatherers (Etkin, 2003). That is why living inside fixed farmer communities where *Anopheles gambiae* were seen naturally would have subjected the Bantu population to year-round falciparum infection. Endemic falciparum malaria has regularly killed their children and first-time mothers. But those who have survived would have enjoyed a lifelong advantage that no outsider could have. The local strain of falciparum to which their immune systems had been primed, could infect them now but no longer kill them. It weakened them occasionally, but no adult from endemic regions would have died from falciparum malaria. On the other hand, the not-so-fortunate nomadic hunter-gatherers who did not live around *A. gambiae* habitats became vulnerable to the fatal fever. As the malaria-proof farmers multiplied and spread with steadfast growth, the unprotected nomadic hunter-gatherer tribes would have to limit their distribution in certain fringe parts of Bantu tribe settlements. Following the historical records, fragmented survivors of once wide-ranging African nomadic hunter-gatherer communities like the Khoisan, Pygmy, and Cushite groups were forced to move to the periphery of the continent, where they remain even at present (Webb, 2005). In nutshell, historically the African Bantu linguistic tribes were supported by effective immunity from the *P. falciparum* infection, which helped them to expand their agro-based territory remarkably.

Malaria in the New World and the Infamous Slave Business

It is quite evident now that the Americas (both North and South, new world continents) were free from malaria for thousands of years before Europeans arrived and colonized in the 1600s. The European colonizers had malaria parasites probably both vivax and falciparum, and they encountered a land swarming with potential vector mosquitoes and extensive wetlands. Huge beaver-made dams and lodges (submerged habitat of beavers) flooded even more land every year and that temperate mosquito wonderland was connected by a narrow isthmus to the tropical South American continent, rich with dense mosquito-friendly impenetrable rain forests and river basins.

An array of different Anopheles species was present in the new worlds like *Anopheles crucians*, *An. punctipennis* and *An. quadrimaculatus* (King and Bradley, 1941), etc. Due to the unique ability to find a suitable vector from the mosquito community, the parasite *P. vivax* immediately spreads infection like wildfire. The southern USA states suffered badly due to many diverse mosquito breeding pools, which remain intact in the absence of freezing winter. Southern states, despite having a large malaria death toll, possess rich black soil that was hot beds for growing crops like maize, Sugarcane, tobacco and cotton, which need a huge number of manual laborers to maintain. To manage the large gap in the workforce in the farming sector and to plug depleting trend of European settlers, this crisis and the crisis management was the very beginning of the slave trade and forced labor industry in the plantations of southern states. In the era of the new world, a new business economy boomed on slave trading when the European colonizers had to depend on the African slaves to withstand the loss of workforce in fields and mines (Curtin, 1968). The slaves, who are majorly from the sub-Saharan region, are genetically Duffy -ve and GPC6 mutant, which gave them a natural immunity against vivax malaria, which in turn became beneficial for the European plantation owners. Around six million slaves were brought from Africa to America within a century, from 1700 to 1800 (Klein, 1999). Though a new problem appeared readily as with humans from sub-Saharan Africa, *P. falciparum* also reached America and altered the course of American history as well as its black people forever.

THE RISE AND FALL OF ANTI-MALARIAL DRUGS

From the dawn of *Homo sapiens* or even before, we as primates consumed plant matters to exploit the pharmacological potency of their natural chemicals. In a picture-perfect world, malaria parasites that lurk inside vectors and host animals generally should not be affected by these botanical bio-weapons. But luckily for humans, some of those compounds in the leaves and branches of plants can inflict severe damage to malaria parasites even when they hide in the liver or RBCs (Shah, 2010). Perhaps it's just a coincidental side effect, or maybe, another example of the evolutionary arms race, the red queen hypothesis.

In central Africa, both humans and chimpanzees often apply a specific shrub locally known as Mululuza (*Vernonia amygdalina D.*) that provides relief from malarial infection (Tenywa, 2007). It suggests the curious idea that our collective knowledge of ethnobotanical malaria medicines may have survived the perilous journey from ape to human (Helitzer *et al.,* 1993). Secondary metabolite-rich edible plants help destroy malaria parasites by exposing infected cells to the armies of free radicals. They may explain why people around the globe gathered and added these nearly nutritionally empty products to their regular diets throughout history.

One of the best potent botanical anti-malarial components can be found thousands of miles away from malaria's African cradle, in the inner bark of the cinchona, *Cinchona officinalis L.*, a neotropical tree that grows along the slopes of the Andes. The cinchona trees possess complex alkaloids for their defence against many pathogens and herbivores (Crozier *et al.,* 2006). One of the notable compounds from the bark is quinine, which has a significant effect on the Plasmodium parasite. The compound interferes with the parasite's main food source, haemoglobin, and get them poisoned the undigested toxic residue of its meal (Warrell & Gilles, 2002). The application of quinine partially has made it possible to prevent and mitigate malaria and its harmful effects successfully. It was this wonder drug that helped the British colonize even the highly malaria-endemic regions of continent Africa and Asia during the 19th century. To satiate the rising demand and supply many European nations started to import cinchona seed to start their plantations in their colonized lands in Asia, such as the Dutch government in Java (Indonesia) and the British in India (Nilgiris and Darjeeling mostly) (Shah, 2010). During Second World War, Germany simply looted the Dutch stash of quinine (Rocco, 2003), and Japan invaded Indonesia and captured all cinchona plantations; thus world's quinine hub had fallen into the hands of axis forces (Duran-Reynals, 1946). As the quinine flow dried up, Plasmodium flourished among the non-immune troops and sickened 60% of Allied troops in Southeast Asia (Harrison, 1996).

To control this mishap, scientists tried their best and discovered powerful synthetic anti-malaria chemicals like quinacrine and chloroquine. Chloroquine proved a success, whereas quinine failed and lasted longer with insignificant side effects; but the best part is the industrial mass production of Chloroquine. This cemented the downfall of heritage botanical medicine, quinine. Chloroquine kills the malaria parasite by concentrating inside its food vacuoles. But the scenario changed when the pfmdr1 gene of the parasite started to reject chloroquine out of their food vacuoles at a fifty times rapid rate (Institute of Medicine, 2004). Not just chloroquine, parasites endowed with pfmdr1 started resisting other antimalarial drugs, including amodiaquine, mefloquine, halofantrine and quinidine in a very short time. Researchers again started searching for a fast-acting stronger alternative to the quinine family and accidentally found medicinal properties of *Artemisia annua L.*, or sweet wormwood tree (Qinghaosu, 1979), from which a delicate compound called artemisinin is harvested, which can kill malaria parasites, by exerting oxidative stress on their cell membranes (Hien & White, 1993). Artemisinin proved to be a natural killer that repels malaria parasites faster, even in gametocyte stages, with lesser toxicity. Despite the promising drug's discovery, confusion, neglect, and misunderstanding delayed artemisinin's widespread distribution for decades, while drug-resistant parasites slaughtered scores. Though artemisinin-induced medicines became a mainstream medicine for eradication of malarial parasites not before 2004, they failed to ensure proper recovery again due to unequal distribution and exposure of long-term medicine in low doses, distribution of outdated medicines in underdeveloped malaria-prone areas and fake or adulterated medicines (Newton *et al.,* 2008). The ill consequences were clear, within 2007, in parts of SE Asia, artemisinin combination drugs were failing by up to 30 percent, because of artemisinin-resistant malaria. This failure reflects that the parasite's genome was mutating rapidly to counter the stresses of artemisinin's assault. In contrast, some of the mutated parasites exhibited a decreased sensitivity to artemisinin-laced compounds (Shah, 2010). Thus, under nature's observant gaze, the silent war goes on, with the allegory of the Red Queen hypothesis.

CONCLUSION

Like every higher organism (extant and extinct) on planet Earth, humans also have their own or shared predator, parasite and pathogen group. The Plasmodium belongs to the last two categories, always loitering between these two colossal ecological and epidemiological terms. It circuits human hepatocyte and midgut of mosquito in the same life cycle, but in different stages. The evolution of Plasmodium inside giant apes and our genetic makeup postulates their ever-presence around us. New candidates for expanding human malaria also imply the unnatural stresses and interventions that we offer to the close-knitted ecosystems. Thus, fine host-parasite tuning often distorts and causes unspeakable mayhem, and malaria's historical and geopolitical ballads become even longer. The unfathomable liaison of poverty and inequality with malaria is quite prominent in developing nations, reflecting the underdevelopment of the suitable schemes and their proper way of execution. The everlasting war of arms between cutting-edge medical interventions and a tiny protozoa's will to survive under undesirable conditions has shown us that panacea doesn't exist. The longest road of ongoing systemic management against malaria is our only ray of hope.

REFERENCES

Adsul, B. B., Laad, P. S., Howal, P. V., & Chaturvedi, R. M. (2011). Health problems among migrant construction workers: A unique public-private partnership project. *Indian Journal of Occupational and Environmental Medicine*, *15*(1), 29. doi:10.4103/0019-5278.83001 PMID:21808498

Allison, A. C. (1965). Population genetics of abnormal haemoglobins and glucose-6-phosphate dehydrogenase deficiency. International Journal for HP Jonxis.

Bassat, Q., & Alonso, P. L. (2011). Defying malaria: Fathoming severe *Plasmodium vivax* disease. *Nature Medicine*, *17*(1), 48–49. doi:10.1038/nm0111-48 PMID:21217683

Beet, E. A. (1946). Sickle cell disease in the Balovale district of Northern Rhodesia. *East African Medical Journal*, *23*, 75–86. PMID:21027890

Bronzan, R. N., Taylor, T. E., Mwenechanya, J., Tembo, M., Kayira, K., Bwanaisa, L., Njobvu, A., Kondowe, W., Chalira, C., Walsh, A. L., Phiri, A., Wilson, L. K., Molyneux, M. E., & Graham, S. M. (2007). Bacteremia in Malawian children with severe malaria: Prevalence, etiology, HIV coinfection, and outcome. *The Journal of Infectious Diseases*, *195*(6), 895–904. doi:10.1086/511437 PMID:17299721

Carme, B., Bouquety, J. C., & Plassart, H. (1993). Mortality and sequelae due to cerebral malaria in African children in Brazzaville. *The American Journal of Tropical Medicine and Hygiene*, *48*(2), 216–221. doi:10.4269/ajtmh.1993.48.216 PMID:8447526

Cavalli-Sforza, L. L., Menozzi, P., & Piazza, A. (1994). *The history and geography of human genes*. Princeton University Press.

Chitnis, C. E., & Miller, L. H. (1994). Identification of the erythrocyte binding domains of *Plasmodium vivax* and *Plasmodium knowlesi* proteins involved in erythrocyte invasion. *The Journal of Experimental Medicine*, *180*(2), 497–506. doi:10.1084/jem.180.2.497 PMID:8046329

Coatney G. R. & National Institute of Allergy and Infectious Diseases (U.S.). (1971). The primate malarias. U.S. Department of Health Education and Welfare National Institutes of Health National Institute of Allergy and Infectious Diseases.

Coban, C., Ishii, K. J., Kawai, T., Hemmi, H., Sato, S., Uematsu, S., Yamamoto, M., Takeuchi, O., Itagaki, S., Kumar, N., Horii, T., & Akira, S. (2005). Toll-like receptor 9 mediates innate immune activation by the malaria pigment hemozoin. *The Journal of Experimental Medicine*, *201*(1), 19–25. doi:10.1084/jem.20041836 PMID:15630134

Collins, W. E. (2012). *Plasmodium knowlesi*: A malaria parasite of monkeys and humans. *Annual Review of Entomology*, *57*(1), 107–121. doi:10.1146/annurev-ento-121510-133540 PMID:22149265

Crozier, A., Clifford, M. N., & Ashihara, H. (Eds.). (2008). *Plant secondary metabolites: occurrence, structure and role in the human diet*. John Wiley & Sons.

Curtin, P. D. (1968). Epidemiology and the Slave Trade. *Political Science Quarterly*, *83*(2), 190–216. doi:10.2307/2147089 PMID:11632739

D'Ombrain, M. C., Voss, T. S., Maier, A. G., Pearce, J. A., Hansen, D. S., Cowman, A. F., & Schofield, L. (2007). Plasmodium falciparum erythrocyte membrane protein-1 specifically suppresses early production of host interferon-γ. *Cell Host & Microbe*, *2*(2), 130–138. doi:10.1016/j.chom.2007.06.012 PMID:18005727

del Portillo, H. A., Fernandez-Becerra, C., Bowman, S., Oliver, K., Preuss, M., Sanchez, C. P., Schneider, N. K., Villalobos, J. M., Rajandream, M. A., Harris, D., da Silva, L. H. P., Barrell, B., & Lanzer, M. (2001). A superfamily of variant genes encoded in the subtelomeric region of Plasmodium vivax. *Nature*, *410*(6830), 839–842. doi:10.1038/35071118 PMID:11298455

Dev, V., & Sharma, V. P. (2013). The dominant mosquito vectors of human malaria in India. In *Anopheles mosquitoes-New insights into malaria vectors*. IntechOpen. doi:10.5772/55215

Dhingra, N., Jha, P., Sharma, V. P., Cohen, A. A., Jotkar, R. M., Rodriguez, P. S., Bassani, D. G., Suraweera, W., Laxminarayan, R., & Pet, R. (2010). Adult and child malaria mortality in India: A nationally representative mortality study. *Lancet*, *376*(9754), 1768–1774. doi:10.1016/S0140-6736(10)60831-8 PMID:20970179

Dondorp, A. M., Kager, P. A., Vreeken, J., & White, N. J. (2000). Abnormal blood flow and red blood cell deformability in severe malaria. *Parasitology Today (Personal Ed.)*, *16*(6), 228–232. doi:10.1016/S0169-4758(00)01666-5 PMID:10827427

Duran-Reynals. (1946). *The Fever Bark Tree*. Academic Press.

Etkin, N. L. (2003). The Co-evolution of People, Plants, and Parasites: Biological and Cultural Adaptations to Malaria. *The Proceedings of the Nutrition Society*, *62*(2), 311–317. doi:10.1079/pns2003244 PMID:14506878

Fan, J., & Malik, A. B. (2003). Toll-like receptor-4 (TLR4) signaling augments chemokine-induced neutrophil migration by modulating cell surface expression of chemokine receptors. *Nature Medicine*, *9*(3), 315–321. doi:10.1038/nm832 PMID:12592402

Franklin, B. S., Parroche, P., Ataíde, M. A., Lauw, F., Ropert, C., de Oliveira, R. B., Pereira, D., Tada, M. S., Nogueira, P., da Silva, L. H. P., Bjorkbacka, H., Golenbock, D. T., & Gazzinelli, R. T. (2009). Malaria primes the innate immune response due to interferon-γ induced enhancement of toll-like receptor expression and function. *Proceedings of the National Academy of Sciences of the United States of America*, *106*(14), 5789–5794. doi:10.1073/pnas.0809742106 PMID:19297619

Gardner, M. J., Hall, N., Fung, E., White, O., Berriman, M., Hyman, R. W., Carlton, J. M., Pain, A., Nelson, K. E., Bowman, S., Paulsen, I. T., James, K., Eisen, J. A., Rutherford, K., Salzberg, S. L., Craig, A., Kyes, S., Chan, M.-S., Nene, V., ... Barrell, B. (2002). Genome sequence of the human malaria parasite Plasmodium falciparum. *Nature*, *419*(6906), 498–511. doi:10.1038/nature01097 PMID:12368864

Genton, B., Al-Yaman, F., Mgone, C. S., Alexander, N., Paniu, M. M., & Alpers, M. P. (1995). Ovalocytosis and cerebral malaria. *Nature*, *378*(6557), 564–565. doi:10.1038/378564a0 PMID:8524388

Haldane, J. B. S. (1948). The rate of mutation of human genes. *Hereditas*, *35*(S1), 267–273. doi:10.1111/j.1601-5223.1949.tb03339.x

Hamblin, M. T., & Di Rienzo, A. (2000). Detection of the signature of natural selection in humans: Evidence from the Duffy blood group locus. *American Journal of Human Genetics*, *66*(5), 1669–1679. doi:10.1086/302879 PMID:10762551

Harrison, M. (1996). Medicine and the Culture of Command: The Case of Malaria Control in the British Army during the Two World Wars. *Medical History*, *40*(4), 437–452. doi:10.1017/S0025727300061688 PMID:8952289

Hien, T. T., & White, N. J. (1993). Qinghaosu. *Lancet*, *341*(8845), 603–608. doi:10.1016/0140-6736(93)90362-K PMID:8094838

Institute of Medicine. (2004). Saving Lives, Buying Time: Economics of Malaria Drugs in an Age of Resistance. National Academies Press.

Kamat, V. (2000). Resurgence of Malaria in Bombay. *Parasitologia*, *42*, 135–148. PMID:11234324

Kermack, W. O., & McKendrick, A. G. (1922). Contribution to the mathematical theory to epidemics. *Proceedings of the Royal Society of London. Series A, Containing Papers of a Mathematical and Physical Character*, *115*, 100–121.

Klein, H. S. (1999). The Atlantic Slave Trade. Cambridge University Press.

Krief, S., Escalante, A. A., Pacheco, M. A., Mugisha, L., André, C., Halbwax, M., Fischer, A., Krief, J. M., Kasenene, J. M., Crandfield, M., Cornejo, O. E., Chavatte, J.-M., Lin, C., Letourneur, F., Grüner, A. C., McCutchan, T. F., Rénia, L., & Snounou, G. (2010). On the diversity of malaria parasites in African apes and the origin of Plasmodium falciparum from Bonobos. *PLoS Pathogens*, *6*(2), e1000765. doi:10.1371/journal.ppat.1000765 PMID:20169187

Krishnamoorthy, K., Jambulingam, P., Natarajan, R., Shriram, A. N., Das, P. K., & Sehgal, S. C. (2005). Altered environment and risk of malaria outbreak in South Andaman, Andaman & Nicobar Islands, India affected by tsunami disaster. *Malaria Journal*, *4*(1), 32. doi:10.1186/1475-2875-4-32 PMID:16029514

Krupnick, J. G., & Benovic, J. L. (1998). The role of receptor kinases and arrestins in G protein-coupled receptor regulation. *Annual Review of Pharmacology and Toxicology*, *38*(1), 289–319. doi:10.1146/annurev.pharmtox.38.1.289 PMID:9597157

Langkabel, P., Zwirner, J., & Oppermann, M. (1999). Ligand-induced phosphorylation of anaphylatoxin receptors C3aR and C5aR is mediated by G protein-coupled receptor kinases. *European Journal of Immunology*, *29*(9), 3035–3046. doi:10.1002/(SICI)1521-4141(199909)29:09<3035::AID-IMMU3035>3.0.CO;2-Z PMID:10508278

Leech, J. H., Barnwell, J. W., Miller, L. H., & Howard, R. J. (1984). Identification of a strain-specific malarial antigen exposed on the surface of *Plasmodium falciparum*-infected erythrocytes. *The Journal of Experimental Medicine*, *159*(6), 1567–1575. doi:10.1084/jem.159.6.1567 PMID:6374009

Leoratti, F. M. D. S., Trevelin, S. C., Cunha, F. Q., Rocha, B. C., Costa, P. A. C., Gravina, H. D., Tada, M. S., Pereira, D. B., Golenbock, D. T., do Valle Antonelli, L. R., & Gazzinelli, R. T. (2012). Neutrophil paralysis in Plasmodium vivax malaria. *PLoS Neglected Tropical Diseases*, *6*(6), e1710. doi:10.1371/journal.pntd.0001710 PMID:22745844

Loy, D. E., Liu, W., Li, Y., Learn, G. H., Plenderleith, L. J., Sundararaman, S. A., Sharp, P. M., & Hahn, B. H. (2017). Out of Africa: Origins and evolution of the human malaria parasites *Plasmodium falciparum* and *Plasmodium vivax*. *International Journal for Parasitology*, *47*(2-3), 87–97. doi:10.1016/j.ijpara.2016.05.008 PMID:27381764

Lucius, R., Loos-Frank, B., Lane, R., Poulin, R., Roberts, C., & Grencis. R. (2017). The Biology of Parasites. Wiley.

Mandal, S., Sarkar, R. R., & Sinha, S. (2011). Mathematical models of malaria -a review. *Malaria Journal*, *10*(202), 1–19. PMID:21777413

Manimunda, S. P., Sugunan, A. P., Sha, W. A., Singh, S. S., Shriram, A. N., & Vijayachari, P. (2011). Tsunami, post-tsunami malaria situation in Nancowry group of islands, Nicobar district, Andaman and Nicobar Islands. *The Indian Journal of Medical Research*, *133*(1), 76–82. PMID:21321423

McFadden, G. I., & Yeh, E. (2017). The apicoplast: Now you see it, now you don't. *International Journal for Parasitology*, *47*(2–3), 137–144. doi:10.1016/j.ijpara.2016.08.005 PMID:27773518

Miller, L. H., Mason, S. J., Clyde, D. F., & McGinniss, M. H. (1976). The resistance factor to *Plasmodium vivax* in blacks: The Duffy blood- group genotype, FyFy. *The New England Journal of Medicine*, *295*(6), 302–304. doi:10.1056/NEJM197608052950602 PMID:778616

Nagpal, B. N., & Sharma, V. P. (1995). *Indian Anophelines*. Oxford & IBH Publishing Co. Pvt. LTD.

Napier, L. E., & Campbell, H. (1932). Observations on a Plasmodium Infection Which Causes Hæmoglobinuria in Certain Species of Monkey. *Indian Medical Gazette*, *67*(5), 246–249. PMID:29011028

Narain, J. P. (2008). Malaria in the South-East Asia: Myth & the reality. *The Indian Journal of Medical Research, 128*, 1–3. PMID:18820349

National Vector Borne Disease Control Programme. (n.d.). Government of India. http://www.nvbdcp.gov.in

Newton, P. N., Fernández, F. M., Plançon, A., Mildenhall, D. C., Green, M. D., Ziyong, L., Christophel, E. M., Phanouvong, S., Howells, S., McIntosh, E., Laurin, P., Blum, N., Hampton, C. Y., Faure, K., Nyadong, L., Soong, C. W. R., Santoso, B., Zhiguang, W., Newton, J., & Palmer, K. (2008). A collaborative epidemiological investigation into the criminal fake artesunate trade in South East Asia. *PLoS Medicine, 5*(2), e32. doi:10.1371/journal.pmed.0050032 PMID:18271620

Qinghaosu Antimalaria Coordinating Research Group. (1979). Antimalaria Studies on Qinghaosu. *Chinese Medical Journal, 92*, 811–816. PMID:117984

Quammen, D. (2013). *Spillover: Animal Infections and the Next Human Pandemic.* Vintage.

Rich, S. M., Leendertz, F. H., Xu, G., LeBreton, M., Djoko, C. F., Aminake, M. N., Takang, E. E., Diffo, J. L., Pike, B. L., Rosenthal, B. M., Formenty, P., Boesch, C., Ayala, F. J., & Wolfe, N. D. (2009). The origin of malignant malaria. *Proceedings of the National Academy of Sciences of the United States of America, 106*(35), 14902–14907. doi:10.1073/pnas.0907740106 PMID:19666593

Rocco, F. (2003). The Miraculous Fever-Tree: Malaria and the Quest for a Cure That Changed the World. New York: HarperCollins.

Ross, R. (1915). Some a priori pathometric equations. *British Medical Journal, 1*(2830), 546–447. doi:10.1136/bmj.1.2830.546 PMID:20767555

Sekhsaria, P. (2021, Mar. 20). NITI Ayog vision for Great Nicobar ignores tribal, ecological concerns. *The Hindu.*

Shah, S. (2010). *The Fever: How Malaria has ruled humankind for 500,000 years.* Academic Press.

Sharma, S., Chattopadhyay, R., Srivastava, V. K., Das, B. S., Pati, S. S., Mahanty, S., Tyagi, P. K., Chitnis, C., Misra, S. K., Adak, T., & Chakrabarti, K. (2004). Epidemiology of Malaria Transmission and Development of Natural Immunity in a Malaria endemic Village, San Dulakudra, in Orissa State, India. *The American Journal of Tropical Medicine and Hygiene, 71*(4), 457–465. doi:10.4269/ajtmh.2004.71.457

Sharma, S. N., Srivastava, P. K., Singh, S., Sharma, R. S., Sonal, G. S., & Dhariwal, A. C. (2014). Urban malaria scheme—Past, present and future. *The Journal of Communicable Diseases*, *46*(2), 77–84.

Sharma, S. N., Subbarao, S. K., Choudhury, D. S., & Pandey, K. C. (1993). Role of *An. culicifacies* and *An. stephensi* in malaria transmission in urban Delhi. *Indian Journal of Malariology*, *30*, 155–168. PMID:8131883

Shivalli, S., Pai, S., Akshaya, K. M., & D'Souza, N. (2016). Construction site workers' malaria knowledge and treatment-seeking pattern in a highly endemic urban area of India. *Malaria Journal*, *15*(1), 1–10. PMID:27530110

Singh, B., Lee, K. S., Asmad, M., Anand, R., Sunita, S., & Shamsul, G. (2004). A Large Focus of Naturally Acquired *Plasmodium knowlesi* Infections in Human Beings. *Lancet*, *363*(9414), 1017–1024. doi:10.1016/S0140-6736(04)15836-4 PMID:15051281

Singh, R. K., Mittal, P. K., Gourshettiwar, M. P., Pande, S. J., & Dhiman, R. C. (2012). Susceptibility of malaria vectors to insecticides in Gadchiroli district (Maharashtra), India. *Journal of Vector Borne Diseases*, *49*, 42–44. PMID:22585244

Subbarao, S. K. (1998). Anopheline Species Complexes in South-East Asia. *World Health Organization. Technical Publication, SEARO*, *18*, 82.

Subbarao, S. K., Vasantha, K., & Sharma, V. P. (1998). Response of *Anopheles culicifacies* sibling species A and B to DDT and HCH in India: Implications in malaria control. *Medical and Veterinary Entomology*, *2*(3), 219–223. doi:10.1111/j.1365-2915.1988.tb00186.x PMID:2485170

Sunish, I. P., Shriram, A. N., De, A., & Vijayachari, P. (2015). Malaria in the Andaman and Nicobar Islands: Challenges and opportunities for elimination. *Asian Pacific Journal of Tropical Disease*, *5*(10), 837–840. doi:10.1016/S2222-1808(15)60942-2

Tenywa G. (2007). *Chimps Eat Herbs to Cure Malaria*. AllAfrica.com

Tishkoff, S. A., Varkonyi, R., Cahinhinan, N., Abbes, S., Argyropoulos, G., Destro-Bisol, G., Drousiotou, A., Dangerfield, B., Lefranc, G., Loiselet, J., Piro, A., Stoneking, M., Tagarelli, A., Tagarelli, G., Touma, E. H., Williams, S. M., & Clark, A. G. (2001). Haplotype diversity and linkage disequilibrium at human *G6PD*: Recent origin of alleles that confer malarial resistance. *Science*, *293*(5529), 455–462. doi:10.1126cience.1061573 PMID:11423617

Tyagi, R. K., Das, M. K., Singh, S. S., & Sharma, Y. D. (2013). Discordance in drug resistance-associated mutation patterns in marker genes of *Plasmodium falciparum* and *Plasmodium knowlesi* during coinfections. *The Journal of Antimicrobial Chemotherapy, 68*(5), 1081–1088. doi:10.1093/jac/dks508 PMID:23292346

Warrell and Gilles. (2002). *Essential Malariology* (4th ed.). CRC.

Webb, J. L. A. (2005). Malaria and the Peopling of Early Tropical Africa. *Journal of World History, 16*(3), 269–291. doi:10.1353/jwh.2006.0011

World Health Organization. Report of an Informal Consultation. (2007). *Containment of Malaria Multi-drug Resistance on the Cambodia-Thailand Border*. WHO.

World malaria report 2021. (2021). Geneva: World Health Organization.

Chapter 6
Physical School Environment and Infectious Diseases:
A Case of Primary School Context in Bangladesh

Afzal Hossain
Institute of Education and Research, University of Rajshahi, Bangladesh

Emaj Uddin
https://orcid.org/0000-0002-6721-6856
University of Rajshahi, Bangladesh

ABSTRACT

The mechanisms underlying relationships between unhealthy physical environment in schools and infectious disease transmission among students are understudied. Drawing from Bronfenbrenner's ecological systems theory, this chapter analyzes behavioral patterns as mechanisms by which an unhealthy physical school environment transmits infectious diseases to primary students in Bangladesh. For the study, the authors randomly selected 300 students from grade-C primary schools located in disadvantaged areas in Rajshahi City, Bangladesh; over 12 months, they collected data on the physical school environment, children's behavioral patterns, and their infections from infectious diseases in the primary school context. The study found that 20-45% of the students suffered from various infectious diseases during the study period. The results from structural equation modeling suggested that poor qualities of the physical school environment via unhealthy behavioral patterns transmitted infectious diseases, affecting the students' physical health in the primary schools.

DOI: 10.4018/978-1-7998-9414-8.ch006

INTRODUCTION

Children are part and parcel of the school environment for healthy development and achievement (Ferguson et al., 2013). Ecological systems perspective and research suggests that school as a microsystem plays an important part in promoting children's healthy physical growth and psychosocial development, and performance (Ferguson et al., 2009, 2013; Kagan et al., 2002; Uddin, 2017a). The physical school environment, in which children spend about 4 to 6 hours a day, 180 days yearly, influences their health and success (Wigle, 2003). While a healthy physical school environment positively influences children's health, education, and well-being, the unhealthy school environment negatively influences their health and achievement (Aprihatin et al., 2020; Kuzik et al., 2022; Rohmah et al., 2022; WHO, 2004, 2016, 2020).

In Bangladesh, growing research reveals that unhealthy physical school environments, in which students from low economic backgrounds enroll and learn, transmit various infectious diseases (Asadullah & Chaudhury, 2011; Dreibelbis et al., 2016). Further studies indicate that students in over-crowded classrooms with poor lighting and ventilation face poor physical school (Bangladesh Bureau of Statistics, BBS, 2018, 2020; Jamal et al., 2013). In turn, poor qualities of the physical school environment frequently infect students with transmitted diseases in the primary school context (Afroz et al., 2019; Uddin, 2017b).

Although Rajshahi is a clean and green city, periodical research reveals that due to population growth and urban expansion, the low-quality of the primary schools' physical environment in the city is deteriorating (BBS, 2020). Compared to primary schools of high-quality education, low-quality schools (40-60%) with over-class size have poor physical environment conditions (BBS, 2018). Although poor qualities of the physical school environment in impoverished areas severely affect students' physical health with infectious diseases and increase morbidity and mortality by 10-15%, only a few studies have focused on the city (Karim et al., 2019; Khan et al., 2021; Wadood et al., 2021; Uddin & Ferdous, 2009; 2015). Based on Urie Bronfenbrenner's (1974, 1977) ecological system perspective, this chapter focuses on how the physical school environment affects the spreading of infectious diseases via behavioral patterns among primary school students in Bangladesh. The first part of the chapter mentions Bronfenbrenner's (1974, 1977) ecological systems perspective on which we develop a conceptual framework and its mechanisms by which the physical school environment transmits infectious diseases into primary school students. The second part of the chapter focuses on methodology. The third and last one presents the results and its discussion.

Ecological Systems Theory

Ecological systems theory by Urie Bronfenbrenner (1974, 1977) is a framework to understand child development and academic achievement across time and space. The theory is divided man-environment interaction into five different systems: the microsystem, the mesosystem, the exosystem, the macrosystem, and the chronosystem ((Bronfenbrenner, 1977, 1995). The microsystem and mesosystem of the theory are the most fundamental, immediate, environmental settings that contain interaction between child, family, and school (Bronfenbrenner & Ceci, 1994; Bronfenbrenner & Evans, 2000). The theory asserts that man-environment interaction across the different system levels produces positive or negative outcomes among children over time (Bronfenbrenner, 1974, 1977, 1994). This basic assertion of the theory is widely applied in social and behavioral sciences to understand children's psychosocial development, health, and academic achievement (Aturupane et al., 2013; Hayes et al., 2017; Stumm et al., 2021; Toyinbo et al., 2016). For example, Paat (2013) applied Bronfenbrenner's theory to understand the development of immigrant children and found that their culture and experience in the various ecological systems influence their development. Using ecological systems theory, Lippard et al. (2018) investigated teacher-child relationships through teacher's reports and classroom observations and found that supportive relationships between teacher and student significantly improve children's development and school achievement. Wilson et al., (2002) also found that a positive school environment developing through ethos and values positively influences student relationships, growth, and achievement. Langford et al., (2014) found that whole-school approaches to the health curriculum positively improve achievement and enhance student well-being. Kelly and Coughlan (2019) used the constructivist grounded theory analysis to develop a theoretical framework for youth mental health recovery and found that their theory, consistent with Bronfenbrenner's ecological systems theory, sufficiently explains influential relationships of ecological contexts such as peers, family, and school for youth mental recovery and development. Jamal et al. (2013) and Palinkas (1991) found that unhealthy physical school factors increase students' chronic diseases (e.g., stress, blood pressure, and respiratory problems), after accounting for socioeconomic background. Although Bronfenbrenner's ecological systems theory and early literature contributes to the relationships between family or school physical factors and children's chronic diseases, few studies have directly analyzed how the physical school environment transmits infectious diseases to students via their behavioral patterns in the primary school context (Jamal et al., 2013; Kagan et al., 2002; Palinkas, 1991).

Conceptual Framework

The unhealthy physical environment reflects the emergence and spread of infectious diseases in a school context (Ceddia et al., 2013; WHO, 2004, 2016). WHO (2004) broadly defines physical school environment as the "physical qualities of schools, encompassing "the school building and all its contents, including physical structures, infrastructure, furniture, and the use and presence of chemicals and biological agents; the site on which a school is located; and the surrounding environment including the air, water, and materials with which children may come into contact, as well as nearby land uses, roadways and other hazards." The term, 'infectious disease' is defined as a physical illness associated with pathogens or toxic products transmitted from infected persons, animals or contaminated inanimate objects to the susceptible hosts (Seventer & Hochberg, 2017). Ecological systems framework suggests that contamination of infectious diseases is determined by the dynamic interaction between unhealthy physical school objects and humans over time (Batterman et al., 2009; Jones et al., 2007; Kutsyuruba et al., 2015; Wilcox & Colwell, 2005). Based on the assertion of ecological systems theory we next review the literature to describe relationships between unhealthy physical school environment and infectious disease infection via behavioral patterns among students in the primary school context (see, Figure 1).

Figure 1. Conceptual framework

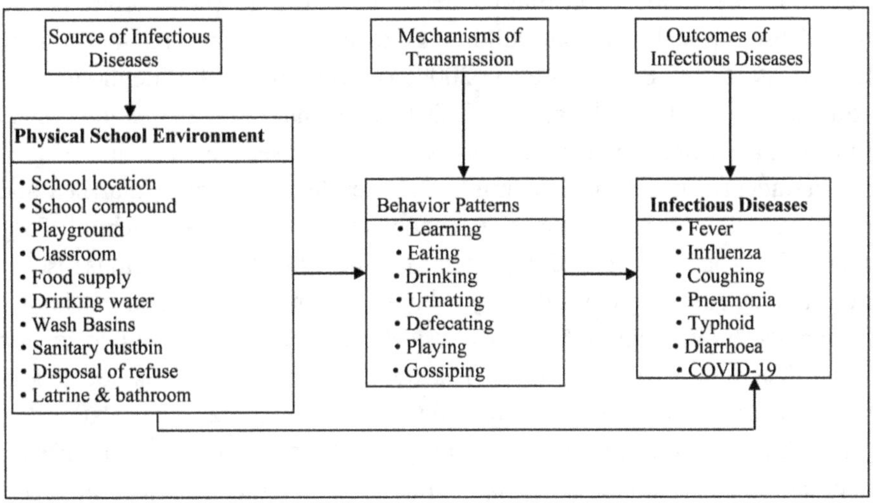

Physical School Environment and Infectious Diseases

The physical school environment may vary from one society to another. Based on Florence Nightingale's (1859) environmental theory, several studies have shown that unhealthy physical school environment characterized by an old or dilapidated school building, unsafe lavatory or bathroom, playground, basin, water, and food supply, poor sanitation, and irregular removal of refusal is more likely to infect students with infectious diseases (e.g., fever, influenza, pneumonia, diarrhea, bacteria, viruses, fungi or parasites; see, Ade et al., 2020; Chard et al., 2019; Ogunfowokan et al., 2017; Palinkas, 1991; Skretkowicz, 1992). Further, other physical factors such as high noise levels, low indoor air quality, high heat, and low access to clean drinking water and food also negatively influence the pupils' physical health (Jamal et al., 2013; McMichael et al., 2019; Njambi et al., 2020). Kutsyuruba et al. (2015) found that dilapidated school buildings affect students' poor mental and physical health. Overcrowding and heavy workloads create stressful working conditions for students and lead them to higher absenteeism from school (Chard et al., 2019). Crowded classroom conditions negatively influence students to engage in learning and stay in the class.

In tropical countries, several systematic studies have found that children are at higher risks due to drainage, open school compounds, unsafe playgrounds, water and food supply, poor sanitation, and hygiene that transmit infectious diseases (e.g., helminth, trachoma, diarrhea) into young children in school context (Merisa et al., 2014; McMichael, 2019; Strunz et al., 2014; Stocks et al., 2014). For example, in developed and developing countries, several studies have found that an unhealthy physical school environment transmits pandemic influenza and water-borne and soil-transmitted diseases that accelerate children's infection, morbidity, and mortality in the school context (see, Brouwers et al., 2009; McMichael, 2019; Ridenhour et al., 2011; Sadique et al., 2008; Yoldascan et al., 2010). Freeman et al. (2013a) intervened over time to analyze and compare the effects of school-based water, sanitation, and hygiene on students' absenteeism, helminth, diarrhea, fever, influenza, diarrhea, and respiratory infection between beneficiary school and non-beneficiary school. They found that school-based intervention programs reduce 40-50% of infectious disease infections among students in beneficiary schools than in non-beneficiary schools.

In Bangladesh, although unhealthy physical school environments and related infectious diseases are increasing in low-quality primary schools located in disadvantaged areas, there are no systematic studies to showcase the relationship between unhealthy school physical environments and the spread of infectious disease results in morbidity or mortality among primary students (Asadullah et al., 2011; Dreibelbis et al., 2016; Haque et al., 2016). Some descriptive studies have revealed that about 19-45% of primary schools have no safe drinking water, sanitation, latrine, and hygiene facilities, resulting in 3 to 9 million school children being at higher risk for infectious diseases (Haque et al., 2016). In the study context, most primary schools (25) studied have no healthy school buildings and boundaries. In addition, clean drinking water, food supply, and WASH facilities (e.g., toilets, basins, playgrounds, dustbins, and soap) are not available in schools (Akther et al., 2021). Lack of the WASH facilities in impoverished schools has a greater probability of infecting students with several infectious diseases in school settings (Sultana et al., 2017). Linkon et al. (2015) studied school environment and infection of contagious diseases among students in a primary school setting and found that children (N=84) enrolled in low-quality of primary school were infected more by scabies (31%) than influenza (21%), measles (15%), whooping cough (11%), chicken pox (10%), typhoid (8%), and pneumonia (6%) respectively. Consequently, more than 50 thousand children die each year associated with infection of several infectious diseases and poor hygiene in school settings (Rahman et al., 2019). Like intervention studies in developing countries, research in Bangladesh also has found that best practices of personal hygiene and improving physical qualities of the school environment reduce infection and morbidity from infectious diseases among primary school children (Akther et al., 2021; Haque et al., 2016; Rahman et al., 2019).

Mechanisms of Infectious Disease Transmission

Although the relationship between an unhealthy physical environment and infectious disease is strong, less is known about how an unhealthy physical school environment transmits infectious diseases in young students in primary school settings (Eisenberg et al., 2007). Behavioral patterns are important pathways by which an unhealthy physical school environment might transmit infectious diseases to students (Batterman et al., 2009; McLaren & Hawe, 2005; Wilcox et al., 2005; WHO, 2004). Despite this, these pathways are understudied (WHO, 2004, 2015, 2016). This study focuses on the behavioral patterns by which infectious diseases are transmitted into the human body in the primary school setting (see Figure 1). Children spend about ¼ times each working day at primary school. Meanwhile, they sit on the benches and read books from the school library. They eat foods and drink water at school. In their leisure time, they play on the school playground. Sometimes, they use toilet or bathroom when necessary (WHO, 2004).

Research shows infectious diseases transmission among students is significantly associated with their day-to-day behavioral patterns, including eating, drinking, playing, learning, urinating, and defecating during working hours in the school setting (Pinter-Wollman et al., 2018; WHO, 2004, 2016). WHO (2004, 2020) argued that children aged 5-10 years lack experience in the risks of infectious disease exposure in unhealthy physical school environments and its properties used. In developed and less-developed countries, several studies have found that 40-60% of children place their hands on their mouths, nose, eyes, and ears in poor schools (Pinter-Wollman et al., 2018). They do not wash their hands with soap or use hand wash liquid before eating, drinking or after using the toilet or bathroom. In their leisure time, they play on the dirty playground without shoes. They do not wear masks during school periods. Consequently, 40-50% of the students infected by infectious diseases, including fever, influenza, diarrhea, pneumonia, and COVID-19 (Caruso et al., 2014; Hunter et al., 2014; Ismail et al., 2021; Rosen et al., 2006; Talaat et al., 2011). In developing countries, some studies have found that intervention in water, food, handwashing, and hand sanitizer reduces students' infection by infectious diseases in schools than in non-intervention schools (Dreibelbis et al., 2016; Pickering et al., 2013). In Bangladesh, few studies have focused on children's behavior patterns as mechanisms by which unhealthy school physical environments transmit infectious diseases into their human body and affect their physical health and well-being ((Akther et al., 2021; Haque et al., 2016; Rahman et al., 2019; Sultana et al., 2017). Sultana et al. (2017) and their colleagues (Akther et al., 2021; Biswas et al., 2019) found that intervention in water, food, sanitation, and changes in unhealthy behavioral patterns reduces 30-60% of infections by infectious diseases in the primary school setting.

Relationships between physical school environment, behavior patterns, and infection disease infection among primary students may vary by school quality, socioeconomic status and age pattern (Heyneman & Loxley, 1983; Stumm et al., 2021). Aturupane et al. (2013) and their colleagues (Engin-Demir, 2009) found similar findings in developing and less developed countries. Likely, relationships between the physical school environment and infectious diseases via behavior patterns may also vary by students' age and socioeconomic status (SES). Research has shown that young children with low SES are less conscious about unhealthy physical school environments and infectious disease transmission in school settings (Rutter, 1983). Consequently, they follow less health sanitation and hygiene than the students with older in age and higher in SES (Dreibelbis et al., 2016). As these factors are related to a physical school environment and infection via behavioral patterns, this study will control the factors to understand the relations between the physical school environment and infectious diseases via students' behavior patterns in the Bangladesh context (Hossain & Uddin, 2021).

The current study hypothesizes that indicators of an unhealthy physical school environment are significantly associated with infectious diseases in a primary school context. Second, the study also hypothesizes that unhealthy behavior patterns of children through which unhealthy physical school environments transmit infectious diseases to young students in the primary school context. These two hypotheses were tested in an urban Bangladesh context where impoverished primary schools are available. In so doing, background factors such as age, SES and school quality were controlled to understand the relationships between unhealthy physical school environment and infectious disease transmission through children's behavior patterns in primary school.

METHOD

Sample

The present study to examine how the physical school environment infects students with infectious diseases via their behavior patterns was conducted in Rajshahi city. First we visited the primary education office and primary schools in Rajshahi city. We found 25 primary schools with three categories: A Grade = 5, B Grade = 10, C Grade = 10. Based on school environment, teacher-student ratio, class assessment and results, the Ministry of Primary and Mass Education has classified the primary schools into the three categories. For this study, we treated these three types of schools as high-quality of primary school, medium-quality of primary school, and low-quality of primary school respectively. In these schools, there were about 1000 students. Of the students, 300 students were randomly selected from impoverished schools for the study and then the samples were intensively studied for this research purpose.

Table 1. Physical school environment characteristics of the students (N = 300), Rajshahi City, Bangladesh, 2022

Physical School Environment Characteristics	Frequency (Yes/ No)	Percent
1. Do you have well school buildings and location (yes/no)?	110/ 190	57.89
2. Do you have well school boundary walls (yes/no)?	80/220	36.36
3. Do you have clean drinking water supply at school (yes/no)?	75/225	33.33
4. Do you have clean toilets at school (yes/no)?	50/250	20.00
5. Do you have a working tube-well at school (Yes/ No)?	40/260	15.38
6. Do you have clean playgrounds at school (Yes/No)?	70/230	30.43
7. Do you have sanitary dustbins at school (Yes/No)?	55/245	22.45
8. Do you have available electricity and lighting at school (Yes/No)?	75/225	33.33
9. Do you have a rich library at school (Yes/ No)?	35/265	13.21
10. Do you have regular disposals of refuses at school (Yes/ No)?	80/220	36.36
11. Do you have available handwashing facilities at school (Yes/No)?	40/260	15.38
12. Do you have available foods/tiffin at school (Yes/ No)?	35/265	13.21
13. Do you have comfortable classrooms at school (Yes/No)?	45/255	17.65

Variable and Measure

This study used four types of variables - predictor, mediator, outcome, and control- to analyze the research hypotheses mentioned above. These variables were defined and measured in the following ways:

Physical school environment characteristics. There are a lot of physical environment characteristics of a school. The current study included school building, boundary walls, furniture (chairs, tables, benches, blackboards), availability of electricity and lighting, library, clean drinking water supply, available clean toilets, handwashing, tube-well, open playground/space, and school tiffin. The respondents were asked about whether their school had a well-school building, boundary wall, furniture (chair, table, bench, blackboard), availability of electricity and lighting, library, clean drinking water supply, clean toilets, basins, handwashing facility, tube-well, open playground, and space (WHO, 2004). These variables of the physical school environment were nominally measured (0= no, 1= yes). The findings of the physical school environment indicators are presented in Table 1.

Students' behavior pattern characteristics. When students spend time at school, they engage in different behaviors/activities, including participation in class, gossiping with the classmate, playing on the field and interacting with the school staff. In so doing, they contact different types of school physical objects and people through which infectious diseases may transmit into their human body. To measure students' behavioral patterns, seven items: learning/reading books, eating, drinking, urinating, defecating, playing and gossiping were included in the study (WHO, 2004). The respondents were asked whether they washed their hands with soap or hand liquid before or after reading books or participating in class, eating food, drinking water, urinating, defecating, playing with wearing shoes/cates and gossiping with wearing masks. These items were measured with Likert scaling, ranging from 1 = Never, 2= sometimes, and 3= always. These items were applied twice, from: - January to June 2022 and July to December 2022. The summary results of the scale on children's behavior patterns are presented in Table 2.

Table 2. Measuring indicators for behavior pattern among the students (N = 300), Rajshahi City, Bangladesh, 2022

Items of Behavioral Patterns	Range	January-June 2022	July-December 2022
		Mean (SD)	Mean (SD)
1. Washing hands with soap after reading a book at school	1-3	1.09 (0.71)	1.10 (0.72)
2. Washing hands before eating food at school	1-3	1.22 (0.82)	1.25 (0.85)
3. Washing glass before drinking at school	1-3	1.31 (0.55)	1.34 (0.59)
4. Washing hands after using toilet (defecating) at school	1-3	2.01 (1.00)	2.11 (1.02)
5. Washing hands after using bath room (urinating)	1-3	1.50 (0.99)	1.56 (1.00)
6. Playing with wearing shoes/cates on the school field	1-3	1.02 (0.45)	1.32 (0.55)
7. Wearing mask when gossiping with others at school	1-3	1.04 (0.52)	1.34 (0.72)
Average grand total	1-3	1.31 (0.72)	1.43 (0.78)

Note: SD = Standard Deviation

Infectious disease (symptoms) characteristics. There are several types of infectious diseases such as water-borne, food-borne, air-borne, soil-borne, vector-borne and others (Eisenberg et al., 2007; WHO, 2004, 2015). To measure infectious diseases, this study included fever, influenza, coughing, typhoid, Diarrhoea, pneumonia, and COVID-19. These variables of infectious diseases were numerically measured. The respondents were particularly asked how many times they were infected by fever, influenza, coughing, typhoid, Diarrhoea, pneumonia, and COVID-19 in the last six months (e.g., January to June 2022). This measure was repeated after six months (e.g., July to December 2022). The summary results on the indicators of infectious diseases are presented in Table 3.

Table 3. Infectious disease characteristics among the students (N = 300), Rajshahi City, Bangladesh, 2022

Physical Disease Characteristics	Range	January–June 2022	July-December 2022
		Mean (SD)	Mean (SD)
1. Number of fever infected in the last six months	0-12	4.40 (1.00)	4.48 (1.20)
2. Number of influenza infected in the last six months	0-12	3.22 (0.99)	3.29 (0.99)
3. Number of coughing infected in the last six months	0-5	1.24 (0.25)	1.28 (0.25)
4. Number of typhoid infected in the last six months	0-1	0.50 (0.08)	0.50 (0.08)
5. Number of Diarrhoea infected in the last six months	0-12	6.20 (1.12)	8.25 (1.18)
6. Number of pneumonia infected in the last six months	0-6	2.25 (0.78)	3.28 (1.78)
7. Number of COVID-19 infected in the last six months	0-1	0.50 (0.01)	0.50 (0.01)
Average grand total		2.62 (0.60)	3.08 (0.78)

Note: SD = Standard Deviation

Background characteristics. Past research shows that background characteristics such as age, SES and school quality influence variations in physical school environment, behavior patterns, and infectious disease infection; hence we used the children's age, SES and school quality as control variables. Particularly, the age of the students was numerically measured (Mean= 11.25, SD= 2.09). The variable SES was categorically measured and coded as 1= low-SES (60%), 2= medium-SES (30%) and 3= high-SES (10%). The quality of primary schools was measured and coded as 1= low-quality school (70.25%), 2= medium-quality schools (20.20%), and high-quality schools (9.55%). These background variables were used as control variables or covariates in the model.

Procedure

This study used semi-longitudinal survey design to analyze relationships between the physical school environment, behavior patterns, and infectious diseases among primary school students in urban Bangladesh. A questionnaire and scaling method was designed for data collection. Based on the questionnaire, each student was asked for relevant background questions. Before final data collection data collectors visited the respondent's homes and primary schools. Fieldwork for this research was conducted from January to December, 2022. To collect data using the questionnaire and scaling methods, researchers built rapport with the students and school head teacher to create consciousness about the research purposes and objectives and encourage them to actively participate in the research process. During the first month (January, 2022) of data collection, we built rapport with the selected students and school teachers, especially head of the schools. Then the researchers collected first-wave (W1) data on background and physical school environment indicators in February, 2022. Four months later, the researchers collected first-wave (W1) data on children's behavior patterns and infectious diseases in June, 2022 in school setting. Five months later (December, 2022), the researchers again collected second-wave (W2) data on children's behavior patterns and infectious diseases in the school setting. The researchers maintained privacy, mutuality, and confidentiality in data collection.

Data Analysis Plan

The data analysis was carried out with descriptive (see, Table 1, 2, 3) and bivariate correlation (see, Table 4) techniques for all variables. The mechanisms of infectious diseases were tested from W1 physical school environment through W1-W2 children's behavior patterns at school to W1-W2 infectious diseases in a primary school context. Testing the pathways, structural equation modeling (SEM) with Amos (version 20.0) was applied (Byrne, 2016). The study used physical school environment indicators, children's behavior patterns, and infectious diseases as three latent constructs. The analysis of latent variable modeling can decompose the extent to which physical school environment indicators and children's behavior patterns directly affect infectious diseases in school. The latent variable analysis also has indirect effects of physical school environment indicators through children's behavior patterns on later their infectious disease transmission.

In the SEM analysis, there are two types of models: Measurement model and structural model. In this study, the measurement model described model identification and factors loadings of the latent constructs: Physical school environment, children's behavior patterns, and infectious diseases. The structural model outlined the coefficients of latent relations, including direct and indirect pathways (shown in Figure 1) through which the W1 physical school environment relates to W1-W2 children's behavior patterns to W1-W2 infectious diseases transmission among primary students. Before testing mediation, the current study conducted confirmatory factor analysis (CFA) to identify the model, including factor loadings of the latent constructs, particularly children's behavior patterns (see, Table 5). The results showed that the model was fully identified and the factor loadings were high (see, Table 5). As the first requirements in the measurement model were fulfilled, the effects of children's behavior patterns on the relationships between the physical school environment and infectious diseases were tested. In doing so, a full mediation model to identify the effects of children's behavior patterns on the association between physical school environment and infectious diseases was tested (see, Figure 2). The model included background variables (e.g., age, SES, and school quality) to determine whether these control variables affect the focal model (Uddin, 2022). The model fit was checked, using chi-square, goodness-of-fit index (GFI), comparative fit index (CFI), and the root mean square error of approximation (e.g., RMSEA; Byrne, 2016).

RESULTS

Bivariate Correlation

Descriptive findings presented in Table 1 showed that unhealthy physical school environments were the major sources of infectious diseases. Consequently, 50-60% of the students were frequently (2 to 5 times) affected by several infectious diseases via their behavioral patterns in the primary school setting. The results of bivariate correlation revealed that the unhealthy physical school environment was significantly ($p<0.05$) associated with infectious diseases via children's behavior patterns (see Table 4). Particularly, unhealthy physical school environment was significantly associated with W1-W2 children's unhealthy behavior patterns ($r=$

0.57, 0.52, p<0.05) and W1-W2 infectious diseases (r = 0.46, 0.49, p<0.05). W1-W2 children's unhealthy behavior pattern was significantly (r = 0.31 to 0.56, p<0.05) associated with W1-W2 infectious diseases(r= 0.52, p<0.05). The background factors included in the bivariate model also were moderately and positively related to the unhealthy physical school environment and children's behavior patterns and infectious diseases (r = 0.15 to 0.20, p<0.01).

Table 4. Result of bivariate correlation for all major variables

Variable	1	2	3	4	5	6	7	8
1 W1Physical school environment	-							
2 W1 Children's behavior pattern	0.57*	-						
3 W2 Children's behavior pattern	0.52*	0.50*	-					
4 W1 Infectious diseases	0.46*	0.56*	0.52*	-				
5 W2 Infectious diseases	0.49*	0.31*	0.40*	0.48*	-			
6 Age	0.15**	0.20*	0.14*	0.39*	0.46*	-		
7 SES	0.16**	0.22*	0.13*	0.28*	0.38*	0.25*	-	
8 School quality	0.20**	0.18**	0.10*	0.13*	0.23*	0.17*	0.29*	-

Note: * p<0.05, ** p<0.01

Measurement Model

Standardized factor loadings from CFA conducted for children's behavior patterns at school are shown in Table 5. The results indicated that the measurement model was fully identified. The convergent validity for children's behavior indicators onto their seven respective factors (the standardized factor loadings of the latent mediators) were relatively high from .591 to .780; CFA model fitting with the data was adequate at χ^2 (112, N=300, df= 15) = .128; other fit indices were well (GFI=.911; TLI= .915; CFI= .932; RMSEA= .034).

Table 5. Factor loadings of children's behavior patterns

Items of Children's Behavior Patterns at Primary School	Standardized Loading	SE
1. Learning/reading	.651	.012
2. Eating	.712	.013
3. Drinking	.764	.024
4. Urinating	.653	.051
5. Defecating	.780	.050
6. Playing	.631	.042
7. Masking	.591	.052

Structural Model

Direct Effects

The SEM model was found to have adequate model fit (χ^2 (122) = 116.99, p = .152; GFI = .925; TLI = .948; CFI = .924; RMSEA = .031; see, Figure 2). The standardized coefficients (*βs*) of the direct effects of the SEM model are also described in the text. The results of the model that explained 23 - 38% variance in the infectious disease outcomes indicated positive and significant direct relations between physical school environment, children's behavior patterns, and infectious disease in primary school, after accounting for background characteristics. The present study found that an unhealthy physical school environment was directly and significantly associated with W1-W2 infectious disease outcomes at primary school level (β = .38, SE= .02, $p < .05$, β = .44, SE= .01, $p < .05$). The unhealthy physical school environment was directly and significantly associated with W1 children's behavior patterns (β = .34, SE= .03, $p < .05$); W1 children's behavior patterns were directly and significantly associated with W2 children's behavior patterns (β = .35, SE= .04, $p < .05$) and that were associated with W1 infectious diseases at primary school (β = .47, SE= .04, $p < .001$); further W1 infectious disease outcomes were directly and significantly associated with W2 infectious diseases at primary school level (β = .48, SE= .04, $p < .001$). Overall, the study's findings suggested that unhealthy physical school environment directly affected infectious disease outcomes and children's behavior patterns. Likely, children's behavior patterns directly affected infectious disease outcomes in the primary school context.

Indirect and Total Effects

Like direct effects, the current study also found indirect effects of W1-W2 children's behavior problems on the relationship between an unhealthy physical school environment and infectious disease outcomes. Specifically, when children's behavior problems were included in the baseline model with control variables, significant direct effects of unhealthy physical school environment on infectious disease outcomes were substantially reduced(β= .09, SE= .03, $p< .001$) for W1 infectious disease outcomes and (β= .07, SE= .02, $p< .001$), but the association between unhealthy physical school environment and infectious disease outcomes were still significant. The findings suggested that children's behavior patterns partially explain the associations between an unhealthy physical school environment and infectious disease outcomes among primary school students. Finally, the total effects indicated that an unhealthy physical school environment had a marginal total effect on infectious disease outcomes within the sample and was also significant in urban Bangladesh, $\beta = .51$, SE= .04, $p < .001$.

DISCUSSION

The aim of the study was to examine how unhealthy physical school environment (e.g., unhealthy school location, unclean school building, classroom, toilet, basin, washroom, playground) via unhealthy behavior patterns (e.g., learning/reading, eating, drinking, urinating, and defecating before or after handwashing, playing without shoes, and unmasking) transmits infectious diseases (e.g., fever, influenza, pneumonia, coughing, Diarrhoea and typhoid) in a sample (N=300) of primary school children in urban Bangladesh. The results from structural equation modeling indicated that children's unhealthy behavior patterns partially explained 28-35% variance in the relations between unhealthy physical school environment and children's infectious disease outcomes. The findings are consistent with previous studies in developed and developing countries (Brouwers et al., 2009; Caruso et al., 2014; Freeman et al., 2013a, 2013b; Hunter et al., 2014; WHO, 2004, 2016).

*Figure 2. Full mediating model for physical school environment predicting infectious disease transmission at primary school. Standardized path coefficients are shown. The pathways with or without control variables are significant at * $p < .05$ and *** $p < .001$. χ^2 (122) = 116.99, p = .152; GFI = .925; TLI = .948; CFI = .924; RMSEA = .031*

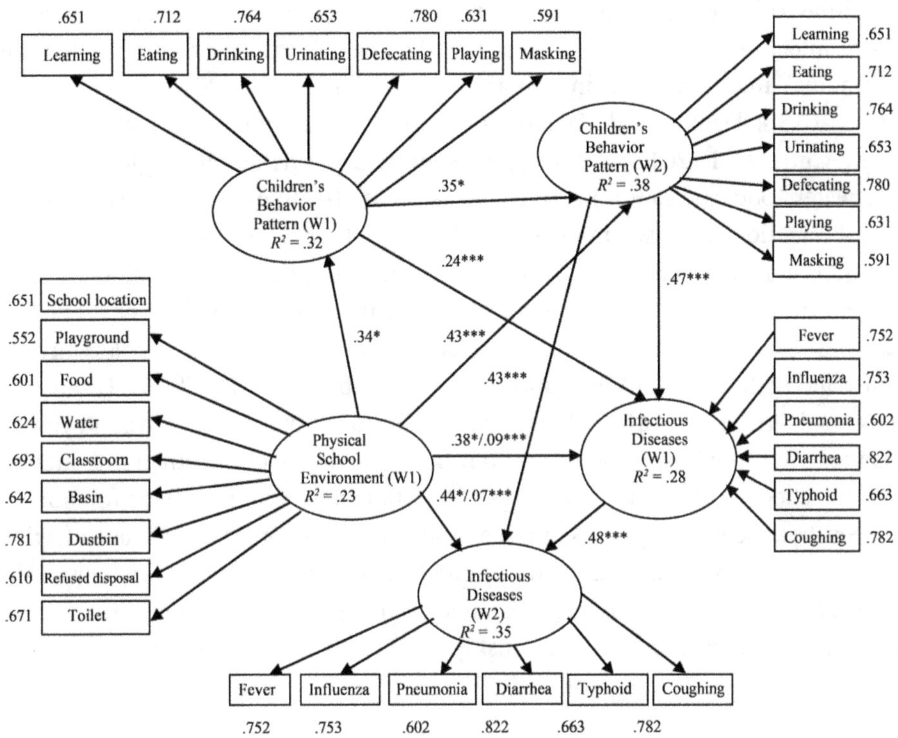

Although the present study found direct relations between the unhealthy physical school environment and the spread of infectious diseases, some of the relations accounted for children's unhealthy behavior patterns. Particularly, this study found that unhealthy physical environments such as dilapidated school buildings, unclean classrooms, toilets, bathrooms, basins, open dustbins, irregular waste disposal, and playground were the main sources of infectious diseases that were transmitted to students via their unhealthy behavior patterns. The study revealed that most children could not wash their hands with soap or hand liquid before eating and drinking. They could not wash their hands with soap after reading a book, class participation, urinating or defecating. They would play or communicate without wearing shoes and masks. These findings of the study are replicated in previous studies (Ferguson et al., 2013; Hunter et al., 2014; Ogunfowokan et al., 2017; Trinies et al., 2016).

These studies have found that unhealthy physical school environments spread fever, influenza, diarrhea, and typhoid among students in the primary school setting.

WHO (2004, 2016) found that young students than older ones were more affected by their behavior patterns (e.g., learning, eating, drinking, reading, class participation, and playing)in spreading diseases. In developing countries, studies have found that school-based intervention programs on water, sanitation, and hygiene reduce 40-50% of infectious diseases (e.g., helminth, diarrhea, fever, influenza, diarrhea, and pneumonia) among students in primary schools (Curtis et al., 2011; Freeman et al., 2013a, 2013b; Trinies et al., 2016). For example, based on a systematic literature review, Curtis et al. (2011) in developing countries found that safe hygiene practices (e.g., water, food, and sanitation) are the most cost-effective measures to prevent and control infectious diseases in primary schools. Rahman et al. (2019) and Ghosh et al. (2020) indicated that knowledge and best practices of personal hygiene protect children from infectious diseases. Further positive changes in children's behavior patterns (e.g., using handwashing, sanitizer, mask, and shoes) may improve physical health and well-being in the school setting (Dreibelbis et al., 2016). Biran et al. (2014) and Dreibelbis et al. (2016) in Bangladesh found that changes in handwashing behavior after toilet use protect children from diarrhea and intestinal infections in primary school. Although the findings of the study help understand the relationship between the unhealthy physical school environment, behavioral patterns, and infectious diseases transmission among primary students, causal direction of the relationship is obscure. Future longitudinal research should collect and analyze three or more wave data to establish causal relationships among the major variables.

CONCLUSION AND IMPLICATION

Based on ecological system theory, the current study followed Bangladeshi children over twelve months to examine how an unhealthy physical school environment, behavior patterns, and interaction transmits infectious diseases and affects them. The descriptive findings indicated that unhealthy physical school environments were significantly associated with spreading of infectious diseases in primary school. The results from the structural equation modeling suggested that unhealthy behavior patterns as mechanisms by which unhealthy school physical environments transmit infectious diseases to unaware young students in Bangladesh. In light of the school intervention programs, best practices of school hygiene knowledge and infrastructural development, the current findings may have implications for primary school promotion policy. In doing so, educationists and policy-makers should take the initiatives to improve the physical environment, hygiene facilities and change behavior patterns.

REFERENCES

Ade, A. D., Nagaraj, K., & Vallepalii, C. (2020). School environment and sanitation: A comparative study. *National Journal of Community Medicine, 11*(4), 6–12.

Afroz, S., Debsarma, S., Dutta, S., Rhaman, M. M., & Mohsena, M. (2019). Prevalence of helminthic infestations among Bangladeshi rural children and its trend since mid-seventies. *IMC J Med Sci, 13*(1), 1–8. doi:10.3329/imcjms.v13i1.42038

Akther, N., Begum, M., Islam, F., Khan, S. S., Doha, M., & Chowdhury, S., S. (2021). Practices of hygiene among primary school children in a selected semi urban area of Bangladesh. *Journal of Z H Sikder Women's Medical College, 3*(2), 16-20. doi:10.47648/zhswmcj.2021.v0302.04

Aprihatin, Y., Barlian, S., Fatimah, S., Yanti, E., & Armaita, A. (2020). Impact of environmental sanitation and infection disease as a determining stunting factor for children. *Sumatra Journal of Disaster. Geography and Geography Education, 4*(2), 209–211. doi:10.24036jdgge.v4i2.336

Asadullah, M. N., & Chaudhury, N. (2011). Poisoning the mind: Arsenic contamination of drinking water wells and children's educational achievement in rural Bangladesh. *Economics of Education Review, 30*(5), 873–888. doi:10.1016/j.econedurev.2011.05.001

Aturupane, H., Glewwe, P., & Wisniewski, S. (2013). The impact of school quality, socioeconomic factors, and child health on students' academic performance: Evidence from Sri Lankan primary schools. *Education Economics, 21*(1), 2–37. doi:10.1080/09645292.2010.511852

Bangladesh Bureau of Statistics. (2018). *Bangladesh demographic profile*. Bangladesh Bureau of Statistics.

Bangladesh Bureau of Statistics. (2020). *Bangladesh demographic profile*. Bangladesh Bureau of Statistics.

Batterman, S., Eisenberg, J., Hardin, R., Kruk, M. E., Lemos, M. C., Michalak, A. M., Mukherjee, B., Renne, E., Stein, H., Watkins, C., & Wilson, M. L. (2009). Sustainable control of water-related infectious diseases: A review and proposal for interdisciplinary health-based systems research. *Environmental Health Perspectives, 117*(7), 1023–1032. doi:10.1289/ehp.0800423 PMID:19654908

Biran, A., Schmidt, W.-P., Varadharajan, K. S., Rajaraman, D., Kumar, R., Greenland, K., Gopalan, B., Aunger, R., & Curtis, V. (2014). Effect of a behavior-change intervention on handwashing with soap in India (SuperAmma): A cluster-randomized trial. *The Lancet. Global Health, 2*(3), e145–e154. doi:10.1016/S2214-109X(13)70160-8 PMID:25102847

Biswas, D., Ahmed, M., Roguski, K., Ghosh, P. K., Parveen, S., Nizame, F. A., Rahman, M. Z., Chowdhury, F., Rahman, M., Luby, S. P., Sturm-Ramirez, K., & Iuliano, A. D. (2019). Effectiveness of a behavior change intervention with hand sanitizer use and respiratory hygiene in reducing laboratory-confirmed influenza among school children in Bangladesh: A cluster randomized controlled trial. *The American Journal of Tropical Medicine and Hygiene, 00*(0), 1–10. doi:10.4269/ajtmh.19-0376 PMID:31701861

Bronfenbrenner, U. (1974). Developmental research, public policy, and the ecology of childhood. *Child Development, 45*(1), 1–5. doi:10.2307/1127743

Bronfenbrenner, U. (1977). Toward an experimental ecology of human development. *The American Psychologist, 32*(7), 513–531. doi:10.1037/0003-066X.32.7.513

Bronfenbrenner, U. (1995). *Developmental ecology through space and time: A future perspective.* Academic Press.

Bronfenbrenner, U., & Ceci, S. J. (1994). Nature-nurture reconceptualized: A bio-ecological model. *Psychological Review, 10*(4), 568–586. doi:10.1037/0033-295X.101.4.568 PMID:7984707

Bronfenbrenner, U., & Evans, G. W. (2000). Developmental science in the 21st century: Emerging questions, theoretical models, research designs and empirical findings. *Social Development, 9*(1), 115–125. doi:10.1111/1467-9507.00114

Brouwers, L., Cakici, B., Camitz, M., Tegnell, A., & Boman, M. (2009). Economic consequences to society of H1N1 influenza 2009-prelimianry results for Sweden. *Eurosurveillance, 15*, 5–11.

Byrne, B. B. (2016). *Structural equation modeling with Amos: Basic concepts, applications, and programming* (3rd ed.). Rutledge. doi:10.4324/9781315757421

Caruso, B. A., Freeman, M. C., Garn, J. V., Dreibelbis, R., Saboori, S., Muga, R., & Rheingans, R. (2014). Assessing the impact of a school-based latrine cleaning and handwashing program on pupil absence in Nyanza Province, Kenya: A cluster-randomized Trial. *Tropical Medicine & International Health, 19*(10), 1185–1197. doi:10.1111/tmi.12360 PMID:25055716

Ceddia, M. G., Bardsley, N. O., Goodwin, R., Holloway, G. J., Nocella, G., & Stasi, A. (2013). A complex system perspective on the emergence and spread of infectious diseases: Integrating economic and ecological aspects. *Ecological Economics, 90*, 124–131. doi:10.1016/j.ecolecon.2013.03.013

Chard, A. N., Garn, J. V., Chang, H. H., Clasen, T., & Freeman, M. C. (2019). Impact of a school-based water, sanitation, and hygiene intervention on school absence, diarrhea, respiratory function, and soil-transmitted helminths: Results from the WASH HELPS cluster-randomized trial. *Journal of Global Health, 9*(2), 1–14. doi:10.7189/jogh.09.020402 PMID:31360445

Curtis, V., Schmidt, W., Luby, S., Florez, R., Touré, O., & Biran, A. (2011). Hygiene: New hopes, new horizons. *The Lancet. Infectious Diseases, 11*(4), 312–321. doi:10.1016/S1473-3099(10)70224-3 PMID:21453872

Dreibelbis, R., Kroeger, A., Hossain, K., Venkatesh, M., & Ram, P. K. (2016). Behavior change without behavior change communication: Nudging handwashing among primary school students in Bangladesh. *International Journal of Environmental and Residential Public Health, 13*.

Eisenberg, J. N. S., Desai, M. N., Levy, K., Bates, S. J., Liang, S., Naumoff, K., & Scott, J. C. (2007). Environmental determinants of infectious diseases: A framework for tracking causal links and guiding public health research. *Environmental Health Perspectives, 115*(8), 1216–1223. doi:10.1289/ehp.9806 PMID:17687450

Engin-Demir, C. (2009). Factors influencing the academic achievement of the Turkish urban poor. *International Journal of Educational Development, 29*(1), 17–29. doi:10.1016/j.ijedudev.2008.03.003

Ferguson, K. T., Cassells, R. C., MacAllister, J. W., & Evans, G. W. (2013). The physical environment and child development: An international review. *International Journal of Psychology, 48*(4), 437–468. doi:10.1080/00207594.2013.804190 PMID:23808797

Ferguson, K. T., Kim, P., Dunn, J. R., & Evans, G. W. (2009). An ecological model of urban child health. In N. Freudenberg, S. Klitzman, & S. Saegert (Eds.), *Urban health and society: Interdisciplinary approaches to research and practice*. Jossey Bass.

Freeman, M. C., Clasen, T., Brooker, S. J., Akoko, D. O., & Rheingans, R. (2013b). The impact of a school-based hygiene, water quality and sanitation intervention on soil-transmitted helminth reinfection: A cluster-randomized trial. *The American Journal of Tropical Medicine and Hygiene, 89*(5), 875–883. doi:10.4269/ajtmh.13-0237 PMID:24019429

Freeman, M. C., Clasen, T., Dreibelbis, R., Saboori, S., Greene, L. E., Brumback, B., & Rheingans, R. (2013a). The impact of a school-based water supply and treatment, hygiene, and sanitation programme on pupil diarrhea: A cluster-randomized trial. *Epidemiology and Infection, 142*(2), 340–351. doi:10.1017/S0950268813001118 PMID:23702047

Ghosh, S., Kabir, M. R., Khan, M. M. I., Shill, L. C., & Alam, S. S. (2020). Knowledge, attitude, and practice (KAP) regarding personal hygiene among primary school going children in Sadar Upazilla, Noakhali District, Bangladesh. *Indian Journal of Public Health Research & Development, 11*(6), 1464–1469.

Haque, F., Sturm-Ramirez, K., Homaira, N., Gurley, E. S., Hossain, M. J., Hasan, S. M. M., Chowdhury, S., Sarkar, S., Khan, A. K. M. D., Rahman, M., Rahman, M., & Luby, S. P. (2016). Influenza B virus outbreak at a religious residential school for boys in northern Bangladesh, 2011. *Influenza and Other Respiratory Viruses, 11*(2), 165–169. doi:10.1111/irv.12430 PMID:27603154

Hayes, N., O'Toole, L., & Halpenny, A. M. (2017). *Introducing Bronfenbrenner: A guide for practitioners and students in early years education*. Taylor & Francis. doi:10.4324/9781315646206

Heyneman, S. P., & Loxley, W. A. (1983). The effect of primary-school quality on academic achievement across twenty-nine high- and low-income countries. *American Journal of Sociology, 88*(6), 1162–1194. doi:10.1086/227799 PMID:6614303

Hossain, M. A., & Uddin, M. E. (2021). Education and health: Mediating role of socioeconomic status and health lifestyle among Muslim, Hindu, and Santal men in rural Bangladesh. *The Journal of School Health, 4*(2), 13–29.

Hunter, P. R., Risebro, H., Yen, M., Lefebvre, H., Lo, C., Hartemann, P., Longuet, C., & Jaquenoud, F. (2014). Impact of the provision of safe drinking water on school absence rates in Cambodia: A quasi-experimental study. *PLoS One, 9*(3), e91847. doi:10.1371/journal.pone.0091847 PMID:24632573

Ismail, S. A., Saliba, V., Bernal, J. L., Ramsay, M. E., & Ladhani, S. N. (2021). SARS-CoV-2 infection and transmission in educational settings: A prospective, cross-sectional analysis of infection clusters and outbreaks in England. *The Lancet. Infectious Diseases, 21*(3), 344–353. doi:10.1016/S1473-3099(20)30882-3 PMID:33306981

Jamal, F., Fletcher, A., Harden, A., Wells, H., Thomas, J., & Bonell, C. (2013). The school environment and student health: A systematic review and meta-ethnography of qualitative research. *BMC Public Health, 13*(1), 1–11. doi:10.1186/1471-2458-13-798 PMID:24007211

Jones, S., Axelrad, R., & Wattigney, W. (2007). Healthy and safe school environment, part II, physical school environment: Results from the school health policies and programs study 2006. *The Journal of School Health*, *77*(8), 544–556. doi:10.1111/j.1746-1561.2007.00234.x PMID:17908107

Kagan, L. J., Aiello, A. E., & Larson, E. (2002). The role of the home environment in the transmission of infectious diseases. *Journal of Community Health*, *27*(4), 247–267. doi:10.1023/A:1016378226861 PMID:12190054

Karim, M. R., Mondal, M. N. I., Haque, M. J., Aik, S., Chakrabarty, S., & Hossain, M. G. (2019). Preventive knowledge on seven infectious diseases in rural Rajshahi District of Bangladesh: A cross-sectional study. *Human Biology Review*, *8*(1), 9–24.

Kelly, M., & Coughlan, B. (2019). A theory of youth mental health recovery from a parental perspective. *Child and Adolescent Mental Health*, *24*(2), 161–169. doi:10.1111/camh.12300 PMID:32677184

Khan, S., Haque, M. N., & Hossen, M. S. (2021). Nutritional status of autistic school children: An evidence from Rajshahi city in Bangladesh. *JSM Nutritional Disorders*, *3*(1). Advance online publication. doi:10.47739/2578-3203/1010 PMID:29881374

Kutsyuruba, B., Klinger, D. A., & Hussain, A. (2015). Relationships among school climate, school safety, and student achievement and well-being: A review of literature. *Review of Education*, *3*(2), 103–135. doi:10.1002/rev3.3043

Langford, R., Bonell, C. P., Jones, H. E., Pouliou, T., Murphy, S. M., Waters, E., Komro, A. A., Gibbs, L. F., Magnus, D., & Campbell, R. (2014). The WHO health promoting school framework for improving the health and well-being of students and their academic achievement. *Cochrane Database of Systematic Reviews*, 4. doi:10.1002/14651858.CD008958.pub2 PMID:24737131

Linkon, M. R., Prodhan, U. K., Liton, M., Islam, M. K., Lisa, L. A., & Paul, D. K. (2015). Prevalence of contagious diseases of school going children in Bangladesh. *Journal of Biosciences and Medicines*, *3*(7), 17–24. doi:10.4236/jbm.2015.37003

Lippard, C. N., La Paro, K. M., Rouse, H. L., & Crosby, D. A. (2018). A closer look at teacher–child relationships and classroom emotional context in preschool. *Child and Youth Care Forum*, *47*(1), 1–21. doi:10.100710566-017-9414-1

McLaren, L., & Hawe, P. (2005). Ecological perspectives in health research. *Journal of Epidemiology and Community Health*, *59*(1), 6–14. doi:10.1136/jech.2003.018044 PMID:15598720

McMichael, C. (2019). Water, sanitation, and hygiene (WASH) in schools in low-income countries: A review of evidence of impact. *International Journal of Environmental Research and Public Health*, *16*(3), 1–21. doi:10.3390/ijerph16030359 PMID:30696023

Merisa, D., Soeyoko, S., & Sutomo, A. H. (2014). Sanitation of house and school, personal hygiene and infection of soil transmitted helminths among elementary school students. *International Journal of Public Health Sciences*, *3*(1), 43–50.

Njambi, E., Magu, D., Masaku, J., Okoyo, C., & Njenga, S. M. (2020). Prevalence of intestinal parasitic infections and associated water, sanitation, and hygiene risk factors among school children in Mwea irrigation scheme, Kirinyaga County, Kenya. *Journal of Tropical Medicine*, *2020*, 1–9. doi:10.1155/2020/3974156 PMID:32454837

Ogunfowokan, A. A., Esan, D. T., Salau, O. R., Bamidele, Y. R., & Ojo, D. E. (2017). Assessment of the physical school environment of selected private high schools in Ile-Ife, Nigeria. *British Journal of School Nurshing*, *11*(10), 498–505.

Paat, Y. F. (2013). Working with immigrant children and their families: An application of Bronfenbrenner's ecological systems theory. *Journal of Human Behavior in the Social Environment*, *23*(8), 954–966. doi:10.1080/10911359.2013.800007

Palinkas, L. A. (1991). Effects of physical and social environments on the health and well-being of Antarctic Winter-over personnel. *Environment and Behavior*, *23*(6), 782–799. doi:10.1177/0013916591236008

Pickering, A. J., Davis, J., Blum, A. G., Scalmanini, J., Oyier, B., ... Ram, P. K. (2013). Access to waterless hand sanitizer improves student hand hygiene behavior in primary schools in Nairobi, Kenya. *The American Journal of Tropical Medicine and Hygiene*, *89*(3), 411–418. doi:10.4269/ajtmh.13-0008 PMID:23836575

Pinter-Wollman, N., Jelic, A., & Wells, N. M. (2018). The impact of the built environment on health behaviors and disease transmission in social systems. *Philosophical Transaction of the Royal Society B*, *373*, 20170245. doi:10.1098/rstb.2017.0245

Rahman, M. M., Alil, M. A., Parvez, A., Shahriar, M., & Rahman, T., & Shameem, D. (2019). Knowledge and practice of personal hygiene among primary school students in Dhaka, Bangladesh. *Journal of Pharmaceutical Sciences & Research*, *11*(9), 3140–3144.

Ridenhour, B. J., Braun, A., Teyrasse, T., & Goldsman, D. (2011). Controlling the spread of disease in schools. *PLoS One*, *6*(12), e29640. doi:10.1371/journal.pone.0029640 PMID:22242138

Rohmah, F. A., Setiawan, R., Adriyani, R., & Sham, S. (2022). Personal hygiene as a risk factor of helminthiasis among primary school students in Asia and Africa: A literature review. *Journal of Environmental Health*, *14*(3), 139–152. Advance online publication. doi:10.20473/jkl.v14i3.2022.139-152

Rosen, L., Manor, O., Engelhard, D., Brody, D., Rosen, B., Peleg, H., Meir, M., & Zucker, D. (2006). Can a handwashing intervention make a difference? Results from a randomized controlled trial in Jerusalem preschools. *Preventive Medicine*, *42*(1), 27–32. doi:10.1016/j.ypmed.2005.09.012 PMID:16300823

Rutter, M. (1983). School effects on pupil progress: Research findings and policy implications. *Child Development*, *54*(1), 1–29. doi:10.2307/1129857

Sadique, M. Z., Adams, E. J., & Edmunds, W. J. (2008). Estimating the costs of school closure for mitigating an influenza pandemic. *BMC Public Health*, *8*(1), 8. doi:10.1186/1471-2458-8-135 PMID:18435855

Skretkowicz, V. (1992). Florence Nightingale's notes on nursing. London: Scutari Press.

Stocks, M. E., Ogden, S., Haddad, D., Addiss, D. G., McGuire, C., & Freeman, M. C. (2014). Effect of water, sanitation, and hygiene on the prevention of trachoma: A systematic review and meta-analysis. *PLoS Medicine*, *11*(2), e1001605. doi:10.1371/journal.pmed.1001605 PMID:24586120

Strunz, E. C., Addiss, D. G., Stocks, M. E., Ogden, S., Utzinger, J., & Freeman, M. C. (2014). Water, sanitation, hygiene, and soil-transmitted helminth infection: A systematic review and meta-analysis. *PLoS Medicine*, *11*(3), e1001620. doi:10.1371/journal.pmed.1001620 PMID:24667810

Stumm, S., von, Smith-Woolley, E., & Cheesman, R. (2021). School quality ratings are weak predictors of students' achievement and well-being. *Journal of Child Psychology and Psychiatry, and Allied Disciplines*, *62*(3), 339–348. doi:10.1111/jcpp.13276 PMID:32488912

Sultana, F., Nizame, F. A., Southern, D. L., Unicomb, L., Winch, P. J., & Luby, S. P. (2017). Pilot of an elementary school cough etiquette intervention: Acceptability, feasibility, and potential for sustainability. *The American Journal of Tropical Medicine and Hygiene*, *97*(6), 1876–1885. doi:10.4269/ajtmh.16-0914 PMID:29141703

Talaat, M., Afiffi, S., Dueger, E., El-Ashry, N., Marfin, A., ... El-Sayed, N. (2011). Effects of hand hygiene campaigns on incidence of laboratory-confirmed influenza and absenteeism in school children, Cairo, Egypt. *Emerging Infectious Diseases*, *17*(4), 619–625. doi:10.3201/eid1704.101353 PMID:21470450

Toyinbo, O., Matilainen, M., Turunen, M., Putus, T., Shaughnessy, R., & Haverinen-Shaughnessy, U. (2016). Modeling associations between principals' reported indoor environmental quality and students' self-reported respiratory health outcomes using GLMM and ZIP models. *International Journal of Environmental Research and Public Health, 13*(4), 1–12. doi:10.3390/ijerph13040385 PMID:27043595

Trinies, V., Garn, J. V., Chang, H. H., & Freeman, M. C. (2016). The impact of a school-based water, sanitation, and hygiene program on absenteeism, diarrhea, and respiratory infection: A matched-control trial in Mali. *The American Journal of Tropical Medicine and Hygiene, 94*(6), 1418–1425. doi:10.4269/ajtmh.15-0757 PMID:27114292

Uddin, M. E. (2017a). Ecological framework for primary school attainment in the tri-ethnic communities in rural Bangladesh. *Child Indicators Research, 10*(3), 693–713. doi:10.100712187-016-9401-3

Uddin, M. E. (2017b). Family demographic mechanisms linking of socioeconomic status to subjective physical health in rural Bangladesh. *Social Indicators Research, 130*(3), 1263–1279. doi:10.100711205-015-1209-x

Uddin, M. E. (2022). Parental investment mediates associations between lower family income and primary math achievement. *Journal of Applied Developmental Psychology, 81*(4), 1–11. doi:10.1016/j.appdev.2022.101422

Uddin, M. E., & Ferdous, Z. (2009). Gender of birth control methods practice in sexual behavior patterns in Bangladesh: A literatures review. *Antrocom: Online Journal of Anthropology, 5*(2), 87–91.

Uddin, M. E., & Ferdous, Z. (2015). Men's condom use or non-use and health risk: A comparative study on premarital, marital and extramarital men at Rajshahi city, Bangladesh. *Advances in Applied Psychology, 1*(2), 128–134.

van Seventer, J. M., & Hochberg, N. S. (2017). Principles of infectious diseases: Transmission, diagnosis, prevention, and control. International Encyclopedia of Public Health, 6, 22-39. doi:10.1016/B978-0-12-803678-5.00516-6

Wadood, M. A., Mamun, A. S. M. A., Rafi, M. A., Islam, M. K., Mohd, S., Lee, L. L., & Hossain, M. G. (2021). *Knowledge, attitude, practice and perception regarding COVID-19 among students in Bangladesh: Survey in Rajshahi University.* SciMedicine Journal. doi:10.28991/SciMedJ-2021-03-SI-9

Wigle, D. T. (2003). *Child health and the environment.* Oxford University Press.

Wilcox, B. A., & Colwell, R. R. (2005). Emerging and reemerging infectious Diseases: Biocomplexity as an interdisciplinary paradigm. *EcoHealth*, *2*(4), 244–257. doi:10.100710393-005-8961-3

Wilson, P., Atkinson, M., Hornby, G., Thompson, M., Cooper, M., Hooper, C. M., & Southall, A. (2002). *Young minds in our schools-a guide for teachers and others working in schools*. Young Minds.

World Health Organization. (2004). *The physical school environment: An essential elements of a health promoting school*. World Health Organization.

World Health Organization. (2015). *School environment: policies and current status*. WHO Regional Office for Europe.

World Health Organization. (2016). *School and youth health-global school health initiative*. https://www.who.int/school_youth_health/gshi/en

World Health Organization. (2020). *Soil-transmitted helminth infections*. World Health Organization.

Yoldascan, E., Kurtaran, B., Koyunchu, M., & Koyunchu, E. (2010). Modeling the economic impact of pandemic influenza: A case study in Turkey. *Journal of Medical Systems*, *34*(2), 139–145. doi:10.100710916-008-9225-x PMID:20433052

Chapter 7
Post-Antibiotic Era:
A Lurking Global Threat

Shafa S. Contractor
Karnatak University, India

Suresh B. Arakera
https://orcid.org/0000-0002-6234-666X
Karnatak University, India

ABSTRACT

The discovery of antibiotics was a breakthrough in the field of medicine in the late 1920s. Antibiotics have saved millions of lives ever since and even increased life expectancy. But due to their irrational use in humans, animals, and agriculture, there is a global concern about antibiotic resistance. The 'magic bullets' are no longer effective against 'superbugs', posing difficulties for the medical practitioners in treatment. The pathogens have evolved different resistance strategies for their survival: enzymatic inactivation of antibiotics, altering the movement of antibiotics across the cell, or modifying target sites that interact with the antibiotics. As the efficacy of the critical drugs wanes and the discovery of new antibiotics stagnates, exploration and use of alternative therapies could offer a much-needed solution. Researchers worldwide are constantly striving to find a solution to combat antibiotic resistance. Some of the alternative therapies and the regulatory measures to deal with antibiotic resistance are discussed in this chapter.

DOI: 10.4018/978-1-7998-9414-8.ch007

INTRODUCTION

The discovery of antibiotics by the Scottish physician and microbiologist Alexander Fleming in 1928 is one of the breakthroughs in human history and the field of modern medicine. Until then, infectious diseases were considered the primary reason for human death. A study by Yoneyama and Katsumata (2006) inferred that contagious diseases were the leading cause of death worldwide in 1900. It is widely known that before the discovery of antibiotics, most Americans died, before reaching their forties, due to infectious bacterial diseases (Lin et al., 2017). The advent of antibiotics led to a significant decrease in morbidity and mortality (Ventola, 2015). Antibiotics were widely used during Worldwar-II against bacterial infections. It was known as the miracle drug, bringing a revolution in treating bacterial infections, and that marked the beginning of the 'Antibiotic era'. Antibiotics selectively target pathogens and were believed to be without must negative effect on the host. They work efficiently in low concentrations and on a wide range of pathogens. Hence, they were also known as the 'wonder drug' and 'magic bullets' (Aminov, 2010). According to a report, since the use of antibiotics in the treatment, the life expectancy of men and women increased from 47 in the 1900s to 74 and 80 years in the year 2000 (Lederberg, 2000). According to the world health organization, antimicrobial drug is a collective term for medicines such as antibiotics, antivirals, antifungals, and antiparasitics used to treat bacteria, viruses, fungi, and parasites.

Antibiotics are compounds of low molecular weight mainly natural products, produced by microorganisms (e.g., Penicillin), or derivatives of natural products (e.g., Ampicillin). Some antibiotics are chemically synthesised, e.g., the Sulpha drugs. Some are even derived from the distillation of plant products, e.g., quinolones (Martens & Demain, 2017). Antibiotics have either a bacteriostatic or a bactericidal mode of action. They target the conserved pathways crucial for the viability of bacteria, such as DNA replication, cell wall synthesis, protein synthesis, and metabolism creating strong selective pressure for their survival (Kapoor et al., 2021). As a result, bacteria have evolved an array of defence mechanisms, making them resistant to antibiotics. These include modification of the antibiotic target, reduced permeability of the outer membrane, efflux pumps, and producing enzymes that can degrade antibiotics (Gordillo Altamirano & Barr, 2019). The ability of the bacteria to grow even in the presence of therapeutic levels of a drug is called antibiotic resistance. Prophylactic use of antibiotics in animal husbandry or self, incomplete antibiotic treatment, and easy availability of antibiotics without any medical prescription have led to the emergence of resistant strains in alarming numbers (Zaman et al., 2017). Moreover, the broad-spectrum activity of the antibiotics destroys the normal microbiome of the patients. The resident microflora shares a commensal relationship with the living organisms, playing vital roles in digestion and in developing immunity. They compete with the pathogenic microorganisms by occupying space in the gut and utilising the nutrients. In the absence of such resident microflora due to broad-spectrum antibiotics, the patient becomes susceptible to opportunistic pathogens and related secondary infections (Stach & Good, 2011).

In 2013, The Centers for Disease Control and Prevention (CDC) announced that the "post-antibiotic era" had commenced. The world health organisation (WHO) later confirmed that antibiotic resistance is one of the most significant global health challenges of modern times (Michael et al., 2014). In the USA alone, over 35000 deaths and 2.8 million infections occur annually due to antibiotic resistance (CDC, 2019).

Medical interventions involving prophylactic usage of antibiotics in critical and lifesaving surgeries pose extreme risks due to the rise and prevalence of resistant pathogens (Ventola, 2015). It is reported that antibiotic-resistant infections will lead to the death of approximately 10 million people annually by the year 2050, exceeding the current toll of annual deaths due to cancer (O'Neill et al., 2016).

The slow and time-consuming process of clinical development of new antimicrobials fails to keep pace with the rapidly emerging antibiotic resistance (WHO, 2017). Only very few new antimicrobials are developing despite the collective and collaborative efforts of researchers worldwide. It takes nearly 15 years and costs as much as two billion dollars to develop a new antibiotic, clear the clinical trials and come out in the market (Streicher, 2021). As a result, several pharmaceutical companies are becoming increasingly unwilling to manufacture antibiotics, worsening the situation. Pharmaceutical corporations are increasingly shifting their focus away from antibiotic development to sectors with significantly higher financial returns. Even the government has less public research funds, just 1.2% for Antimicrobial Resistance (AMR) related research between 2009 and 2014, compared to 18.6% for cancer research (O'Neill et al., 2016).

Bold and innovative measures must be taken to resolve the challenge of antibiotic resistance. There is a dire need to conserve antibiotics and find novel antimicrobials. Following a more hygienic lifestyle, antibiotic stewardship - considerate use of antibiotics and organic farming - agriculture and farming without antibiotics (Streicher, 2021) could also play a crucial role in controlling the issue. Researchers are now looking into novel therapy and procedures less prone to resistance development, particularly strategies without side and harmful effects. Many other lesser-known therapies, such as phage therapy and faecal microbiome transplantation (FMT) have also regained the interest of researchers. Other recent and promising treatments include lysins, bacteriocins, probiotics, monoclonal antibodies, nanoparticles, and combinatorial therapies, to name a few (Martens & Demain, 2017).

ANTIBIOTIC RESISTANCE:

The Emergence

Within a few years of the discovery of antibiotics, the evolution of antibiotic-resistant strains of bacteria was also observed (Davies & Davies, 2010). AMR is the acquired ability of the bacteria to grow in the presence of even therapeutic levels of a drug; it occurs when bacteria, fungi, viruses, and parasites evolve, no longer responding to antimicrobials. This makes treatment difficult and increases the risk of infection, disease transmission, severe illness, and even the death of the infected (Streicher, 2021). Bacteria become resistant by way of rare random mutations that occur while they divide. Most mutations may not have any effect, but very few may capacitate the bacteria to resist a particular antibiotic and allows the growth of such mutant bacteria even in the presence of the antibiotic. As more and more antibiotics increase in circulation, more resistant mutations occur, and because of natural selection superbugs arise. Another way of acquiring resistance is via horizontal gene transfer, where mutant genes from neighbouring resistant species are incorporated into the host bacterial genome. Figure 1 exhibits the chronological emergence of antibiotic resistance for the respective antibiotics.

Figure 1. Chronological representation of the onset of antibiotic resistance for the respective antibiotics

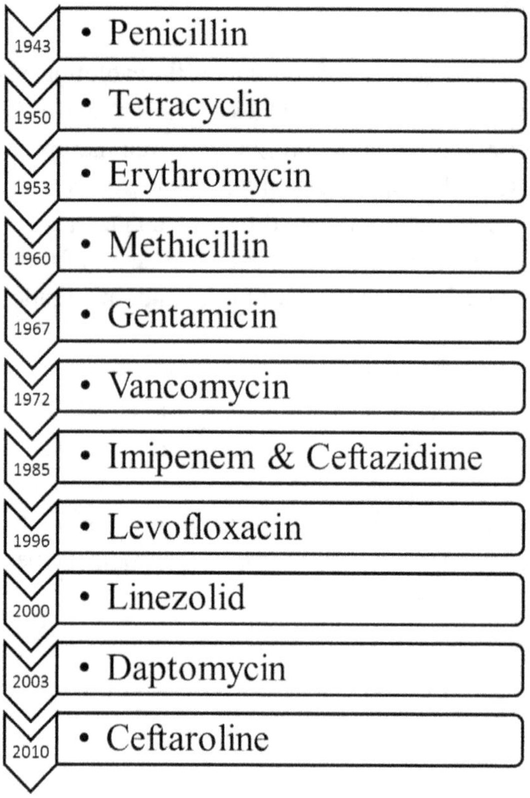

The three basic mechanisms of antimicrobial resistance are (1) enzymatic inactivation of the antibiotics (e.g., by the production of β-lactamase enzyme), (2) modification of target bacterial proteins that reduce the binding of an antibiotic, and (3) alteration in membrane permeability to antibiotics i.e, increased efflux of the antibiotic out of the cell or decreased uptake of the antibiotic inside the cell. Genes responsible for antibiotic resistance can be present on the plasmid or the bacterial chromosome (Dever, 1991). Overuse, misuse, and abuse of antibiotics have resulted in the emergence of some of the extremely drug-resistant superbugs that are resistant to almost every drug under the sleeve (Lowy, 2003).

Present-Day Crisis

About 17 million people are reportedly dying yearly from bacterial infections worldwide (Butler & Buss, 2006), making it the second leading killer in the world (Demain and Martens, 2017). The emergence and prevalence of multidrug-resistant (MDR) strains increasing globally pose great difficulties in treating common diseases. Bacteria that have developed resistance to every known antibiotic are known as 'Superbugs'. According to a recent study published in the Lancet, it is estimated that about five million human deaths in 2019 were closely associated with AMR; of this, 1.27 million deaths were directly attributable to drug-resistant infections (Murray et al., 2022). This is the most comprehensive study on the global burden of antimicrobial resistance based on data from 204 countries and territories. The CDC, in its recent report, mentioned a threat list of top Superbugs in the United States as 'urgent', 'serious', and 'concerned' based on levels of concern. The 'urgent' threats include Carbapenem-resistant *Acinetobacter, Clostridioides difficile, Candida auris,* Carbapenem-resistant Enterobacterales, and drug-resistant *Neisseria gonorrhoeae.* 'Serious' threats include drug-resistant *Campylobacter,* ESBL-producing Enterobacterales, drug-resistant *Candida,* multidrug-resistant *Pseudomonas aeruginosa, v*ancomycin-resistant *Enterococci* (VRE), drug-resistant *Salmonella* serotype Typhi, drug-resistant nontyphoidal *Salmonella,* drug-resistant *Shigella,* drug-resistant Tuberculosis, methicillin-resistant *Staphylococcus aureus* (MRSA), and drug-resistant *Streptococcus pneumonia*. Bacteria that are of 'concerned' threats include Clindamycin-resistant Group-B *Streptococcus* and Erythromycin-Resistant Group-A *Streptococcus*, Azole-resistant *Aspergillus fumigatus,* drug-resistant *Mycoplasma genitalium*, and drug-resistant *Bordetella pertussis* (CDC, 2021)

A lack of access to quality antimicrobials has led to the Antibiotic crisis, which is related to the emergence rate of MDR pathogens and the delay in the discovery and development of antimicrobials in recent years. On average, it takes over a decade for a new drug to make it to the market shelves from its initial discovery and through a battery of clinical trials. However, it takes only about a couple of years to have significant clinical resistance to the newly introduced antibiotics creating an unsustainable situation. This has drawn the attention of scientific societies, public health agencies, and political governing bodies. Thus, the unavailability of new antibiotics is impacting the development of nations' healthcare systems. If people continue using antibiotics injudiciously, every new antibiotic will have the same fate as the previous ones and will turn ineffective over time.

Some of the reasons proposed for the inability to discover novel antibiotics are (a) enzymes vital for the survival of the pathogens are not 'druggable' easily (i.e. lack of binding sites for the inhibitors or restriction of entry of compounds into the cell), (b) *In vitro* screening is not accessible for the targets (nascent molecules of peptidoglycan or ribosomal), and (c) unavailability of detailed information on the complex nature of antibiotics in the chemical libraries. Pharmaceutical companies alone cannot be held responsible for finding new antibiotics and for that, a partnership of academic medical research with the biotechnological and pharmaceutical industries will be required (Demain and Martens, 2017). It is estimated that failure in addressing the problem of AMR could result in 10 million deaths per year by 2050, with a $100 trillion economic loss worldwide (O'Neill et al., 2016).

Tackling the Issue

The emergence of antibiotic resistance has compelled researchers to find alternative therapies against resistant pathogens. Few of the alternatives are as follows: Antibiotics hybrids- antibiotics in combination or covalently linked with other antibiotics or with adjuvants, Phage therapy, use of lytic enzymes, antimicrobial peptides, nanoparticles, photodynamic therapy, and vaccines to boost immunity are all currently in use or under trial as alternative therapies (Mulani et al., 2019) as shown in table 1.

Table 1. Alternative products to tackle infections

	Alternative Products to Tackle Infections	
1.	Antibiotics in combination	The synergistic effect of combined antibiotics is greater than the effect of each antibiotic alone.
2.	Bacteriophage therapy	Bacterial viruses with broad or narrow host ranges kill the bacteria.
3.	Bacteriophage lytic enzymes	Enzymes are produced by bacteriophages that have a lytic effect on bacteria.
4.	Peptides	Non-mammalian animals' natural defences against infection
5.	Photodynamic Light Therapy	Results in a photo-oxidative stress response leading to microbial death
6.	Silver Nanoparticles in Therapy	The metal ions disrupt the electron transport or signal transduction pathways or lead to the generation of reactive oxygen species (ROS), causing damage to critical biomolecules, including cell walls and cell membranes of the pathogens.
7.	Faecal Microbiota Transplantation	Replenishing the gut microflora of a recipient by faecal extract transfer from a healthy donor
8.	Probiotics	Enriching the gut with commensal microflora to prevent opportunistic pathogen colonisation
9.	Vaccines	Helps patients develop natural immunity against pathogens.

Combinatorial approaches are recommended for effective treatment as it reduces the chances of the emergence of potential resistant bacteria, thereby minimising the limitations of monotherapies. It comprises the use of two or more mono-therapies in combination (co-administered, functionalised, or conjugated).

Antibiotics in Combination

Several researchers, after testing, have found the use of antibiotics in combination to be better than a single antibiotic. It is less likely for a pathogen to acquire resistance against a combination of two drugs than just one. Also, the cumulative effect of two antibiotics is higher and hence more effective. Drugs in combination also increase the range of coverage, proving effective in treating severe infections involving multiple pathogens (Baronia & Ahmed, 2014).

When certain compounds such as the adjuvants are coupled with antibiotics, a previously ineffective medicine becomes effective. These adjuvants lack antimicrobial activity of their own. They work to break the resistance mechanism by blocking efflux pumps, increasing the permeability of the bacterial cell membrane for antibiotics, and disrupting the integrity of the biofilm into planktonic forms (Bueno, 2016). Quorum quenchers can be used with antibiotics for more excellent antimicrobial activity against infections *in vitro*. Quorum quenchers are chemicals that disrupt biofilm formation by disrupting quorum sensing and interrupting bacterial communication. Compared to *in vivo*, with only a few studies, it has not been so far found very advantageous; hence still under research (Mulani et al., 2019).

Bacteriophage Therapy

Phages are the oldest nemesis of bacteria known to be naturally occurring and have antibacterial activity. They were used to treat bacterial infections for more than a century in certain parts of the world before the discovery of antibiotics. The dramatic rise in antimicrobial resistance in recent years has sharpened the focus on phage therapy. Lytic phages against bacterial pathogens can be isolated from their natural reservoirs such as natural water bodies, hospitals, domestic wastewater, and soil facilitating their accessibility as therapeutic agents. Bacteriophages employed in treatment offer several advantages, including excellent host specificity, ease of usage, low cost, low treatment dosages, auto dosing, and rapid proliferation inside the host bacteria without disturbing the normal flora or the eukaryotic cells, and so on (Latz et al., 2016). As phages have their genetic material, they co-evolve along with their bacterial host making them ideal candidates to fight the constantly evolving pathogens.

Phages are highly specific in action; their specificity can be beneficial and a drawback. The therapeutic phages are selected after their *in vitro* interaction with the host. Use of phage cocktail - a mixture of more than one type of phage is considered more effective than monophage therapy as it also helps prevent phage-resistant bacteria's emergence. According to researchers, an optimal phage cocktail consists of a mixture of phages belonging to different families or groups with the ability for good adsorption and a broad host range (Chan et al., 2013, Rohde et al., 2018). As demonstrated by several experts, the genomic characterisation of phages is critical for predicting their "safety" in therapeutic applications. Phages through transduction can transfer virulence genes among bacteria. Ideal therapeutic phages should not contain antibiotic resistance or virulence genes, site-specific recombinases, as well as integrases, or lytic cycle repressors. They should be fully characterised and not produce any allergic or immunological reaction. They should remain stable at various pH and temperatures (Mulani et al., 2019).

The bacterial strains utilised to make phages should be devoid of functioning prophages as they have repressed the lytic cycle and hence prove useless in phage therapy. The ability of phages to reach the target location of the action and the stability of phages has also been mentioned as a constraint. Phages for treatment can be applied topically, inhaled nasally, or ingested orally. Entrapment of phage with liposomes has been demonstrated to boost phage effectiveness in studies. Phage powder formulations can also be administered directly to the infection site for target-specific therapy (Chang et al., 2018).

Coupling phage therapy with antibiotics proves the synergistic impact more effective than monotherapies. Antibiotic treatment in combination with phages has also been shown to reduce the production of bacterial biofilms (Jo et al., 2016; Chaudhry et al., 2017). An interesting study has found that when bacteria develop phage resistance while being attacked by phage, the efflux pump mechanism is altered, increasing the organism's susceptibility to antibiotics. Bacteriophages though relatively harmless to humans, their use for therapy is still under trial in the rest of the world, while in Russia, Georgia, and Poland this has been in practice for over a century.

Antibacterial phage-derived products, such as endolysin or ectolysin, are enzymes encoded by the bacteriophages, which help them penetrate the bacterial cell and lyse the cell inside out. These enzymes can be employed alone or in combination with other antimicrobials for better efficiency in killing bacteria (Rios et al., 2016).

Antimicrobial Peptides in Therapy

Antimicrobial peptides (AMPs) comprise less than 100 amino acids. They are found in all domains of life and play an important role in innate immunity. As plants and insects lack adaptive immune systems, antimicrobial peptides comprise their primary defence against pathogenic microorganisms. Their role varies from defending environmental niche by the bacteria and other microorganisms to a more diverse part of regulating the immune pathways in the higher organisms. They have a broad range of antimicrobial effects, and unlike traditional antibiotics, AMPs cause physical damage to the cells by cell lysis due to electrostatic interactions (Berglund et al., 2015). This makes it impossible for the bacteria to gain resistance to them.

Various natural and bioengineered AMPs have antibacterial, anti-biofilm, anti-inflammatory, and wound healing activities *in vitro* and *in vivo* with minimal cytotoxicity. As a result, AMPs demonstrate promise in *in vivo* action, similar to their extraordinary *in vitro* features (Mulani et al., 2019). The future of peptide-based commercial medication formulations is bright, given the substantial research on various AMPs against various infectious pathogens.

Photodynamic Light Therapy

Photodynamic therapy (PDT) involves destroying the cells, usually cancerous, precancerous, or even bacterial cells, after interacting light energy with a drug (photosensitiser). The photosensitiser (PS) is designed to be specifically taken up by the target cells. Later upon activation by a light source of a certain wavelength, they are selectively targeted for destruction (Tomb et al., 2017). The PSs, activated only in the presence of light energy, are nontoxic otherwise to the cells in their inactive state (National Cancer Institute, 2022; Abrahamse & Hamblin, 2016).

An array of photosensitiser drugs is available to treat skin diseases, cancers, and infections from fungal, viral, and bacterial pathogens. It is also known that PDT can trigger an immune response in the body. An ideal photosensitiser used in treating bacterial infections should have high permeability for bacterial cell walls while causing negligible damage to the host cells/ tissue. The PSs should not be removed by the micro-efflux processes of the pathogens (Sueoka et al., 2018).

PDT has a promising future in treating antimicrobial resistance, particularly effective treatments involving topical application. Nevertheless, due to the presence of some side effects, there is a need for studies on *in vivo* models to examine the toxicity and biocompatibility of diverse combinations before implementing them into clinical practice (Mulani et al., 2019).

Silver Nanoparticles in Therapy

Nanomedicine is another recent branch of medicine used in treating drug-resistant pathogens. Metal nanoparticles have a wide range of biological uses as antibacterial agents because of their unique physio-chemical properties (Dakal et al., 2016). Amongst metal nanoparticles produced by physical, chemical, or natural means, silver nanoparticles (AgNPs) exhibit better antibacterial activity due to the reduced probability of resistance to their multi-targeted action strategy. AgNPs impair electron transport or signal transduction pathways by releasing Ag^+ ions or lead to the generation of reactive oxygen species (ROS), causing damage to critical biomolecules, including cell walls and cell membranes. AgNPs work by inhibiting or disrupting planktonic cells and biofilms of MDR pathogens. Despite early findings suggesting that AgNPs are cytotoxic (Mohanty et al., 2012), they have recently been proven safe both in *in vitro* and *in vivo* tests (Qayyum et al., 2017). One of the well-studied uses is using AgNPs in hydrogels or dressings for treating topical infected wounds. Medical equipment such as catheters are coated with a thin layer of silver nanoparticles, and implants are made of polymer-based nanomaterials to prevent infections.

After repeated exposure to sub-inhibitory concentrations, bacterial pathogens may develop resistance to AgNPs. To overcome this constraint, combining AgNPs with antibiotics has been proposed to boost the therapeutic efficacy of both while lowering the dosage and hence toxicity. AgNPs are seen as an effective therapeutic agent yet the *in vivo* interactions such as their toxicity, effectiveness, pharmacokinetics, and immunomodulatory response need to be studied in detail. AgNPs can be used in wound dressings or medical devices after further research in well-defined studies and clinical trials (Mulani et al., 2019).

Faecal Microbiota Transplantation (FMT)

It is the administration of a faecal extract from a healthy donor into the intestinal tract of a recipient to directly alter the microbiome of the recipient and confer a health benefit. Opportunistic pathogens colonise the gut in the absence of normal commensal flora. Instead of using antibiotics, a better option is replenishing the gut with commensal flora. Hence, the aim of FMT is the restoration of healthy bacterial flora in the gut (Gupta et al., 2015).

FMT was practised in ancient china during the 4th century to treat various gastrointestinal conditions. While earlier days, a filtered solution of healthy donor faeces was administered orally. Now dried faecal extracts are administered through colonoscopy, enema, or orally in capsules, also popularly known as poop pills or capsules. FMT has shown to be 85-90% effective in people with CD (*Clostridioides difficile*) infections, in patients with recurrent infections, and where antibiotics were ineffective. It has therapeutic potential for conditions such as obesity, gastrointestinal diseases, and even certain neurological disorders such as Parkinson's and multiple sclerosis (Gupta et al., 2015).

Other alternatives include the use of probiotics and vaccines. Probiotics help maintain the healthy gut flora and restrict opportunistic pathogens' colonisation. Of the six leading pathogens, only one *Streptococcus pneumoniae* has been the target of a major vaccination program. Vaccines for other pathogens are still under the discovery and trial phase.

Regulatory Measures

To tackle the global issue of drug resistance, the UK government and the Wellcome Trust commissioned and sponsored an Antimicrobial Resistance (AMR) review in 2014. Economist Jim O'Neill was assigned to analyse the situation globally and propose future action plans internationally. From an economic and social standpoint, the study consulted international stakeholders to have a better view of the global condition and offer concrete solutions against AMR. It produced its final report and recommendations (O'Neill et al., 2016), according to which the following measures (shown in figure 2) are presented.

1. Global awareness in public: There is a need to spread awareness of antimicrobial resistance among medical practitioners and veterinarians not to prescribe antibiotics on demand. Patients and farmers also should not force clinicians to prescribe antibiotics. According to a study in Belgium, campaigns to reduce drugs during the winter flu season resulted in a reduction in prescriptions by up to 36% (O'Neill et al., 2016). Awareness should be customised, appealing, and spread to reach the masses.
2. Improvement of hygiene: This can be achieved by improving the quality of drinking water and by providing proper sanitation. Teaching the importance of proper hand washing, better sanitation, and hygiene helps control infectious diseases and reduce antibiotic use. It reduces the chances of the emergence of antimicrobial-resistant strains of pathogens.
3. Reducing the irrational use of antimicrobials in farming and agriculture: In agriculture, several antimicrobials are used to prevent crop damage and avoid losses due to infections. Even the livestock is fed an antibiotic-rich diet to prevent bacterial infections as they are reared close by under poor hygienic conditions raising the risks for the spread of infection. Earlier antibiotics were also injected as growth promoters in farm animals, which were later labelled illegal (O'Neill et al., 2016). These antibiotics enter humans through their food chain. To avoid this, the use of antibiotics in agriculture should be restricted and replaced by the use of natural pesticides. Highly critical antibiotics that are usually used as last-line drugs in treating human infections should be prohibited for use in livestock.

Antibiotics disseminate in nature through animal and human waste, the effluents of pharmaceutical companies, and the water from agricultural fields, encouraging the development of resistance among pathogens. Hence, proper treatment of the wastewater and effluents should be managed before letting them into the water bodies (Shankar, 2016).

4. Global surveillance of consumption of antibiotics and development of resistance in humans and animals: Global Antibiotic Resistance Surveillance System (GLASS) is initiated by the World Health Organization (WHO) to take necessary actions with the governments to keep AMR in check. Governments of different countries have introduced acts, policies, and antibiotic stewardship programs to tackle the issue. By reducing the unnecessary consumption of antibiotics, we can combat resistance. It should be mandatory to test for antibiotic susceptibility of the pathogen before prescribing antibiotics. By using existing drugs judiciously, the rate of demand for new drugs can be slowed down. Imposing a ban on 'over-the-counter sale' of antimicrobials, i.e., purchase of medicines without prescription from a registered medical practitioner would help in the same. Decisive action must be taken for restrictions on the sale of antimicrobials through online pharmacies that cater to orders without prescriptions.
5. Use of Rapid diagnostic tools: The use of rapid diagnostic tools to infer the disease and the causative agent makes treatment precise and specifically the use of antibiotics. Rapid diagnostic tests (RDTs) for infectious diseases detect pathogens within a couple of hours, reducing treatment time compared to conventional susceptibility tests. These are compact and quick; help improve patient care and infection prevention in hospitals and critical care units (Bouzid et al., 2021).
6. Use of vaccines and their development: Vaccines boost natural immunity, thereby preventing infections. This reduces the use of antibiotics, in turn controlling antimicrobial resistance.
7. Incentives for research in discovering and developing new drugs and improving existing drugs: search for novel antibiotics in nature, modification of existing antibiotics, and antibiotics in combination are some strategies. Novel drug discovery requires high investments in terms of both time and money. Another reason for the abandonment of antibiotic research by the manufacturing companies is that medicines for heart disease or high blood pressure are usually taken on a chronic basis by patients and that offer more profits and higher return on investment compared to the antibiotics that are taken only for short durations (Demain and Martens, 2017).

Figure 2. Regulatory measures proposed to tackle AMR

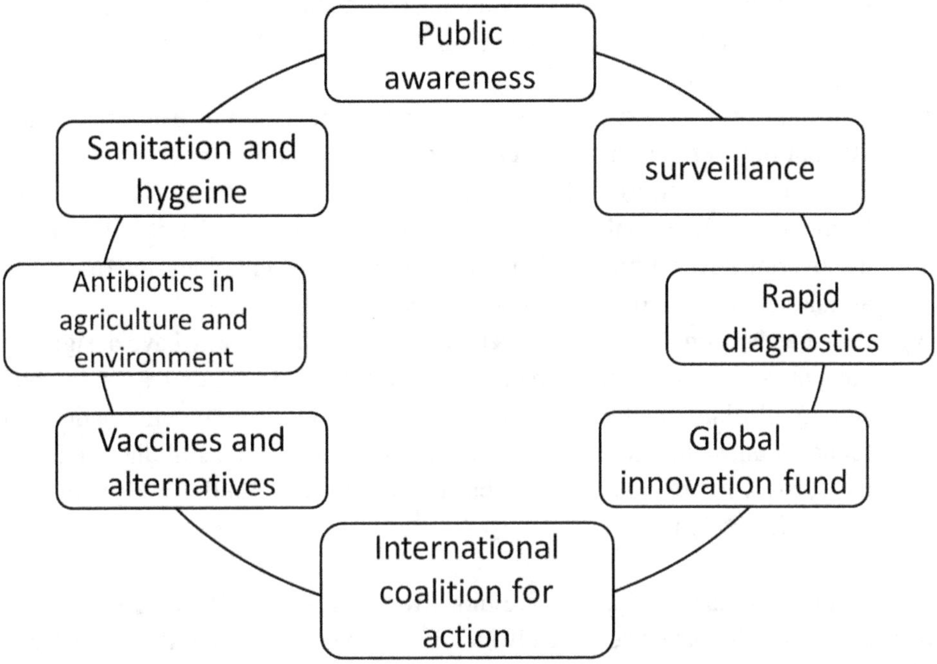

New generation antibiotics are desperately needed worldwide. Efforts like free-sharing information in the form of compound libraries regarding drugs among students and companies worldwide, and providing financial incentives by national and international governments for the early development of new antibiotics required for preclinical trials can prove beneficial.

FUTURE RESEARCH DIRECTIONS

Apart from the remedies discussed above, the following are some other possible remedies and prospective solutions to deal with the menace of antimicrobial resistance:

i) New medications against multi-drug resistant diseases in the critical category of the CDC threat list should be given a fast-track designation.
ii) Genome mining uses advanced approaches to find new natural compounds. Genome mining has discovered a slew of new 'silent' secondary metabolites. Germicidins, coelichelin, and mycothiol from *Streptomyces coelicolor* and geosmin and epi-isozizaene from *Streptomyces avermitilis* are among them.

Other mining microorganisms have also yielded a plethora of new chemicals. Genome mining has a lot of potential for medication development in the future, especially after sequencing hundreds of microbial genomes (Martens & Demain, 2017).

iii) The number of bacterial species in nature is estimated to be between 10^7 and 10^9, but only a fraction of that can be cultured in laboratory conditions. The isolation chip (i-chip) with native soil nutrients in its many compartments facilitates 50-60% of the bacteria to survive, increasing the possibilities of discovering new antibiotics (Jabr, 2022). Using this technology, the first novel antibiotic, Teixobactin, was found in decades (Borel, 2022).

iv) Researchers from the Massachusetts Institute of Technology have designed a computer-based model that screens chemical compounds for their antimicrobial activity. Within a matter of days, it can screen 100 million combinations. It found an unusually potent new antibiotic, Halisin, which, still under clinical trials, can kill 35 kinds of deadly bacteria ("Artificial intelligence yields new antibiotic", 2022).

About 1.1 million fungal species are known to exist in nature. Of this, about 14,000 micro-fungi species have been already identified. So far, only 30 of the 61 bacterial phyla have been cultured. A metagenomic method is advised for discovering novel medications from such sources. It is claimed that the lack of antibiotics can be met by genome mining, isolation of marine actinomycetes, and large-scale fermenters (Martens & Demain, 2017; Katz & Baltz, 2016).

CONCLUSION

Resistant bacteria are here to stay, constantly evolving to thwart the utility of new antibiotics. The question is not in beating superbugs but in containing them. This is possible only if the antibiotics are valued and not overused and abused. The emergence of the superbugs and the increasing ineffectiveness of the antibiotics seem to be taking us to a 'post-antibiotic era', where none of the antibiotics would be effective for treatments. AMR in bacterial pathogens is a major challenge mainly due to the lack of effective therapies, absence of prevention measures, and lack of new antibiotics. It is found that treating gram-positive and Gram-negative bacteria with multidrug-resistant qualities with standard antibiotics is difficult, and at times, they become untreatable. These challenges associated with bacterial infection are demanding in terms of the development of alternative antimicrobial therapies and novel treatment strategies.

Antibiotic stewardship and the spread of awareness regarding the crisis of antibiotic resistance in healthcare, agriculture, and livestock is the need of the hour. We must urgently replenish our antimicrobial arsenal to remain ahead of the ever-increasing drug-resistant bacteria. In addition to the pipeline of new medications, there is a shortage of effective antibiotic combinations. Antibiotics and adjuvants have been utilised to target critical metabolic systems and pathways that lead to antibiotic resistance (lactamase inhibitors, permeabilisers, quorum sensing inhibitors, toxin inhibitors, efflux pump inhibitors, etc.). The scientific community worldwide has come up together and united to find alternative ways to combat the menace of antibiotic resistance. Failure to do so will lead to an apocalyptic situation in the future, where even simple infections will lead to high mortality rates. Researchers strive in every possible niche to bring alternatives to deal with superbugs, from getting back to century-old therapies to modern technologies including artificial intelligence. As the world has become more aware of the possible threats in the future, and with the help of political aid, the end seems promising if joint efforts are employed globally to fight the resistance from microbes.

REFERENCES

Abrahamse, H., & Hamblin, M. (2016). New photosensitisers for photodynamic therapy. *The Biochemical Journal, 473*(4), 347–364. doi:10.1042/BJ20150942 PMID:26862179

Aminov, R. (2010). A Brief History of the Antibiotic Era: Lessons Learned and Challenges for the Future. *Frontiers in Microbiology, 1*. Advance online publication. doi:10.3389/fmicb.2010.00134 PMID:21687759

Artificial intelligence yields new antibiotics. (2022). MIT News, Massachusetts Institute of Technology. Retrieved on 3 June 2022, from https://news.mit.edu/2020/artificial-intelligence-identifies-new-antibiotic-0220

Baronia, A., & Ahmed, A. (2014). Current concepts in combination antibiotic therapy for critically ill patients. *Indian Journal of Critical Care Medicine : Peer-Reviewed, Official Publication of Indian Society of Critical Care Medicine, 18*(5), 310–314. doi:10.4103/0972-5229.132495 PMID:24914260

Berglund, N., Piggot, T., Jefferies, D., Sessions, R., Bond, P., & Khalid, S. (2015). Interaction of the Antimicrobial Peptide Polymyxin B1 with Both Membranes of E. coli: A Molecular Dynamics Study. *PLoS Computational Biology, 11*(4), e1004180. doi:10.1371/journal.pcbi.1004180 PMID:25885324

Borel, B. (2022). *iChip: The Future of Antibiotic Discovery*. Popular Science. Retrieved on 3 June 2022, from https://www.popsci.com/ichip-new-way-find-antibiotics-and-other-key-drugs/

Bouzid, D., Zanella, M., Kerneis, S., Visseaux, B., May, L., Schrenzel, J., & Cattoir, V. (2021). Rapid diagnostic tests for infectious diseases in the emergency department. *Clinical Microbiology and Infection*, *27*(2), 182–191. doi:10.1016/j.cmi.2020.02.024 PMID:32120036

Bueno, J. (2016). Antimicrobial Adjuvants Drug Discovery, the Challenge of Avoid the Resistance and Recover the Susceptibility of Multidrug-Resistant Strains. *Journal of Microbial & Biochemical Technology*, *8*(3). Advance online publication. doi:10.4172/1948-5948.1000281

Butler, M., & Buss, A. (2006). Natural products - The future scaffolds for novel antibiotics. *Biochemical Pharmacology*, *71*(7), 919–929. doi:10.1016/j.bcp.2005.10.012 PMID:16289393

Centers for Disease Control and Prevention. (2021). *Antibiotic-resistant Germs: New Threats*. Retrieved on 9 October 2021, from https://www.cdc.gov/drugresistance/biggest-threats.html

Centers for Disease Control and Prevention (U.S.) & National Center for Emerging Zoonotic and Infectious Diseases (U.S.). Division of Healthcare Quality Promotion. Antibiotic Resistance Coordination and Strategy Unit. (Eds.). (2019). *Antibiotic resistance threats in the United States, 2019* (CDC:82532). https://stacks.cdc.gov/view/cdc/82532

Chan, B., Abedon, S., & Loc-Carrillo, C. (2013). Phage cocktails and the future of phage therapy. *Future Microbiology*, *8*(6), 769–783. doi:10.2217/fmb.13.47 PMID:23701332

Chang, R., Chen, K., Wang, J., Wallin, M., Britton, W., Morales, S., Kutter, E., Li, J., & Chan, H.-K. (2018). Proof-of-Principle Study in a Murine Lung Infection Model of Antipseudomonal Activity of Phage PEV20 in a Dry-Powder Formulation. *Antimicrobial Agents and Chemotherapy*, *62*(2), e01714-17. Advance online publication. doi:10.1128/AAC.01714-17 PMID:29158280

Dakal, T., Kumar, A., Majumdar, R., & Yadav, V. (2016). Mechanistic Basis of Antimicrobial Actions of Silver Nanoparticles. *Frontiers in Microbiology*, *7*. Advance online publication. doi:10.3389/fmicb.2016.01831 PMID:27899918

Davies, J., & Davies, D. (2010). Origins and Evolution of Antibiotic Resistance. *Microbiology and Molecular Biology Reviews*, *74*(3), 417–433. doi:10.1128/MMBR.00016-10 PMID:20805405

de Kraker, M., Stewardson, A., & Harbarth, S. (2016). Will 10 Million People Die a Year due to Antimicrobial Resistance by 2050? *PLoS Medicine*, *13*(11), e1002184. doi:10.1371/journal.pmed.1002184 PMID:27898664

Dever, L. (1991). Mechanisms of bacterial resistance to antibiotics. *Archives of Internal Medicine*, *151*(5), 886–895. doi:10.1001/archinte.1991.00400050040010 PMID:2025137

Domalaon, R., Idowu, T., Zhanel, G., & Schweizer, F. (2018). Antibiotic Hybrids: The Next Generation of Agents and Adjuvants against Gram-Negative Pathogens? *Clinical Microbiology Reviews*, *31*(2), e00077-17. Advance online publication. doi:10.1128/CMR.00077-17 PMID:29540434

Gomes, B., Augusto, M., Felício, M., Hollmann, A., Franco, O., Gonçalves, S., & Santos, N. (2018). Designing improved active peptides for therapeutic approaches against infectious diseases. *Biotechnology Advances*, *36*(2), 415–429. doi:10.1016/j.biotechadv.2018.01.004 PMID:29330093

Gordillo Altamirano, F., & Barr, J. (2019). Phage Therapy in the Postantibiotic Era. *Clinical Microbiology Reviews*, *32*(2), e00066-18. Advance online publication. doi:10.1128/CMR.00066-18 PMID:30651225

Gupta, S., Allen-Vercoe, E., & Petrof, E. (2015). Fecal microbiota transplantation: In perspective. *Therapeutic Advances in Gastroenterology*, *9*(2), 229–239. doi:10.1177/1756283X15607414 PMID:26929784

Jabr, F. (2022). *The Age of Infection*. Foreign Policy. Retrieved on 3 June 2022, from https://foreignpolicy.com/2015/09/30/the-age-of-infection-antibiotics-microbes-germ-ichip/

Jo, A., Ding, T., & Ahn, J. (2016). Synergistic antimicrobial activity of bacteriophages and antibiotics against Staphylococcus aureus. *Food Science and Biotechnology*, *25*(3), 935–940. doi:10.100710068-016-0153-0 PMID:30263357

Kapoor, G., Saigal, S., & Elongavan, A. (2021). *Action and resistance mechanisms of antibiotics: A guide for clinicians.* Academic Press.

Katz, L., & Baltz, R. (2016). Natural product discovery: Past, present, and future. *Journal of Industrial Microbiology & Biotechnology*, *43*(2-3), 155–176. doi:10.100710295-015-1723-5 PMID:26739136

Kaur, I. (2016). Novel Strategies to Combat Antimicrobial Resistance. *Journal of Infectious Disease and Therapy*, *4*(4). Advance online publication. doi:10.4172/2332-0877.1000292

Latz, S., Wahida, A., Arif, A., Häfner, H., Hoß, M., Ritter, K., & Horz, H. (2016). A preliminary survey of local bacteriophages with lytic activity against multi-drug resistant bacteria. *Journal of Basic Microbiology*, *56*(10), 1117–1123. doi:10.1002/jobm.201600108 PMID:27194637

Lederberg, J. (2000). Infectious History. *Science*, *288*(5464), 287–293. doi:10.1126cience.288.5464.287 PMID:10777411

Lowy, F. (2003). Antimicrobial resistance: The example of Staphylococcus aureus. *The Journal of Clinical Investigation*, *111*(9), 1265–1273. doi:10.1172/JCI18535 PMID:12727914

Mahlapuu, M., Håkansson, J., Ringstad, L., & Björn, C. (2016). Antimicrobial Peptides: An Emerging Category of Therapeutic Agents. *Frontiers in Cellular and Infection Microbiology*, *6*. Advance online publication. doi:10.3389/fcimb.2016.00194 PMID:28083516

Martens, E., & Demain, A. (2017). The antibiotic resistance crisis, with a focus on the United States. *The Journal of Antibiotics*, *70*(5), 520–526. doi:10.1038/ja.2017.30 PMID:28246379

Michael, C. A., Dominey-Howes, D., & Labbate, M. (2014). The antimicrobial resistance crisis: Causes, consequences, and management. *Frontiers in Public Health*, *2*. Advance online publication. doi:10.3389/fpubh.2014.00145 PMID:25279369

Mohanty, S., Mishra, S., Jena, P., Jacob, B., Sarkar, B., & Sonawane, A. (2012). An investigation on the antibacterial, cytotoxic, and antibiofilm efficacy of starch-stabilized silver nanoparticles. *Nanomedicine; Nanotechnology, Biology, and Medicine*, *8*(6), 916–924. doi:10.1016/j.nano.2011.11.007 PMID:22115597

Möhler, J., Sim, W., Blaskovich, M., Cooper, M., & Ziora, Z. (2018). Silver bullets: A new luster on an old antimicrobial agent. *Biotechnology Advances*, *36*(5), 1391–1411. doi:10.1016/j.biotechadv.2018.05.004 PMID:29847770

Mulani, M. S., Kamble, E. E., Kumkar, S. N., Tawre, M. S., & Pardesi, K. R. (2019). Emerging Strategies to Combat ESKAPE Pathogens in the Era of Antimicrobial Resistance: A Review. *Frontiers in Microbiology*, *10*, 539. Advance online publication. doi:10.3389/fmicb.2019.00539 PMID:30988669

National Cancer Institute. (2022). *Photodynamic Therapy to Treat Cancer*. Retrieved 14 June 2022, from https://www.cancer.gov/about-cancer/treatment/types/photodynamic-therapy

O'Neill, J. (2016). *Tackling Drug-Resistant Infections Globally: Final Report and Recommendations*. Review on Antimicrobial Resistance. Welcome Trust and UK Government. https://amr-review.org/sites/default/files/160518_Final%20paper_with%20cover.pdf

Pirnay, J., Verbeken, G., Ceyssens, P., Huys, I., De Vos, D., Ameloot, C., & Fauconnier, A. (2018). The Magistral Phage. *Viruses*, *10*(2), 64. doi:10.3390/v10020064 PMID:29415431

Qayyum, S., Oves, M., & Khan, A. (2017). Obliteration of bacterial growth and biofilm through ROS generation by facilely synthesized green silver nanoparticles. *PLoS One*, *12*(8), e0181363. doi:10.1371/journal.pone.0181363 PMID:28771501

Reinhardt, A., & Neundorf, I. (2016). Design and Application of Antimicrobial Peptide Conjugates. *International Journal of Molecular Sciences*, *17*(5), 701. doi:10.3390/ijms17050701 PMID:27187357

Rios, A., Moutinho, C., Pinto, F., Del Fiol, F., Jozala, A., Chaud, M., Vila, M. M. D. C., Teixeira, J. A., & Balcão, V. M. (2016). Alternatives to overcoming bacterial resistances: State-of-the-art. *Microbiological Research*, *191*, 51–80. doi:10.1016/j.micres.2016.04.008 PMID:27524653

Rohde, C., Resch, G., Pirnay, J., Blasdel, B., Debarbieux, L., Gelman, D., Górski, A., Hazan, R., Huys, I., Kakabadze, E., Łobocka, M., Maestri, A., Almeida, G., Makalatia, K., Malik, D., Mašlaňová, I., Merabishvili, M., Pantucek, R., Rose, T., ... Chanishvili, N. (2018). Expert Opinion on Three Phage Therapy Related Topics: Bacterial Phage Resistance, Phage Training and Prophages in Bacterial Production Strains. *Viruses*, *10*(4), 178. doi:10.3390/v10040178 PMID:29621199

Seo, M., Won, H., Kim, J., Mishig-Ochir, T., & Lee, B. (2012). Antimicrobial Peptides for Therapeutic Applications: A Review. *Molecules (Basel, Switzerland)*, *17*(10), 12276–12286. doi:10.3390/molecules171012276 PMID:23079498

Shankar, P. (2016). Book review: Tackling drug-resistant infections globally. *Archives of Pharmacy Practice (Mumbai)*, *7*(3), 110. doi:10.4103/2045-080X.186181

Stach, J., & Good, L. (2011). Synthetic RNA Silencing in Bacteria – Antimicrobial Discovery and Resistance Breaking. *Frontiers in Microbiology*, *2*, 185. doi:10.3389/fmicb.2011.00185 PMID:21941522

Streicher, L. M. (2021). Exploring the future of infectious disease treatment in a post-antibiotic era: A Comparative Review of Alternative Therapeutics. *Journal of Global Antimicrobial Resistance*, *24*, 285–295. doi:10.1016/j.jgar.2020.12.025 PMID:33484895

Sueoka, K., Chikama, T., Latief, M., Ko, J., Kiuchi, Y., Sakaguchi, T., & Obana, A. (2018). Time-dependent antimicrobial effect of photodynamic therapy with TONS 504 on Pseudomonas aeruginosa. *Lasers in Medical Science*, *33*(7), 1455–1460. doi:10.100710103-018-2490-0 PMID:29589177

Tomb, R., Maclean, M., Coia, J., MacGregor, S., & Anderson, J. (2017). Assessment of the potential for resistance to antimicrobial violet-blue light in Staphylococcus aureus. *Antimicrobial Resistance and Infection Control*, *6*(1), 100. Advance online publication. doi:10.118613756-017-0261-5 PMID:29046782

Ventola, C.L. (2015). The antibiotic resistance crisis: part 1: causes and threats. *P & T: A Peer-Reviewed Journal for Formulary Management, 40*(4), 277-83.

Wang, G., Li, X., & Zasloff, M. (2010). A database view of naturally occurring antimicrobial peptides: nomenclature, classification, and amino acid sequence analysis. *Antimicrobial Peptides: Discovery, Design and Novel Therapeutic Strategies*, 1-21. doi:10.1079/9781845936570.0001

World Health Organization (WHO). (2017). *Antibacterial agents in clinical development: an analysis of the antibacterial clinical development pipeline, including tuberculosis*. Geneva, Switzerland: WHO. https://www.who.int/medicines/areas/rational_use/antibacterial_agents_clinical_development/en/

Yoneyama, H., & Katsumata, R. (2006). Antibiotic Resistance in Bacteria and Its Future for Novel Antibiotic Development. *Bioscience, Biotechnology, and Biochemistry*, *70*(5), 1060–1075. doi:10.1271/bbb.70.1060 PMID:16717405

Zaman, S., Hussain, M., Nye, R., Mehta, V., Mamun, K., & Hossain, N. (2017). A Review on Antibiotic Resistance: Alarm Bells are Ringing. *Cureus*. Advance online publication. doi:10.7759/cureus.1403 PMID:28852600

Chapter 8
The Interplay Between Environmental Pollutants, Gut Microbiota, and Infections:
Current Concepts and Therapies

Bilal El-Mansoury
Faculty of Science, Chouaib Doukkali University, Morocco

Kamal Smimih
Faculty of Sciences and Techniques, Sultan Moulay Slimane University, Morocco

Youssef Ait Hamdan
https://orcid.org/0009-0005-8616-8852
Higher Normal School, Cadi Ayyad University, Morocco

Abdelaati El Khiat
Faculty of Medicine and Pharmacy, Cadi Ayyad University, Morocco & ISPITS, Ouarzazate, Morocco.

Samia Elouali
Higher Normal School, Cadi Ayyad University, Morocco

Hanane El Fatimi
Mohammed VI University of Health Sciences, Morocco

Ahmed Draoui
https://orcid.org/0000-0002-7700-059X
Faculty of Science Semlalia, Cadi Ayyad University, Morocco

Omar El Hiba
Faculty of Sciences, Chouaib Doukkali University, Morocco

Arumugam Radhakrishnan Jayakumar
Miller School of Medicine, University of Miami, USA

ABSTRACT

Environmental pollutants (EPs) have become an increasingly common health hazard. Several studies have sustained the impact of these EPs on human gut microbiota (GM). The human GM is made up of thousands of microbes that play a paramount

DOI: 10.4018/978-1-7998-9414-8.ch008

role to maintain hosts' health by providing many physiological functions, as well as protecting the host from pathogens and aggressions. Considering the crucial role that the host's GM plays in maintaining health, its disruption (gut dysbiosis) by the EPs can increase extraordinary the risk of infections and lead to the development of several conditions. Therefore, interventions aimed to restore the gut microbial population are of great interest. This chapter examines the intricate relationship between EPs and GM, highlighting the profound impact of EPs on human health including the development of infections. Additionally, this chapter discusses the potential role of probiotics, prebiotics, and FMT in mitigating the detrimental effects of EPs on GM.

I. INTRODUCTION

Environmental pollutants (EPs) are causing widespread concern due to their ecotoxicology and potential hazards to human health, particularly in light of the pervasive and severe pollution of biotic and abiotic environmental compartments (Chen, 2021). A growing body of research shows that being exposed to these EPs is one of the many environmental factors that contribute to the emergence of many health conditions (Claus et al., 2016b). Indeed, it has been established that chronic exposure to contaminants can raise the risk of developing metabolic and cancerous disorders(Chen, 2021). The basic role of gut microbiota (GM) in the upkeep and regulation of general host health has been confirmed in recent years by accumulating research. The GM is a complex community of microbes that produces energy from digested food, controls immunological response, and defends against pathogens(Omar et al., 2022). It comprises bacteria and other microbes such as fungi, archaea, viruses, and protozoans (Omar et al., 2022). Indeed, recent research has shown a significant connection between GM and human health. Through dietary channels, which can directly contact the gut ecosystem and come into contact with the microorganisms that naturally reside there, water and food pollution promote exposure to EPs (Chen, 2021).

GM are very sensitive to drugs, diet, and EPs (Jin et al., 2017a). Indeed, the GM system is commonly targeted and driven to dysbiosis by various pollutants (Chen, 2021). It is widely recognized that a balanced commensal microbiota is essential for defending the host against a variety of infections, either directly by eradicating them or indirectly by suppressing them inside or outside the intestine (Rothschild et al., 2018). Indeed, he human GM comprises thousands of microbes, many of which fight pathogens in infectious diseases and either inhibit or stimulate inflammation in various immunological situations (Maciel-Fiuza et al., 2023). Therefore, changes in the composition, abundance, diversity, and metabolism of the gut microbial population will undermine the physiological activity and fitness of the host, as well as create an imbalance in the host's immune system and raise the risk of infections. These have been shown to influence disease susceptibility and contribute to the pathogenesis of several conditions, including diabetes, obesity, and inflammation (Chen, 2021). There is mounting evidence that the gut microbiome influences the development and progression of various inflammatory and infectious diseases. On the other hand, medical concerns about EPs' effects on animals have grown over the past few years (Jin et al., 2010, 2016; Z. Liu et al., 2016; Lv et al., 2019; Piccioni et al., 2021; Ye et al., 2010), and the subsequent gut dysbiosis associated with high risks of infection.

Considering the crucial role of the host's GM in maintaining health, its disruption (gut dysbiosis) by the effects of EPs (Chen, 2021) leads to several conditions' development. As a result, intervention aiming to restore the functions of the gut microbial population and the integrity of the gut epithelial barrier appears to be a straightforward but practical approach to combating the negative effects of EPs (Chen, 2021). Therefore, it is crucial to comprehend the interplay between intestinal microbiota and the adverse effects of EPs on its function to understand the mechanisms involved in different pathologies and search for new treatments. This chapter overviews the link between EPs and GM dysfunction, their association with infections, and some current therapeutic approaches to restore GM function.

II. ROLE OF THE GUT MICROBIOTA IN THE REGULATION OF HOST PHYSIOLOGY

The GM is a dynamic biological system comprising pathogenic, commensal, and symbiotic microbial communities (Evariste et al., 2019). Indeed, the GM is a complex collection of microorganisms that includes bacteria and other microbes such as fungi, archaea, viruses, and protozoans (Omar et al., 2022). The GM performs a broad array of metabolic processes by producing metabolites, peptides, and proteins to interact with the host organism and carry out host metabolic functions. This includes metabolism and energy balance. Indeed, the GM can regulate intestinal motility and secretion, break down food's macromolecular polysaccharides, aid in nutrient digestion and absorption, preserve the integrity of the intestinal epithelial barrier, and support and maintain the immune system's normal growth and functions(Round et al., 2010). Additionally, the human host receives genetic and metabolic support from the bacterial community genome (metagenome), greatly expanding its capability to exploit a variety of dietary sources and improving the manufacture of vitamins, isoprenoids, and amino acids (Barlow et al., 2015). A healthy GM can boost immune function, encourage tissue development, shield the host from infections (Honda &

Littman, 2016; Thaiss et al., 2016) and improve the prevention of infectious diseases (Evariste et al., 2019), thus performing a substantial and indispensable part in determining the general health of the host (Zhong et al., 2022).

The GM contains thousands of different bacterial species, the majority of which fall within one of the following six phyla: Firmicutes, Bacteroidetes, Actinobacteria, Proteobacteria, Fusobacteria, and Verrucomicrobia(Jin et al., 2017; Evariste et al., 2019). Most of these GM members' nutritional composition and metabolic processes largely rely on the availability of carbohydrates, making saccharolytic metabolisms their predominant metabolic pathway(Macfarlane & Macfarlane, 2011; Smith et al., 2011). In contrast to other butyrate producers, such as Faecalibacteriumprausnitzii, which has only a weak correlation with fecal butyrate concentrations, it has been reported that *Roseburia intestinalis* and *Eubacterium rectale* play dominant roles in butyrate synthesis, which is crucial for the preservation and defense of the normal colonic epithelium(Duncan et al., 2007). The characterization of the microbial communities has been made possible by recent developments in metagenomic and bioinformatic techniques as well as next-generation sequencing technologies (Cénit et al., 2014). The human gut microbiome's species and composition are currently better understood than ever, with the variance being usually stratified (Jia et al., 2021). The microbiome genome is over 150 times larger than the human genome, with GM genomes encoding more than 3.3 million genes(Jin et al., 2017a).

III. HEALTH TOXICITY RELATED TO ENVIRONMENTAL POLLUTANTS EXPOSURE

GM are extremely sensitive to medications, nutrition, and contaminants of the environment (Jin et al., 2017a), and many diseases are correlated with altered gut microbiome composition. Indeed, humans are continuously exposed to multiple EPs. For instance, pesticide residues are the most frequent compounds discovered in the natural environment, followed by well-studied chemicals in plastic materials and commercial household products like phthalates and bisphenol A. Over the past few decades, EPs have increased in frequency and are now a significant health risk (Jin et al., 2017a) because soil, water, and air regularly contain EPs and their residuals. They can build up in organisms and negatively impact human health and the natural environment (Jin et al., 2017a). There is an urgent need to address human exposure to EPs, such as persistent organic pollutants, pesticides, phthalates, and heavy metals around the world (Hennig et al., 2018). It must be assumed that some of these EPs' effects on human health cause populations to be incredibly susceptible to disease risks (Figure 1).

1. Heavy Metals

In recent years, environmental heavy metals have significantly increased risk to human health (Liu et al., 2016a). Heavy metals are a prevalent type of EPs and are linked to a variety of harmful effects, including DNA damage, oxidative stress, cancer, and immune system impacts (Figure 1) (Jin et al., 2017a). Some non-essential metals, including mercury, silver, cadmium, and lead, are harmful to microorganisms even in low amounts. Accordingly, previous studies have shown that exposure to arsenic (As) is linked to a multitude of illnesses, including diabetes, cardiovascular disease, and cancers of the skin, bladder, lung, and liver (Hughes et al., 2011; Taylor et al., 2011; Van de Wiele et al., 2010). According to other investigations, mice exposed to mercury chloride at dosages of 0 to 80 mg/L developed intestinal lesions that resulted in dysbiosis of the intestinal microbiota and metabolic abnormalities (Zhao et al., 2020). Similar to this, low-level cadmium exposure altered the intestinal microbiota and increased intestinal permeability, which in turn caused a notable buildup of cadmium in mouse livers (Liu et al., 2020). Moreover, copper may be harmful to gut bacteria. In rats, early copper exposure (0,20 and 1,00 mg/kg) resulted in altered metabolic pathways, dysbiosis of the GM, and liver damage (Dai et al., 2020). Furthermore, studies on 109 workers who had been exposed to nickel (Ni) revealed that the exposed group's serum had a high content of Ni (5,70 ± 3,21 µg/L) and that their intestines had been affected with a high uric acid level (355,95 ± 67,87 µmol/L) (Yang et al., 2023).

2. Persistent Organic Pollutants

Persistent organic pollutants (POPs) are a group of organic chemicals that represent a global concern because of their environmental persistence (Tian et al., 2020). POPs are synthetic compounds that can build up in biological tissues and are very hazardous, long-lasting, and semi-volatile. They include polychlorinated biphenyls (PCBs), organochlorine pesticides, and polycyclic aromatic hydrocarbons (PAHs) (Jin et al., 2017a).In addition to older POPs, there are also emerging POPs, including brominated compounds, flame retardants (FRBs), perfluoroalkyl substances (PFASs) and polybrominated diphenyl ethers (PBDEs) (Popli et al., 2022).POPs have been implicated in human health problems, including immune system disorders (Gascon et al., 2013), cancer (Koual et al., 2019; Lim et al., 2017), and reproductive disorders (Vested et al., 2014). POPs are extremely resistant to biological and environmental deterioration; as a result, they linger in the environment and build up in the food chain. POPs are ligands of the aryl hydrocarbon receptor (AhR)(Claus et al., 2016b), which is a ligand-activated transcription factor that incorporates nutritional, environmental, microorganisms, and metabolic stimuli to control intricate regulatory processes in a ligand-specific, cell-type-specific and context-specific manner(Rothhammer & Quintana, 2019), which mediates the toxicity of these chemicals (Claus et al., 2016b).

3. Pesticides

''Pesticides'' is a broad term that refers to all substances used to eliminate or control pests. (Hassaan & El Nemr, 2020). They are utilized in agriculture as well as in gardens, public spaces, or private residences and are used to get rid of living things like insects (insecticides), fungi (fungicides), or weeds (herbicides) (Dahiri et al., 2021; Kim et al., 2017). Pesticides in water supplies have a negative impact on individuals and ecosystems. Although air, water, and the food chain are the most common ways the general public is exposed to pesticides, dietary consumption accounts for up to 90% of all pesticide exposure(Anderson & Meade, 2014; Yilmaz et al., 2020). Due to their widespread detection, pesticides are a severe and ongoing environmental issue. For many pesticides, acute toxicity is well-established; however, there is little information on chronic low-dose exposure scenarios that may occur in the general population (Buchenauer et al., 2022). Since they include elements that can cause DNA variations, such materials have been considered likely mutagens. The World Health Organization (WHO) estimates that pesticide exposure causes acute poisoning in roughly 100,000 people. The annual death rate ranges from 0.4% to 1.9% (Eddleston, 2020; Z.-Q. Jia et al., 2020; Qiu et al., 2017; Thundiyil et al., 2008).

4. Nanomaterials and Food Additives Nanoparticles

Emerging pollutants (EPs), also known as micropollutants, have become a significant concern for the entire world's population in recent years due to the potential risks they pose to the environment and public health (Arman et al., 2021). Numerous items, electrical components, paint that will not get scratched, cosmetics, sports gear, food containers, and dietary supplements, are made using nanoparticles (NPs) (Chaudhry et al., 2008; Mahmoudi et al., 2011). NPs (new EPs) can enter the human body through inhalation, ingestion, skin penetration, or injection because of their small size. NPs can cause oxidative stress, the most harmful intracellular and in vivo mechanism.

5. Microplastic

Microplastic (MP), a plastic particle with a diameter of less than 5 mm, is regarded as a brand-new environmental issue (Moore, 2008; Thompson et al., 2004); as a result, its effects have received much attention recently (Lu et al., 2019). MPs have developed into a major issue and a dangerous contaminant due to the use of these items and the vast manufacture of plastics (Cole et al., 2011). It may be claimed that MP contamination has spread over the globe because it has been found in coastal waterways (Zhao et al., 2018), seawater (Barnes et al., 2009), coastal waters (Zhao et al., 2018), freshwater systems (Li et al., 2018), soil ecosystems (Chae & An, 2018) and even in polar regions (Obbard et al., 2014). More importantly, MPs have also been detected in drinking water and table salt (L. Lu et al., 2019; Yang et al., 2015) and have been found in various products used by humans, including bivalves (Van Cauwenberghe & Janssen, 2014), and chicken gizzards (Huerta et al., 2017), which comes into immediate contact with the humans.

The biological toxicity of MPs, particularly to aquatic organisms, has been steadily increasing over the past few years (Chae & An, 2017; Ha & Yeo, 2018). Currently, numerous studies have found that because of MPs' small size and inability to degrade, fish and other species, particularly those found in the marine environment, can consume them(do Sul & Costa, 2014), which primarily evaluate the toxicity of MP on some parameters including tissue distribution(Ding et al., 2018), growth rate, oxidative damage, biological enzyme activity, and even reproduction. Therefore, MPs could be dangerous to terrestrial organisms and further endanger human health through the food chain or other routes (Sharma & Chatterjee, 2017; Wright & Kelly, 2017). Until now, the toxicity assessment of MPs in terrestrial mammals is somewhat limited (L. Lu et al., 2019).

Figure 1. Toxicity induced by different kinds of environmental pollutants on human health

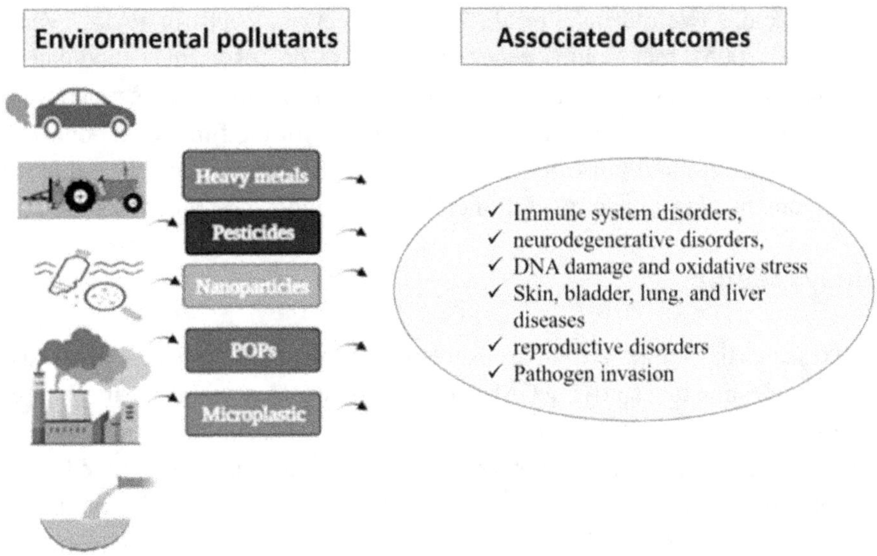

IV. ENVIRONMENTAL POLLUTANTS AND GUT MICROBIOTA ALTERATIONS

Even though most EPs do not specifically target GM, some pollutants can enter the body and interact with GM in various ways. The gut microbiome's extraordinary sensitivity to xenobiotics in the environment has been demonstrated (Jin et al., 2017; Zhong et al., 2022). Previous research has demonstrated that exposure to EPs can change the gut microbiome's composition, resulting in a range of diseases (Jin et al., 2015; S. Zhang et al., 2015). Indeed, growing evidence suggests EPs influence GM composition, leading to dysbiosis (Claus et al., 2016a; Singh & Mullin, 2017).

A number of EPs can influence the composition of GM, which in turn can affect the general homeostasis of the gut microbiome (Claus et al., 2016b; L. Zhang et al., 2015). Recent research, for instance, has indicated that EPs, including POPs, heavy metals, and pesticides, as well as nanomaterials and food additives, could have an impact on the composition of the GM, leading to physiological dysfunction, including host fitness loss, increased susceptibility to pathogens and/or be responsible for metabolic diseases (Carding et al., 2015; DeGruttola et al., 2016; Evariste et al., 2019). Surprisingly, several health conditions are linked to changes in the GM community and/or host functions, like the capacity to process and absorb dietary carbohydrates and complex lipids (Nicholson et al., 2012; Tilg & Moschen, 2014). Because of this, GM is a key indicator for toxicological assessment, and toxicological research should use this indicator as one of the main ones to investigate host health issues (L. Lu et al., 2019).

Although GM has effective biotransformation abilities in the detoxifying of numerous kinds of pollutants, including metals (Diaz-Bone and Van de Wiele, 2010), pesticides (Velmurugan et al., 2017; Yim et al., 2008) or PAHs (Van de Wiele et al., 2005), metabolites generated by these processes may modulate the toxicological pathways and toxic potential toward host organism. Investigations are still needed to fully understand the internal workings of the functional connection between host health and gut flora(Jia et al., 2021). Below are some EPs that have been demonstrated to impact the GI microbiota's composition and function.

1. Heavy Metal

Recent research has suggested that exposure to heavy metals may result in dysbiosis of the GM, suggesting that studying GM offers a fresh way to examine the mechanisms behind heavy metal toxicity (Jin et al., 2017a). For instance, the composition of the mouse microbiome and its metabolomic profiles are markedly altered after 4 weeks of exposure to 10 mg/L of arsenic (As)(Lu et al., 2014). The fluctuation of a variety of microbial co-metabolites, including indole derivatives, found in the urine or feces of exposed animals provided evidence that these taxonomic alterations were connected to a change in the metabolic activity of the gut microbiome (Lu et al., 2014). Moreover, Dheer et al. (2015)observed that long-term exposure to As changed the colon's microbiome's structure and composition and promoted male mouse hosts' metabolisms of amino acids and nitrogen(Dheer et al., 2015). Furthermore, cadmium (Cd), regularly and extensively employed in producing several goods like batteries, metal plating, pigments, and plastics, might cause changes in the GM. For example, after 10 weeks of subchronic exposure of male mice to a low dose of Cd (10 mg/L in drinking water), Bacteroidetes increased in relative abundance in the cecum and feces (as determined by real-time quantitative PCR and 16S RNA gene sequencing) as well as a sharp decrease in the populations of all microbial species was observed(Fazeli et al., 2011). In addition, Breton et al. (2013) found that continuous treatment of mice with cadmium and lead resulted in small but distinct alterations in the family and genus levels of the colonic microbiota composition (Breton et al., 2013). Importantly, heavy metal-consuming animals possessed more Lactobacillaceae and Erysipelotrichaceacae than control animals but less Lachnospiraceae (Breton et al., 2013). More recently, Jin et al. (2017) have suggested that exposure to heavy metals may also cause MG dysbiosis, indicating that studying GM offers a fresh perspective on deciphering the mechanisms underlying heavy metal toxicity (Jin et al., 2017a).

2. Pesticides

Pesticides have the potential to change the gut microbiome because of their antimicrobial activity. For instance, glyphosate is the active component of Roundup, the most widely used herbicide worldwide (Claus et al., 2016b). Due to its target, the enzyme 5-enolpyruvylshikimate-3-phosphate synthase (ESPS), which only exists in plants and not human cells, glyphosate has long been thought to be safe for use on people. However, there are several microorganisms in the human gut that depend on EPSP (Marques et al., 2007). Therefore, glyphosate is supposed to induce alterations in microbiota (Rueda-Ruzafa et al., 2019). Surprisingly, it has been suggested that glyphosate could exert its harmful effects by altering the crosstalk between the gut, brain, and immune systems by inducing dysbiosis of the GM composition (Aitbali et al., 2018; Maddalon et al., 2021). Buchenauer et al. (2022) used 16S rRNA gene profiling of caecum samples to analyze the bacterial community structure of 3-week-old female offspring from glyphosate-exposed dams compared to female offspring from control mice. The variation of microbial communities between the groups was clearly different, even though maternal glyphosate exposure had no impact on the alpha-diversity of the gut microbiome in the F1 mice (Buchenauer et al., 2022). Moreover, glyphosate may disrupt the relationship between specific immune cells and the gut, which could alter the development, progression, or persistence of disease or the immunological response (Buchenauer et al., 2022).

Additionally, pentachlorophenol (PCP) is a widely used pesticide worldwide (Dong et al., 2009). PCP exposure (100 mg/L) for 28 days in goldfish also changed the microbiome's makeup by lowering the Firmicutes/Bacteroidetes ratio. At the genus level, Bacteroides increased in relative abundance while Chryseobacterium, Microbacterium, Arthrobacter, and Legionella had decreased (Kan et al., 2015). Additionally, the broad-spectrum azole fungicide epoxiconazole (EPO) is a frequently used antifungal treatment in agriculture (Paveley et al., 2001; Zarn et al., 2003). Rats fed with large doses of EPO-adulterated meals for 90 days had lower relative abundances of Firmicutes and higher relative abundances of Bacteroidetes and Proteobacteria in their stools. In addition, Lachnospiraceae and Enterobacteriaceae were selectively enriched at the family level in terms of relative abundance. The authors came to the conclusion that these changes in microbiota composition brought on by EPO were related to compromised liver function (Xu et al., 2015). Furthermore, chlorpyrifos (CPF; organophosphate insecticide) was demonstrated to impair the mucosal barrier and microbiome, leading to dysbiosis and raising bacterial translocation (Condette et al., 2015). Chlorpyrifos induced-GM dysbiosis was shown by increased Bacteroides sp. and lower Lactobacillus sp. and Bifidobacterium sp (Claus et al., 2016b). Additionally, it has been noted that the fungicide carbendazim (CBZ) causes GM dysbiosis, resulting in lower relative concentrations of Bacteroidetes and higher relative concentrations of Firmicutes, Proteobacteria, and Actinobacteria at the phylum level. Desulfovibrionaceae, Ruminococcaceae, Lachnospiraceae, and Deferribacteraceae experienced relative abundance increases at the family level, whereas S24-7, Paraprevotellaceae, Rikenellaceae, Prevotellaceae, Bacteroidaceae, Porphyromonadaceae, and Christensenellaceae experienced relative abundance decreases in cecal contents (Jin et al., 2015c). Therefore, fungicide-induced changes to GM can promote inflammation and metabolic problems (Lu et al., 2019), and the possible impacts of pesticides on GM cannot be disregarded in the future due to their widespread use.

3. Nanoparticles

Nanoparticles (NPs) are thought to have a potential role in biodistribution and antimicrobial activity (Hadrup & Lam, 2014). It is conceivable for NPs to affect the GM both orally and non-orally. However, little research is looking at how NPs affect gut flora. According to a recent study, giving mice Ag NPs (1 mg/mL in 2 mM citrate buffer or 100 mg/mL polyvinylpyrrolidone in water for 28 days) did not change the gut microbiome's structure, diversity, or composition (Wilding et al., 2016). However, a different study found that 13 weeks of exposure to Ag NPs

(9, 18, and 36 mg/kg body weight) disrupted the expression of immunomodulatory genes in the rat ileum and reduced the number of Firmicutes at the phylum level (Williams et al., 2015). Additionally, Ag NPs harm fruit fly larvae by reducing the variety of the microbiota in their guts (Jin et al., 2017a).

4. Emulsifiers

Emulsifiers are additives in bread and other processed foods to extend shelf life and maintain freshness. The integrity of intestinal tissue and the GM may be significantly impacted by the detergent-like effects of emulsifiers, according to a recent study (Chassaing et al., 2015). Indeed, changes in GM have been reported in mice given carboxymethylcellulose or polysorbate 80 (emulsifiers) for 12 weeks. Moreover, the microbiota of mice treated with emulsifiers might be used to transmit these changes to germ-free animals. Emulsifiers may promote robust colitis in susceptible hosts and cause low-grade inflammation in others, according to the scientists' observation that they induced robust colitis in mice lacking the Toll-like receptor 5 (TLR5) and IL-10 (Chassaing et al., 2015). Additionally, a recent report showed that chronically feeding noncaloric artificial sweeteners (NAS, including saccharin, sucralose, and aspartame) to mice was associated with an increased abundance of bacteria belonging to the genus Bacteroides and order Clostridiales in the gut and resulted in much higher glucose intolerance (Jin et al., 2017a). This increase in glucose intolerance was GM-dependent (Suez et al., 2014). Furthermore, Saccharin and sucralose, two noncaloric artificial sweeteners, were found to disrupt the GM community, increase intestinal permeability, trigger systemic inflammation, and result in nonalcoholic fatty liver disease (NAFLD) in mice (Shi et al., 2021). In contrast, fructooligosaccharide supplementation significantly stimulates the reproduction of Akkermansiamuciniphila in sucralose-consuming mice, thereby alleviating NAFLD (Shi et al., 2021).

5. Microplastic

The toxicity of microplastics (MPs) to species, particularly aquatic organisms, has been thoroughly researched as a novel type of environmental pollutant (Rezania et al., 2018; Ha & Yeo, 2018). There is, however, little research on the impact of MPs on microbes, particularly the GM. The fact that MPs can directly enter the intestine and accumulate there is the key factor contributing to how MPs can impact the GM (Deng et al., 2017; Jin et al., 2019). The polystyrene MPs may have the ability to cause metabolic problems, intestinal barrier failure, and dysbiosis of the GM in mice (Lu et al., 2019). Additionally, it has been hypothesized that exposure to polystyrene MPs of various sizes could cause GM dysbiosis in mice, which would primarily manifest as a change in the diversity and composition of the GM and could have adverse effects on the host, such as abnormal physiological indicators and metabolic disorders (Jin et al., 2019; Lu et al., 2018). Indeed, in previous research, it has been found that polystyrene MPs have been shown to alter the composition of the GM in both adult and larval zebrafish, resulting in gut inflammation and metabolism disorder, respectively. Hence, MPs can alter the GM of aquatic organisms and have various potential effects (Lu et al., 2019). According to numerous studies, exposure to polystyrene MPs may change the make-up of the GM, causing inflammation in the zebrafish gut, a disturbance of hepatic lipid metabolism, and disruption of the intestinal barrier in mice (Jin et al., 2018; Lu et al., 2018). MPs may impact animal and human health by altering the GM. Therefore, there is a need for studies on the impact of these MPs on animal and human health (Eerkes-Medrano et al., 2015; Horton et al., 2017), which is a novel area of toxicological research in order to analyze the potential health risks to the host and to determine how MPs affect the gut flora.

6. Other Persistent Organic Pollutants

Concerns regarding the effects of POPs on GM have increased in recent years (Claus et al., 2016b; Jin et al., 2017b; Snedeker & Hay, 2012). According to growing experimental animal research, POP exposure may change the makeup of the microbial population and/or its metabolic processes, which could result in host illnesses (Choi et al., 2013; Nichols et al., 2019; L. Zhang et al., 2015). Short-term exposure to POPs has been shown to drastically alter both gene expression and the metabolism of the entire microbial population, in addition to changing the physiology of bacteria (Tian et al., 2020). The reactions of the gut microbiome to four POPs, including 2,3,7,8-tetrachlorodibenzofuran (TCDF), 2, 3, 7, and 8-tetrachlorodibenzo-p-dioxin (TCDD), and two polychlorinated biphenyls (PCBs), using a combination of flow cytometry, 16S rRNA gene sequencing, metatranscriptomics, and metabolomics has been investigated (Figure 1) (Tian et al., 2020). Indeed, the TCDF affects the morphology, population, and composition of GM in mice (Zhang et al., 2015). Moreover, with TCDF exposure for five days at a dose of 24 mg/kg body weight, the ratio of Firmicutes/Bacteroidetes was lowered, Flavobacteria levels were up, and Clostridia levels were decreased at the class level in the cecal content (Zhang et al., 2015). Additionally, PCB injection caused a dysbiosis of the GM in mice, resulting in a drop in Firmicutes and an increase in Bacteroidetes at the phylum level in intestinal contents from the colon(Chi et al., 2019). It has also been shown that PCBs can significantly alter the GM composition within 2 days of exposure in mice (Choi et al., 2013). In more recent studies, Lefever et al. found that TCDD dosing (6 mg/kg body weight) for 26 weeks significantly increased Firmicutes' abundance and decreased Bacteroidetes' abundance. In addition, the levels of the families Lactobacillaceae and Desulfovibrionaceae dramatically rose, and the levels of Prevotellaceae dropped in the feces (Lefever et al., 2016).

Figure 2. Schematic representation of the effects of different types of EPs on GM and their subsequent outcomes. EPs directly affect the profiles of microbiota metabolites and disrupt the gut microbiome's composition. The immune system is typically the first to be impacted by the microbiota and their metabolites. EPs may also have an impact on the endocrine system, the nervous system, the cardiovascular system, and other systems. Exposure to specific EPs will, therefore, induce various hazardous phenotypes.

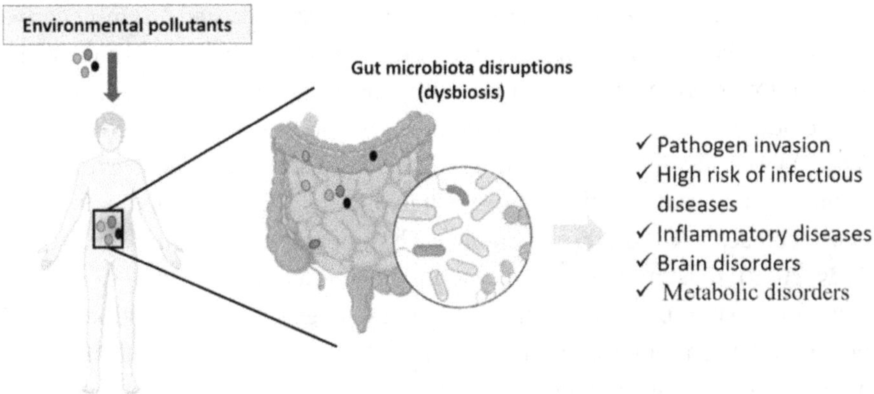

V. THE LINK BETWEEN ENVIRONMENTAL POLLUTANTS, INFECTIONS, AND GUT MICROBIOTA ALTERATIONS

A healthy commensal microbiota is essential for defending the host against several infections, either directly by eradicating them or indirectly by suppressing them inside or outside the intestine (Maciel-Fiuza et al., 2023). However, disruptions to the established GM community's structure and ensuing functionality modify the host and microbiota's overall symbiotic relationship, which alters the host's vulnerability to infections (Libertucci & Young, 2019). Indeed, it has been established, for example, that changes in the GM affect how susceptible the host is to various diseases.

The interplay between host defense mechanisms, the local microbiota, and prospective pathogens determine health and disease. This equilibrium is maintained through mechanisms of colonization resistance, which the microbiota may both directly and indirectly mediate. The host's mucosal immune responses are boosted by the microbiota, which helps to avoid the invasion of foreign bacteria and subsequent infection. In-depth investigations of the relationship between microbial composition and immune function, addressing the interaction between GM and immunological responses, have been made possible by recent technical and computational advancements in combination with ongoing translational research (Van Den Elsen et al., 2017). For instance, innate immune cells can have their activation thresholds adjusted by microbiota signals, increasing the chance of detecting the invasion of pathogens (Van Den Elsen et al., 2017). In addition to regulating systemic immunity in hosts, Gut microbes have some defense mechanisms that they can use to combat diseases brought in by invading pathogens and control systemic immunity in hosts. Consider the secretion of antibacterial compounds by gut microorganisms. It has been determined and sequenced that more than 170 different types of bacteria secrete bacteriocins (Aresti Sanz & El Aidy, 2019). Gram-positive bacteria are responsible for producing 88% of these bacteriocins. Archaea and Gram-negative bacteria produce the remaining 12%. Among them, the more common bacteriocins, such as lantibiotics produced by Gram-positive bacteria, are active against many species, including *Streptococcus pneumoniae*, *Staphylococcus aureus*, *C. difficile*, and *Enterococcus faecium*; microcins produced by Gram-negative bacteria can combat infection caused by other species, including E. coli, Salmonella enterica, Enterobacter cloacae, Klebsiella pneumoniae, Citrobacter, and Shigella (Aresti Sanz & El Aidy, 2019), thus gut dysbiosis could facilitate infections.

On the other hand, prior metagenomic studies suggested that infections could alter the microbial communities' composition and functions, which would affect the physiological aspects of gut homeostasis and the development of innate and adaptive immunity, intestinal immune tolerance, nutrient bioavailability, metabolism, and harvesting of energy (Jandhyala et al., 2015; Lin & Zhang, 2017; Marchesi et al., 2016; Preidis & Versalovic, 2009; Jia et al., 2021).

Previous research has suggested that environmental contaminants may directly alter the GM and alter metabolites and the immune system (Buffie et al., 2015; Candon et al., 2015; S. Zhang et al., 2015). Hence, it is certain that EPs will also impact the host's health, leading to several conditions. Indeed, it is probable that certain diseases, such as metabolic syndrome, undernutrition, atherosclerosis, inflammation, pathogen invasion, nerve damage, and infectious disease susceptibility, would be facilitated as a result of EP exposure (Jin et al., 2017a). Figure 2 presents an overview of the impacts of EPs on the GM and its subsequent implications.

VI. INTERACTIONS BETWEEN GUT MICROBIOTA AND HOSTS AND THEIR ROLE IN INFECTIOUS DISEASES

The microbiota plays a crucial role in colonization resistance to inhibit non-native growth, persistence, and eventual infection (Libertucci & Young, 2019). The study of infectious diseases has traditionally emphasised non-indigenous microorganisms, also known as "pathogens," as being crucial to the etiopathogenesis of infectious diseases (Libertucci & Young, 2019). According to Koch's original model for infectious diseases, infection happens in a linear fashion: a non-indigenous microorganism enters the host, colonizes it, and then produces disease (Libertucci & Young, 2019). Since so many people are affected, infectious diseases are now a major global public health concern. (Mehand et al., 2018). Major infectious diseases, including viral hepatitis and tuberculosis, are challenging to eradicate rapidly (Lv et al., 2019). Infectious diseases continue to pose a serious threat to world health despite significant advancements in prevention and treatment due to the development of antiseptics, vaccines, and antibiotics that significantly decrease infection rates and fatalities. As was already noted, EPs may have a direct impact on the GM, altering metabolites and the immune system leading to dysbiosis that may increase the susceptibility of infectious diseases or make them worse (Figure 2) (Jin et al., 2017a).

The mechanisms by which a pathogen might circumvent host defenses and establish infection have been the subject of investigation for more than a century (Falkow, 2004). A number of theories proposed to account for the increase in such diseases involve alterations to the human microbiome (Moossavi & Bishehsari, 2019). According to some research, GM's balance and health are essential for developing infectious diseases and susceptibility to infections (Lv et al., 2019). For instance, it has been demonstrated that patients with hepatitis B (HVB) had lower levels of Bacteroides than healthy individuals, and numerous studies have revealed that an increase in bacteria from the phyla Proteobacteria Firmicutes and Bacteroidetes can exacerbate dysbiosis caused by viral hepatitis C (HCV) infection (Maciel-Fiuza et al., 2023). As a result, manipulating the GM is a promising therapeutic target for enhancing immune function and preventing infectious diseases (Van Den Elsen et al., 2017). Also, identifying the gut bacteria that are active during the emergence of infectious diseases is crucial as it could help to develop novel approaches for the control and treatment of infectious diseases.

VII. GUT MICROBIOTA MANIPULATION TO MITIGATE THE DETRIMENTAL EFFECTS OF ENVIRONMENTAL POLLUTANTS AND INFECTIONS

GM plays a crucial role in host health, including fighting infections, but EPs could affect GM and lead to GM dysbiosis, which is associated with a high risk of infections. Therefore, interventions aimed at reestablishing GM are of great interest. Indeed, the beneficial modulation of GM is a promising clinical target for many prevalent diseases, including IBD, metabolic abnormalities, low-grade inflammation, allergy and protective immunity against infections (Van Den Elsen et al., 2017).

The enormous cost associated with removing persistent toxic substances from the environment is particularly encouraged by the development of in situ remediation of animal and human health issues related to the long-term stress of EPs. Among these approaches, dietary supplementation with probiotics, prebiotics, and fecal microbiota transplantation are promising.

1. Probiotics and Prebiotics

Probiotics are live microorganisms delivered through food, drugs, supplements, and formula (El-Mansoury et al., 2023). Most of the bacteria in a conventional probiotic are those naturally found in the human stomach. Probiotics may improve the host's health if administered in the right amounts and at the right intervals (Hill et al., 2014; El-Mansoury et al., 2023). Probiotics were described as "live microorganisms that when administered in adequate amounts, confer a health benefit on the host" by an expert panel in 2014(Hill et al., 2014). Most commercially available probiotics include Lactobacillus, Bifidobacterium, or Saccharomyces spp (Andrews & Tan, 2012; Balakrishnan & Floch, 2012). Prebiotics and synbiotics are other terms used in this context (Markowiak & Śliżewska, 2017; Pandey et al., 2015). Prebiotics are mostly indigestible, fibrous food components that support the health of the host by particularly boosting the growth and/or activity of some genera of gastrointestinal microorganisms, primarily lactobacilli and bifidobacterial (Markowiak & Śliżewska, 2017; Pandey et al., 2015). Prebiotics are described as intentionally fermented substances that cause certain alterations in the GM's composition and/or activity and positively affect the host's health (Gibson et al., 2004; Roberfroid, 2007). The use of prebiotics as functional meals to alter the make-up of the GM has seen a sharp rise in interest in recent years(Silk et al., 2009; Vardakou et al., 2008). A variety of dietary fibers, including fructooligosaccharides, mannanoligosaccharides, galactooligosaccharides, and inulin, are the main prebiotics that have attracted widespread recognition. These fibers also shown good effectiveness in reducing the potentially harmful effects of xenobiotics (Chen, 2021). However, from a toxicological standpoint, probiotics' and prebiotics' ability to attenuate or ameliorate the toxicities of pollutants remain mostly unclear (Chen, 2021). Although probiotic and prebiotic applications have been made in the aquaculture sector and for improving human health, it has not yet been sufficiently investigated if these functional ingredients can interact with and decrease the inherent toxicity of environmental contaminants. More combined exposures are anticipated to emphasize the protective effect of probiotics and prebiotics by elucidating their protective mode of action against the toxicities of environmental xenobiotics.

Regarding the toxicities of EPs, probiotic administration has also been shown to yield inhibitory activities (Chen, 2021). For instance, using Lactobacillus plantarum CCFM8610 during treatments effectively sequesters the heavy metal cadmium, reducing bioaccumulation. L. plantarum CCFM8610 can also shield the intestinal barrier and shield mice from the inflammatory signs that cadmium causes (Zhai et al., 2016). Additionally, diethylhexylphthalate poisoning in male mice is reduced by oral supplementation of L. plantarum TW1-1, which also shifts the exposed GM to the control structure and lessens the hyper-permeability of the intestinal barrier brought on by diethylhexylphthalate pollutant (Tian et al., 2019). Moreover, feeding zebrafish probiotic *L. rhamnosus* greatly reduces blood cholesterol buildup. It counteracts the disruption of bile acid metabolism brought on by continuous exposure to perfluorobutane sulfonate, an aquatic contaminant of rising concern (Chen, 2021). Furthermore, inulin feeding of hyperlipidemic mice restores GM, lowers aortic root lesions, lessens disturbance of glucose metabolism, and lessens hepatic lipid accumulation after exposure to a dioxin-like pollutant, PCB (Hoffman et al., 2020).

Some probiotics have shown promise in diagnosing, preventing, and treating several infectious diseases and their consequences. For instance, Hickson et al.'s randomized double-blind placebo-controlled experiment from 2007 found that consuming a probiotic drink containing *L. casei*, *L. bulgaricus*, and *S. thermophilus* can lessen diarrhea caused by *C. difficile* (Hickson et al., 2007). Moreover, oral administration of L. rhamnosus GG to patients with HIV reduced Enterobacteriaceae and inflammation in the gut (Arnbjerg et al., 2018). Additionally, probiotic feeding of Glucose and E. coli had a preventative effect on *V. cholerae* infection in the zebrafish model, indicating a comparable inhibitory effect in patients receiving cholera treatment (Jia et al., 2021). Even though numerous probiotic strains are helpful in preventing and treating many infectious diseases, further clinical trials are required to pinpoint favorable strains and confirm their practical uses.

2. Fecal Microbiota Transplantation

The manipulation of GM by fecal microbiota transplantation (FMT) has recently drawn considerable attention (D'Haens & Jobin, 2019; El-Mansoury et al., 2023). FMT stands for re-establishing a stable microbiota in the GI tract and treating the disease by giving potential patients a fecal suspension (the entire GM) from healthy donors (Choi & Cho, 2016; Cui et al., 2016; Zhang et al., 2018; Zhang et al., 2012). It is important to note that FMT alters the GM and its metabolites by improving its structural integrity, which in turn enhances some negative effects of EPs and infections. Indeed, FMT is a novel approach for treating several gut diseases linked to dysbiosis (El-Mansoury et al., 2023), and plays a significant role in preventing and treating infectious diseases, especially for refractory infections. For instance, recurrent *Clostridium difficile* infection (CDI) can be effectively treated with FMT (Konturek et al., 2016; Mullish et al., 2018; D'Haens & Jobin, 2019), an thus, the Federal Drug Administration (FDA) of the United States has granted it approval (Carlucci et al., 2016). Indeed, FMT is among the most significant methods lately taken into consideration for treating *C. difficile* infection. A recent meta-analysis found that 90% of the trials treated with fecal transplantation for refractory C. difficile infection were effective (Bakken, 2009; Landy et al., 2011) showing that this technique has a potent and reproducible efficacy when broad-spectrum antibiotics, as traditional therapeutic option have failed to treat the disease (Bakken, 2009). Another meta-analysis found that 439 of the 610 CDI patients who received a single FMT treatment across 13 studies were clinically cured. The open-label study's cure rates (82.7% vs. 67.7%) were considerably higher than the randomized trial's cure rates (67.7% vs. 82.7%) (Tariq et al., 2019).

Although FMT shows exciting results in refractory infections and is considered a successful strategy to treat dysbiosis, the relationship between FMT and adverse events such as inflammatory bowel attacks is an issue of concern. It is important to note that FMT is not frequently utilized due to the time needed to find a suitable donor, the chance of introducing harmful microorganisms, and the general distaste of recipients (Borody & Khoruts, 2012). Hence, the development of animal models with many features of FMT in humans with infections, and gut dysbiosis induced by EPs could help to understand the basic mechanisms of successful FMT and develop standardized treatments.

CONCLUSION

The intricate relationship between EPs, GM and infections underscores the complex interplay between human health and the environment. The composition and diversity of human GM are significantly impacted by exposure to several kinds of EPs, leading to dysbiosis, which may result in various health problems, such as inflammation, metabolic abnormalities, weakened immune systems and a high risk of developing infectious diseases. It is worth mentioning that GM plays a pivotal role in the detoxification and metabolization of certain pollutants, thereby acting as a protective barrier against their harmful effects, but these effects are quite limited. The detoxification methods include an effective adherence to the intestinal mucosa, dominating microbial suppression, and a high tolerance for GI fluid. However, the complexity of this interaction necessitates further investigation into the mechanisms underlying the interplay between EPs and GM, emphasizing the need for interdisciplinary research and innovative interventions to mitigate the detrimental effects of these EPs on the gut microbiome and overall well-being. Therefore, a deep understanding of the connections between EPs, GM, and infections could help to develop innovative interventions to improve human health outcomes.

REFERENCES

Aitbali, Y., Ba-M'hamed, S., Elhidar, N., Nafis, A., Soraa, N., & Bennis, M. (2018). Glyphosate based-herbicide exposure affects gut microbiota, anxiety and depression-like behaviors in mice. *Neurotoxicology and Teratology*, *67*, 44–49. doi:10.1016/j.ntt.2018.04.002 PMID:29635013

Anderson, S. E., & Meade, B. J. (2014). Potential health effects associated with dermal exposure to occupational chemicals. *Environmental Health Insights, 8.*

Andrews, J. M., & Tan, M. (2012). Probiotics in luminal gastroenterology: The current state of play. *Internal Medicine Journal*, *42*(12), 1287–1291. doi:10.1111/imj.12015 PMID:23252997

Aresti Sanz, J., & El Aidy, S. (2019). Microbiota and gut neuropeptides: A dual action of antimicrobial activity and neuroimmune response. *Psychopharmacology*, *236*(5), 1597–1609. doi:10.100700213-019-05224-0 PMID:30997526

Arman, N. Z., Salmiati, S., Aris, A., Salim, M. R., Nazifa, T. H., Muhamad, M. S., & Marpongahtun, M. (2021). A review on emerging pollutants in the water environment: Existences, health effects and treatment processes. *Water (Switzerland)*, *13*(22), 1–31. doi:10.3390/w13223258

Arnbjerg, C. J., Vestad, B., Hov, J. R., Pedersen, K. K., Jespersen, S., Johannesen, H. H., Holm, K., Halvorsen, B., Fallentin, E., Hansen, A. E., Lange, T., Kjær, A., Trøseid, M., Fischer, B. M., & Nielsen, S. D. (2018). Effect of Lactobacillus rhamnosus GG supplementation on intestinal inflammation assessed by PET/MRI scans and gut microbiota composition in HIV-infected individuals. *JAIDS Journal of Acquired Immune Deficiency Syndromes*, *78*(4), 450–457. doi:10.1097/QAI.0000000000001693 PMID:29874201

Bakken, J. S. (2009). Fecal bacteriotherapy for recurrent Clostridium difficile infection. *Anaerobe*, *15*(6), 285–289. doi:10.1016/j.anaerobe.2009.09.007 PMID:19778623

Balakrishnan, M., & Floch, M. H. (2012). Prebiotics, probiotics and digestive health. *Current Opinion in Clinical Nutrition and Metabolic Care*, *15*(6), 580–585. doi:10.1097/MCO.0b013e328359684f PMID:23037903

Barlow, G. M., Yu, A., & Mathur, R. (2015). Role of the gut microbiome in obesity and diabetes mellitus. *Nutrition in Clinical Practice*, *30*(6), 787–797. doi:10.1177/0884533615609896 PMID:26452391

Barnes, D. K. A., Galgani, F., Thompson, R. C., & Barlaz, M. (2009). Accumulation and fragmentation of plastic debris in global environments. *Philosophical Transactions of the Royal Society of London. Series B, Biological Sciences*, *364*(1526), 1985–1998. doi:10.1098/rstb.2008.0205 PMID:19528051

Bist, P., & Choudhary, S. (2022). Impact of Heavy Metal Toxicity on the Gut Microbiota and Its Relationship with Metabolites and Future Probiotics Strategy: A Review. *Biological Trace Element Research*, *200*(12), 5328–5350. doi:10.100712011-021-03092-4 PMID:34994948

Borody, T. J., & Khoruts, A. (2012). Fecal microbiota transplantation and emerging applications. *Nature Reviews. Gastroenterology & Hepatology*, *9*(2), 88–96. doi:10.1038/nrgastro.2011.244 PMID:22183182

Breton, J., Massart, S., Vandamme, P., De Brandt, E., Pot, B., & Foligné, B. (2013). Ecotoxicology inside the gut: Impact of heavy metals on the mouse microbiome. *BMC Pharmacology & Toxicology*, *14*(1), 1–11. doi:10.1186/2050-6511-14-62 PMID:24325943

Buchenauer, L., Junge, K. M., Haange, S. B., Simon, J. C., von Bergen, M., Hoh, A. L., Aust, G., Zenclussen, A. C., Stangl, G. I., & Polte, T. (2022). Glyphosate differentially affects the allergic immune response across generations in mice. *The Science of the Total Environment*, *850*(April), 157973. doi:10.1016/j.scitotenv.2022.157973 PMID:35963408

Buffie, C. G., Bucci, V., Stein, R. R., McKenney, P. T., Ling, L., Gobourne, A., No, D., Liu, H., Kinnebrew, M., Viale, A., Littmann, E., van den Brink, M. R. M., Jenq, R. R., Taur, Y., Sander, C., Cross, J. R., Toussaint, N. C., Xavier, J. B., & Pamer, E. G. (2015). Precision microbiome reconstitution restores bile acid mediated resistance to *Clostridium difficile*. *Nature*, *517*(7533), 205–208. doi:10.1038/nature13828 PMID:25337874

Candon, S., Perez-Arroyo, A., Marquet, C., Valette, F., Foray, A.-P., Pelletier, B., Milani, C., Ventura, M., Bach, J.-F., & Chatenoud, L. (2015). Antibiotics in early life alter the gut microbiome and increase disease incidence in a spontaneous mouse model of autoimmune insulin-dependent diabetes. *PLoS One*, *10*(5), e0125448. doi:10.1371/journal.pone.0125448 PMID:25970503

Carding, S., Verbeke, K., Vipond, D. T., Corfe, B. M., & Owen, L. J. (2015). Dysbiosis of the gut microbiota in disease. *Microbial Ecology in Health and Disease*, *26*(1), 26191. doi:10.3402/mehd.v26.26191 PMID:25651997

Carlucci, C., Petrof, E. O., & Allen-Vercoe, E. (2016). Fecal microbiota-based therapeutics for recurrent Clostridium difficile infection, ulcerative colitis and obesity. *EBioMedicine*, *13*, 37–45. doi:10.1016/j.ebiom.2016.09.029 PMID:27720396

Cénit, M. C., Matzaraki, V., Tigchelaar, E. F., & Zhernakova, A. (2014). Rapidly expanding knowledge on the role of the gut microbiome in health and disease. *Biochimica et Biophysica Acta (BBA)-. Molecular Basis of Disease*, *1842*(10), 1981–1992. doi:10.1016/j.bbadis.2014.05.023 PMID:24882755

Chae, Y., & An, Y.-J. (2017). Effects of micro-and nanoplastics on aquatic ecosystems: Current research trends and perspectives. *Marine Pollution Bulletin*, *124*(2), 624–632. doi:10.1016/j.marpolbul.2017.01.070 PMID:28222864

Chae, Y., & An, Y.-J. (2018). Current research trends on plastic pollution and ecological impacts on the soil ecosystem: A review. *Environmental Pollution*, *240*, 387–395. doi:10.1016/j.envpol.2018.05.008 PMID:29753246

Chassaing, B., Koren, O., Goodrich, J. K., Poole, A. C., Srinivasan, S., Ley, R. E., & Gewirtz, A. T. (2015). Dietary emulsifiers impact the mouse gut microbiota promoting colitis and metabolic syndrome. *Nature*, *519*(7541), 92–96. doi:10.1038/nature14232 PMID:25731162

Chen, L. (2021). Gut microbiota manipulation to mitigate the detrimental effects of environmental pollutants. *Toxics*, *9*(6), 2–3. doi:10.3390/toxics9060127 PMID:34205878

Chi, Y., Lin, Y., Lu, Y., Huang, Q., Ye, G., & Dong, S. (2019). Gut microbiota dysbiosis correlates with a low-dose PCB126-induced dyslipidemia and non-alcoholic fatty liver disease. *The Science of the Total Environment*, *653*, 274–282. doi:10.1016/j.scitotenv.2018.10.387 PMID:30412872

Choi, H. H., & Cho, Y.-S. (2016). Fecal microbiota transplantation: Current applications, effectiveness, and future perspectives. *Clinical Endoscopy*, *49*(3), 257–265. doi:10.5946/ce.2015.117 PMID:26956193

Choi, J. J., Eum, S. Y., Rampersaud, E., Daunert, S., Abreu, M. T., & Toborek, M. (2013). Exercise attenuates PCB-induced changes in the mouse gut microbiome. *Environmental Health Perspectives*, *121*(6), 725–730. doi:10.1289/ehp.1306534 PMID:23632211

Claus, S. P., Guillou, H., & Ellero-Simatos, S. (2016a). The gut microbiota: A major player in the toxicity of environmental pollutants? *NPJ Biofilms and Microbiomes*, *2*(1), 1–11. doi:10.1038/npjbiofilms.2016.3 PMID:28721242

Claus, S. P., Guillou, H., & Ellero-Simatos, S. (2016b). The gut microbiota: A major player in the toxicity of environmental pollutants? *NPJ Biofilms and Microbiomes*, *2*(March), 1–12. doi:10.1038/npjbiofilms.2016.3 PMID:28721242

Cole, M., Lindeque, P., Halsband, C., & Galloway, T. S. (2011). Microplastics as contaminants in the marine environment: A review. *Marine Pollution Bulletin*, *62*(12), 2588–2597. doi:10.1016/j.marpolbul.2011.09.025 PMID:22001295

Condette, C. J., Bach, V., Mayeur, C., Gay-Quéheillard, J., & Khorsi-Cauet, H. (2015). Chlorpyrifos exposure during perinatal period affects intestinal microbiota associated with delay of maturation of digestive tract in rats. *Journal of Pediatric Gastroenterology and Nutrition*, *61*(1), 30–40. doi:10.1097/MPG.0000000000000734 PMID:25643018

Cui, B., Li, P., Xu, L., Peng, Z., Xiang, J., He, Z., Zhang, T., Ji, G., Nie, Y., Wu, K., Fan, D., & Zhang, F. (2016). Step-up fecal microbiota transplantation (FMT) strategy. *Gut Microbes*, *7*(4), 323–328. doi:10.1080/19490976.2016.1151608 PMID:26939622

D'Haens, G. R., & Jobin, C. (2019). Fecal microbial transplantation for diseases beyond recurrent clostridium difficile infection. *Gastroenterology*, *157*(3), 624–636. doi:10.1053/j.gastro.2019.04.053 PMID:31220424

Dahiri, B., Martín-Reina, J., Carbonero-Aguilar, P., Aguilera-Velázquez, J. R., Bautista, J., & Moreno, I. (2021). Impact of pesticide exposure among rural and urban female population. An overview. *International Journal of Environmental Research and Public Health*, *18*(18), 9907. doi:10.3390/ijerph18189907 PMID:34574830

Dai, J., Yang, X., Yuan, Y., Jia, Y., Liu, G., Lin, N., Xiao, H., Zhang, L., & Chen, J. (2020). Toxicity, gut microbiota and metabolome effects after copper exposure during early life in SD rats. *Toxicology*, *433*, 152395. doi:10.1016/j.tox.2020.152395 PMID:32027963

DeGruttola, A. K., Low, D., Mizoguchi, A., & Mizoguchi, E. (2016). Current understanding of dysbiosis in disease in human and animal models. *Inflammatory Bowel Diseases*, *22*(5), 1137–1150. doi:10.1097/MIB.0000000000000750 PMID:27070911

Deng, Y., Zhang, Y., Lemos, B., & Ren, H. (2017). Tissue accumulation of microplastics in mice and biomarker responses suggest widespread health risks of exposure. *Scientific Reports*, *7*(1), 46687. doi:10.1038rep46687 PMID:28436478

Dheer, R., Patterson, J., Dudash, M., Stachler, E. N., Bibby, K. J., Stolz, D. B., Shiva, S., Wang, Z., Hazen, S. L., Barchowsky, A., & Stolz, J. F. (2015). Arsenic induces structural and compositional colonic microbiome change and promotes host nitrogen and amino acid metabolism. *Toxicology and Applied Pharmacology*, *289*(3), 397–408. doi:10.1016/j.taap.2015.10.020 PMID:26529668

Ding, J., Zhang, S., Razanajatovo, R. M., Zou, H., & Zhu, W. (2018). Accumulation, tissue distribution, and biochemical effects of polystyrene microplastics in the freshwater fish red tilapia (*Oreochromis niloticus*). *Environmental Pollution*, *238*, 1–9. doi:10.1016/j.envpol.2018.03.001 PMID:29529477

do Sul, J. A. I., & Costa, M. F. (2014). The present and future of microplastic pollution in the marine environment. *Environmental Pollution*, *185*, 352–364. doi:10.1016/j.envpol.2013.10.036 PMID:24275078

Dong, Y.-L., Zhou, P.-J., Jiang, S.-Y., Pan, X.-W., & Zhao, X.-H. (2009). Induction of oxidative stress and apoptosis by pentachlorophenol in primary cultures of Carassius carassius hepatocytes. *Comparative Biochemistry and Physiology. Toxicology & Pharmacology : CBP*, *150*(2), 179–185. doi:10.1016/j.cbpc.2009.04.010 PMID:19410655

Duncan, S. H., Belenguer, A., Holtrop, G., Johnstone, A. M., Flint, H. J., & Lobley, G. E. (2007). Reduced dietary intake of carbohydrates by obese subjects results in decreased concentrations of butyrate and butyrate-producing bacteria in feces. *Applied and Environmental Microbiology*, *73*(4), 1073–1078. doi:10.1128/AEM.02340-06 PMID:17189447

Eddleston, M. (2020). Poisoning by pesticides. *Medicine*, *48*(3), 214–217. doi:10.1016/j.mpmed.2019.12.019

Eerkes-Medrano, D., Thompson, R. C., & Aldridge, D. C. (2015). Microplastics in freshwater systems: A review of the emerging threats, identification of knowledge gaps and prioritisation of research needs. *Water Research*, *75*, 63–82. doi:10.1016/j.watres.2015.02.012 PMID:25746963

El-Mansoury, B., Smimih, K., El Khiat, A., El Fatimi, H., Aimrane, A., El Baz, S., Elmourid, A., Saad, F. E.-Z., El Ghachi, H., & Khanouchi, M. (2023). Potential Role of Probiotics and Fecal Microbiota Transplantation in the Treatment of Parkinson's Disease. In *Experimental and Clinical Evidence of the Neuropathology of Parkinson's Disease* (pp. 296–317). IGI Global. doi:10.4018/978-1-6684-5156-4.ch017

Evariste, L., Barret, M., Mottier, A., Mouchet, F., Gauthier, L., & Pinelli, E. (2019). Gut microbiota of aquatic organisms: A key endpoint for ecotoxicological studies. *Environmental Pollution*, *248*, 989–999. doi:10.1016/j.envpol.2019.02.101 PMID:31091643

Falkow, S. (2004). Molecular Koch's postulates applied to bacterial pathogenicity—A personal recollection 15 years later. *Nature Reviews. Microbiology*, *2*(1), 67–72. doi:10.1038/nrmicro799 PMID:15035010

Fazeli, M., Hassanzadeh, P., & Alaei, S. (2011). Cadmium chloride exhibits a profound toxic effect on bacterial microflora of the mice gastrointestinal tract. *Human and Experimental Toxicology*, *30*(2), 152–159. doi:10.1177/0960327110369821 PMID:20488844

Gascon, M., Morales, E., Sunyer, J., & Vrijheid, M. (2013). Effects of persistent organic pollutants on the developing respiratory and immune systems: A systematic review. *Environment International*, *52*, 51–65. doi:10.1016/j.envint.2012.11.005 PMID:23291098

Gibson, G. R., Probert, H. M., Van Loo, J., Rastall, R. A., & Roberfroid, M. B. (2004). Dietary modulation of the human colonic microbiota: Updating the concept of prebiotics. *Nutrition Research Reviews*, *17*(2), 259–275. doi:10.1079/NRR200479 PMID:19079930

Ha, J., & Yeo, M.-K. (2018). The environmental effects of microplastics on aquatic ecosystems. *Molecular & Cellular Toxicology*, *14*(4), 353–359. doi:10.100713273-018-0039-8

Hadrup, N., & Lam, H. R. (2014). Oral toxicity of silver ions, silver nanoparticles and colloidal silver–a review. *Regulatory Toxicology and Pharmacology*, *68*(1), 1–7. doi:10.1016/j.yrtph.2013.11.002 PMID:24231525

Hassaan, M. A., & El Nemr, A. (2020). Pesticides pollution: Classifications, human health impact, extraction and treatment techniques. *Egyptian Journal of Aquatic Research*, *46*(3), 207–220. doi:10.1016/j.ejar.2020.08.007

Hennig, B., Petriello, M. C., Gamble, M. V., Surh, Y. J., Kresty, L. A., Frank, N., Rangkadilok, N., Ruchirawat, M., & Suk, W. A. (2018). The role of nutrition in influencing mechanisms involved in environmentally mediated diseases. *Reviews on Environmental Health*, *33*(1), 87–97. doi:10.1515/reveh-2017-0038 PMID:29381475

Hickson, M., D'Souza, A. L., Muthu, N., Rogers, T. R., Want, S., Rajkumar, C., & Bulpitt, C. J. (2007). Use of probiotic Lactobacillus preparation to prevent diarrhoea associated with antibiotics: Randomised double blind placebo controlled trial. *BMJ (Clinical Research Ed.)*, *335*(7610), 80. doi:10.1136/bmj.39231.599815.55 PMID:17604300

Hill, C., Guarner, F., Reid, G., Gibson, G. R., Merenstein, D. J., Pot, B., Morelli, L., Canani, R. B., Flint, H. J., Salminen, S., Calder, P. C., & Sanders, M. E. (2014). The International Scientific Association for Probiotics and Prebiotics consensus statement on the scope and appropriate use of the term probiotic. *Nature Reviews. Gastroenterology & Hepatology*, *11*(8), 506–514. doi:10.1038/nrgastro.2014.66 PMID:24912386

Hoffman, J. B., Petriello, M. C., Morris, A. J., Mottaleb, M. A., Sui, Y., Zhou, C., Deng, P., Wang, C., & Hennig, B. (2020). Prebiotic inulin consumption reduces dioxin-like PCB 126-mediated hepatotoxicity and gut dysbiosis in hyperlipidemic Ldlr deficient mice. *Environmental Pollution*, *261*, 114183. doi:10.1016/j.envpol.2020.114183 PMID:32105967

Honda, K., & Littman, D. R. (2016). The microbiota in adaptive immune homeostasis and disease. *Nature*, *535*(7610), 75–84. doi:10.1038/nature18848 PMID:27383982

Horton, A. A., Walton, A., Spurgeon, D. J., Lahive, E., & Svendsen, C. (2017). Microplastics in freshwater and terrestrial environments: Evaluating the current understanding to identify the knowledge gaps and future research priorities. *The Science of the Total Environment*, *586*, 127–141. doi:10.1016/j.scitotenv.2017.01.190 PMID:28169032

Huerta, E., Mendoza Vega, J., Quej, V. K., Chi, J. de los A., Sanchez del Cid, L., Quijano, C., Escalona-Segura, G., Gertsen, H., Salánki, T., & van der Ploeg, M. (2017). Bioaccumulation of microplastics in the terrestrial food chain: An example from home gardens in SE Mexico. *EGU General Assembly Conference Abstracts*, 15847.

Hughes, M. F., Beck, B. D., Chen, Y., Lewis, A. S., & Thomas, D. J. (2011). Arsenic exposure and toxicology: A historical perspective. *Toxicological Sciences*, *123*(2), 305–332. doi:10.1093/toxsci/kfr184 PMID:21750349

Jandhyala, S. M., Talukdar, R., Subramanyam, C., Vuyyuru, H., Sasikala, M., & Reddy, D. N. (2015). Role of the normal gut microbiota. *World Journal of Gastroenterology: WJG*, *21*(29), 8787. doi:10.3748/wjg.v21.i29.8787 PMID:26269668

Jia, P. P., Junaid, M., Wen, P. P., Yang, Y. F., Li, W. G., Yang, X. G., & Pei, D. S. (2021). Role of germ-free animal models in understanding interactions of gut microbiota to host and environmental health: A special reference to zebrafish. *Environmental Pollution, 279*, 116925. doi:10.1016/j.envpol.2021.116925 PMID:33744636

Jia, Z.-Q., Zhang, Y.-C., Huang, Q.-T., Jones, A. K., Han, Z.-J., & Zhao, C.-Q. (2020). Acute toxicity, bioconcentration, elimination, action mode and detoxification metabolism of broflanilide in zebrafish, Danio rerio. *Journal of Hazardous Materials, 394*, 122521. doi:10.1016/j.jhazmat.2020.122521 PMID:32279005

Jin, Y., Chen, R., Liu, W., & Fu, Z. (2010). Effect of endocrine disrupting chemicals on the transcription of genes related to the innate immune system in the early developmental stage of zebrafish (*Danio rerio*). *Fish & Shellfish Immunology, 28*(5–6), 854–861. doi:10.1016/j.fsi.2010.02.009 PMID:20153439

Jin, Y., Lu, L., Tu, W., Luo, T., & Fu, Z. (2019). Impacts of polystyrene microplastic on the gut barrier, microbiota and metabolism of mice. *The Science of the Total Environment, 649*, 308–317. doi:10.1016/j.scitotenv.2018.08.353 PMID:30176444

Jin, Y., Wu, S., Zeng, Z., & Fu, Z. (2017). Effects of environmental pollutants on gut microbiota. *Environmental Pollution, 222*, 1–9. doi:10.1016/j.envpol.2016.11.045 PMID:28086130

Jin, Y., Xia, J., Pan, Z., Yang, J., Wang, W., & Fu, Z. (2018). Polystyrene microplastics induce microbiota dysbiosis and inflammation in the gut of adult zebrafish. *Environmental Pollution, 235*, 322–329. doi:10.1016/j.envpol.2017.12.088 PMID:29304465

Jin, Y., Zeng, Z., Wu, Y., Zhang, S., & Fu, Z. (2015). Oral exposure of mice to carbendazim induces hepatic lipid metabolism disorder and gut microbiota dysbiosis. *Toxicological Sciences, 147*(1), 116–126. doi:10.1093/toxsci/kfv115 PMID:26071454

Jin, Y., Zhu, Z., Wang, Y., Yang, E., Feng, X., & Fu, Z. (2016). The fungicide imazalil induces developmental abnormalities and alters locomotor activity during early developmental stages in zebrafish. *Chemosphere, 153*, 455–461. doi:10.1016/j.chemosphere.2016.03.085 PMID:27035382

Kan, H., Zhao, F., Zhang, X.-X., Ren, H., & Gao, S. (2015). Correlations of gut microbial community shift with hepatic damage and growth inhibition of Carassius auratus induced by pentachlorophenol exposure. *Environmental Science & Technology, 49*(19), 11894–11902. doi:10.1021/acs.est.5b02990 PMID:26378342

Kim, K.-H., Kabir, E., & Jahan, S. A. (2017). Exposure to pesticides and the associated human health effects. *The Science of the Total Environment*, *575*, 525–535. doi:10.1016/j.scitotenv.2016.09.009 PMID:27614863

Konturek, P. C., Koziel, J., Dieterich, W., Haziri, D., Wirtz, S., Glowczyk, I., Konturek, K., Neurath, M. F., & Zopf, Y. (2016). Successful therapy of Clostridium difficile infection with fecal microbiota transplantation. *Journal of Physiology and Pharmacology*, *67*(6), 859–866. PMID:28195066

Koual, M., Cano-Sancho, G., Bats, A.-S., Tomkiewicz, C., Kaddouch-Amar, Y., Douay-Hauser, N., Ngo, C., Bonsang, H., Deloménie, M., Lecuru, F., Le Bizec, B., Marchand, P., Botton, J., Barouki, R., Antignac, J.-P., & Coumoul, X. (2019). Associations between persistent organic pollutants and risk of breast cancer metastasis. *Environment International*, *132*, 105028. doi:10.1016/j.envint.2019.105028 PMID:31382183

Landy, J., Al-Hassi, H. O., Mclaughlin, S. D., Walker, A. W., Ciclitira, P. J., Nicholls, R. J., Clark, S. K., & Hart, A. L. (2011). faecal transplantation therapy for gastrointestinal disease. *Alimentary Pharmacology & Therapeutics*, *34*(4), 409–415. doi:10.1111/j.1365-2036.2011.04737.x PMID:21682755

Lefever, D. E., Xu, J., Chen, Y., Huang, G., Tamas, N., & Guo, T. L. (2016). TCDD modulation of gut microbiome correlated with liver and immune toxicity in streptozotocin (STZ)-induced hyperglycemic mice. *Toxicology and Applied Pharmacology*, *304*, 48–58. doi:10.1016/j.taap.2016.05.016 PMID:27221631

Li, J., Liu, H., & Chen, J. P. (2018). Microplastics in freshwater systems: A review on occurrence, environmental effects, and methods for microplastics detection. *Water Research*, *137*, 362–374. doi:10.1016/j.watres.2017.12.056 PMID:29580559

Libertucci, J., & Young, V. B. (2019). The role of the microbiota in infectious diseases. *Nature Microbiology*, *4*(1), 35–45. doi:10.103841564-018-0278-4 PMID:30546094

Lim, J., Nam, C., Yang, J., Rha, K. H., Lim, K.-M., & Jee, S. H. (2017). Serum persistent organic pollutants (POPs) and prostate cancer risk: A case-cohort study. *International Journal of Hygiene and Environmental Health*, *220*(5), 849–856. doi:10.1016/j.ijheh.2017.03.014 PMID:28420543

Lin, L., & Zhang, J. (2017). Role of intestinal microbiota and metabolites on gut homeostasis and human diseases. *BMC Immunology*, *18*(1), 1–25. doi:10.118612865-016-0187-3 PMID:28061847

Liu, Y., Li, Y., Xia, Y., Liu, K., Ren, L., & Ji, Y. (2020). The dysbiosis of gut microbiota caused by low-dose cadmium aggravate the injury of mice liver through increasing intestinal permeability. *Microorganisms, 8*(2), 211. doi:10.3390/microorganisms8020211 PMID:32033263

Liu, Z., Wang, Y., Zhu, Z., Yang, E., Feng, X., Fu, Z., & Jin, Y. (2016). Atrazine and its main metabolites alter the locomotor activity of larval zebrafish (Danio rerio). *Chemosphere, 148*, 163–170. doi:10.1016/j.chemosphere.2016.01.007 PMID:26803580

Lu, K., Abo, R. P., Schlieper, K. A., Graffam, M. E., Levine, S., Wishnok, J. S., Swenberg, J. A., Tannenbaum, S. R., & Fox, J. G. (2014). Arsenic exposure perturbs the gut microbiome and its metabolic profile in mice: An integrated metagenomics and metabolomics analysis. *Environmental Health Perspectives, 122*(3), 284–291. doi:10.1289/ehp.1307429 PMID:24413286

Lu, L., Luo, T., Zhao, Y., Cai, C., Fu, Z., & Jin, Y. (2019). Interaction between microplastics and microorganism as well as gut microbiota: A consideration on environmental animal and human health. *The Science of the Total Environment, 667*, 94–100. doi:10.1016/j.scitotenv.2019.02.380 PMID:30826685

Lu, L., Wan, Z., Luo, T., Fu, Z., & Jin, Y. (2018). Polystyrene microplastics induce gut microbiota dysbiosis and hepatic lipid metabolism disorder in mice. *The Science of the Total Environment, 631*, 449–458. doi:10.1016/j.scitotenv.2018.03.051 PMID:29529433

Lv, L.-X., Jiang, H.-Y., Yan, R., & Li, L. (2019). Interactions Between Gut Microbiota and Hosts and Their Role in Infectious Diseases. *Infectious Microbes and Diseases, 1*(1), 3–9. doi:10.1097/IM9.0000000000000001

Macfarlane, G. T., & Macfarlane, S. (2011). Fermentation in the human large intestine: Its physiologic consequences and the potential contribution of prebiotics. *Journal of Clinical Gastroenterology, 45*, S120–S127. doi:10.1097/MCG.0b013e31822fecfe PMID:21992950

Maciel-Fiuza, M. F., Muller, G. C., Campos, D. M. S., do Socorro Silva Costa, P., Peruzzo, J., Bonamigo, R. R., Veit, T., & Vianna, F. S. L. (2023). Role of gut microbiota in infectious and inflammatory diseases. *Frontiers in Microbiology, 14*(March), 1098386. Advance online publication. doi:10.3389/fmicb.2023.1098386 PMID:37051522

Maddalon, A., Galbiati, V., Colosio, C., Mandić-Rajčević, S., & Corsini, E. (2021). Glyphosate-based herbicides: Evidence of immune-endocrine alteration. *Toxicology, 459*, 152851. doi:10.1016/j.tox.2021.152851 PMID:34246717

Marchesi, J. R., Adams, D. H., Fava, F., Hermes, G. D. A., Hirschfield, G. M., Hold, G., Quraishi, M. N., Kinross, J., Smidt, H., Tuohy, K. M., Thomas, L. V., Zoetendal, E. G., & Hart, A. (2016). The gut microbiota and host health: A new clinical frontier. *Gut*, *65*(2), 330–339. doi:10.1136/gutjnl-2015-309990 PMID:26338727

Markowiak, P., & Śliżewska, K. (2017). Effects of probiotics, prebiotics, and synbiotics on human health. *Nutrients*, *9*(9), 1021. doi:10.3390/nu9091021 PMID:28914794

Mehand, M. S., Al-Shorbaji, F., Millett, P., & Murgue, B. (2018). The WHO R&D Blueprint: 2018 review of emerging infectious diseases requiring urgent research and development efforts. *Antiviral Research*, *159*, 63–67. doi:10.1016/j.antiviral.2018.09.009 PMID:30261226

Moore, C. J. (2008). Synthetic polymers in the marine environment: A rapidly increasing, long-term threat. *Environmental Research*, *108*(2), 131–139. doi:10.1016/j.envres.2008.07.025 PMID:18949831

Moossavi, S., & Bishehsari, F. (2019). Microbes: Possible link between modern lifestyle transition and the rise of metabolic syndrome. *Obesity Reviews*, *20*(3), 407–419. doi:10.1111/obr.12784 PMID:30548384

Mullish, B. H., Quraishi, M. N., Segal, J. P., McCune, V. L., Baxter, M., Marsden, G. L., Moore, D. J., Colville, A., Bhala, N., Iqbal, T. H., Settle, C., Kontkowski, G., Hart, A. L., Hawkey, P. M., Goldenberg, S. D., & Williams, H. R. T. (2018). The use of faecal microbiota transplant as treatment for recurrent or refractory *Clostridium difficile* infection and other potential indications: Joint British Society of Gastroenterology (BSG) and Healthcare Infection Society (HIS) guidelines. *Gut*, *67*(11), 1920–1941. doi:10.1136/gutjnl-2018-316818 PMID:30154172

Nichols, R. G., Zhang, J., Cai, J., Murray, I. A., Koo, I., Smith, P. B., Perdew, G. H., & Patterson, A. D. (2019). Metatranscriptomic analysis of the mouse gut microbiome response to the persistent organic pollutant 2, 3, 7, 8-tetrachlorodibenzofuran. *Metabolites*, *10*(1), 1. doi:10.3390/metabo10010001 PMID:31861317

Nicholson, J. K., Holmes, E., Kinross, J., Burcelin, R., Gibson, G., Jia, W., & Pettersson, S. (2012). Host-gut microbiota metabolic interactions. *Science*, *336*(6086), 1262–1267. doi:10.1126cience.1223813 PMID:22674330

Obbard, R. W., Sadri, S., Wong, Y. Q., Khitun, A. A., Baker, I., & Thompson, R. C. (2014). Global warming releases microplastic legacy frozen in Arctic Sea ice. *Earth's Future*, *2*(6), 315–320. doi:10.1002/2014EF000240

Omar, E. H., Balzano, T., Abdelmonaim, E. H. M., & Jayakumar, A. R. (2022). The gut microbiota and brain interactions in healthy and pathological circumstances: From understanding to therapy. *Frontiers in Neuroscience*, *16*, 1052350. doi:10.3389/fnins.2022.1052350 PMID:36325481

Pandey, K. R., Naik, S. R., & Vakil, B. V. (2015). Probiotics, prebiotics and synbiotics-a review. *Journal of Food Science and Technology*, *52*(12), 7577–7587. doi:10.100713197-015-1921-1 PMID:26604335

Paveley, N. D., Sylvester-Bradley, R., Scott, R. K., Craigon, J., & Day, W. (2001). Steps in predicting the relationship of yield on fungicide dose. *Phytopathology*, *91*(7), 708–716. doi:10.1094/PHYTO.2001.91.7.708 PMID:18943002

Piccioni, A., de Cunzo, T., Valletta, F., Covino, M., Rinninella, E., Raoul, P., Zanza, C., Mele, M. C., & Franceschi, F. (2021). Gut microbiota and environment in coronary artery disease. *International Journal of Environmental Research and Public Health*, *18*(8), 4242. Advance online publication. doi:10.3390/ijerph18084242 PMID:33923612

Popli, S., Badgujar, P. C., Agarwal, T., Bhushan, B., & Mishra, V. (2022). Persistent organic pollutants in foods, their interplay with gut microbiota and resultant toxicity. *The Science of the Total Environment*, *832*, 155084. doi:10.1016/j.scitotenv.2022.155084 PMID:35395291

Preidis, G. A., & Versalovic, J. (2009). Targeting the human microbiome with antibiotics, probiotics, and prebiotics: Gastroenterology enters the metagenomics era. *Gastroenterology*, *136*(6), 2015–2031. doi:10.1053/j.gastro.2009.01.072 PMID:19462507

Qiu, Y.-W., Zeng, E. Y., Qiu, H., Yu, K., & Cai, S. (2017). Bioconcentration of polybrominated diphenyl ethers and organochlorine pesticides in algae is an important contaminant route to higher trophic levels. *The Science of the Total Environment*, *579*, 1885–1893. doi:10.1016/j.scitotenv.2016.11.192 PMID:27939082

Rezania, S., Park, J., Din, M. F. M., Taib, S. M., Talaiekhozani, A., Yadav, K. K., & Kamyab, H. (2018). Microplastics pollution in different aquatic environments and biota: A review of recent studies. *Marine Pollution Bulletin*, *133*, 191–208. doi:10.1016/j.marpolbul.2018.05.022 PMID:30041307

Ribeiro Marques, M., Henrique Pereira, J., Oliveira, J. S., Augusto Basso, L., Filgueira de Azevedo, W., Santiago Santos, D., & Sergio Palma, M. (2007). The inhibition of 5-enolpyruvylshikimate-3-phosphate synthase as a model for development of novel antimicrobials. *Current Drug Targets*, *8*(3), 445–457. doi:10.2174/138945007780058951 PMID:17348837

Roberfroid, M. (2007). Prebiotics: The concept revisited. *The Journal of Nutrition, 137*(3), 830S–837S. doi:10.1093/jn/137.3.830S PMID:17311983

Rothhammer, V., & Quintana, F. J. (2019). The aryl hydrocarbon receptor: An environmental sensor integrating immune responses in health and disease. *Nature Reviews. Immunology, 19*(3), 184–197. doi:10.103841577-019-0125-8 PMID:30718831

Rothschild, D., Weissbrod, O., Barkan, E., Kurilshikov, A., Korem, T., Zeevi, D., Costea, P. I., Godneva, A., Kalka, I. N., Bar, N., Shilo, S., Lador, D., Vila, A. V., Zmora, N., Pevsner-Fischer, M., Israeli, D., Kosower, N., Malka, G., Wolf, B. C., ... Segal, E. (2018). Environment dominates over host genetics in shaping human gut microbiota. *Nature, 555*(7695), 210–215. doi:10.1038/nature25973 PMID:29489753

Round, J. L., O'Connell, R. M., & Mazmanian, S. K. (2010). Coordination of tolerogenic immune responses by the commensal microbiota. *Journal of Autoimmunity, 34*(3), J220–J225. doi:10.1016/j.jaut.2009.11.007 PMID:19963349

Rueda-Ruzafa, L., Cruz, F., Roman, P., & Cardona, D. (2019). Gut microbiota and neurological effects of glyphosate. *Neurotoxicology, 75*, 1–8. doi:10.1016/j.neuro.2019.08.006 PMID:31442459

Sharma, S., & Chatterjee, S. (2017). Microplastic pollution, a threat to marine ecosystem and human health: A short review. *Environmental Science and Pollution Research International, 24*(27), 21530–21547. doi:10.100711356-017-9910-8 PMID:28815367

Shi, Z., Lei, H., Chen, G., Yuan, P., Cao, Z., Ser, H.-L., Zhu, X., Wu, F., Liu, C., Dong, M., Song, Y., Guo, Y., Chen, C., Hu, K., Zhu, Y., Zeng, X., Zhou, J., Lu, Y., Patterson, A. D., & Zhang, L. (2021). Impaired intestinal *Akkermansia muciniphila* and aryl hydrocarbon receptor ligands contribute to nonalcoholic fatty liver disease in mice. *mSystems, 6*(1), 10–1128. doi:10.1128/mSystems.00985-20 PMID:33622853

Silk, D. B. A., Davis, A., Vulevic, J., Tzortzis, G., & Gibson, G. R. (2009). Clinical trial: The effects of a trans-galactooligosaccharide prebiotic on faecal microbiota and symptoms in irritable bowel syndrome. *Alimentary Pharmacology & Therapeutics, 29*(5), 508–518. doi:10.1111/j.1365-2036.2008.03911.x PMID:19053980

Singh, M. M., & Mullin, G. E. (2017). Diet, environmental chemicals, and the gut microbiome. In *Integrative Environmental Medicine* (pp. 115–140). Oxford University Press New York.

Smith, A. R., Macfarlane, S., Furrie, E., Ahmed, S., Bahrami, B., Reynolds, N., & Macfarlane, G. T. (2011). Microbiological and immunological effects of enteral feeding on the upper gastrointestinal tract. *Journal of Medical Microbiology*, *60*(3), 359–365. doi:10.1099/jmm.0.026401-0 PMID:21088127

Snedeker, S. M., & Hay, A. G. (2012). Do interactions between gut ecology and environmental chemicals contribute to obesity and diabetes? *Environmental Health Perspectives*, *120*(3), 332–339. doi:10.1289/ehp.1104204 PMID:22042266

Suez, J., Korem, T., Zeevi, D., Zilberman-Schapira, G., Thaiss, C. A., Maza, O., Israeli, D., Zmora, N., Gilad, S., Weinberger, A., Kuperman, Y., Harmelin, A., Kolodkin-Gal, I., Shapiro, H., Halpern, Z., Segal, E., & Elinav, E. (2014). Artificial sweeteners induce glucose intolerance by altering the gut microbiota. *Nature*, *514*(7521), 181–186. doi:10.1038/nature13793 PMID:25231862

Tariq, R., Pardi, D. S., Bartlett, M. G., & Khanna, S. (2019). Low cure rates in controlled trials of fecal microbiota transplantation for recurrent Clostridium difficile infection: A systematic review and meta-analysis. *Clinical Infectious Diseases*, *68*(8), 1351–1358. doi:10.1093/cid/ciy721 PMID:30957161

Taylor, T. N., Caudle, W. M., & Miller, G. W. (2011). *VMAT2-Deficient Mice Display Nigral and Extranigral Pathology and Motor and Nonmotor Symptoms of Parkinson's Disease.* doi:10.4061/2011/124165

Thaiss, C. A., Zmora, N., Levy, M., & Elinav, E. (2016). The microbiome and innate immunity. *Nature*, *535*(7610), 65–74. doi:10.1038/nature18847 PMID:27383981

Thompson, R. C., Olsen, Y., Mitchell, R. P., Davis, A., Rowland, S. J., John, A. W. G., McGonigle, D., & Russell, A. E. (2004). Lost at sea: Where is all the plastic? *Science*, *304*(5672), 838. doi:10.1126cience.1094559 PMID:15131299

Thundiyil, J. G., Stober, J., Besbelli, N., & Pronczuk, J. (2008). Acute pesticide poisoning: A proposed classification tool. *Bulletin of the World Health Organization*, *86*(3), 205–209. doi:10.2471/BLT.08.041814 PMID:18368207

Tian, X., Yu, Z., Feng, P., Ye, Z., Li, R., Liu, J., Hu, J., Kakade, A., Liu, P., & Li, X. (2019). Lactobacillus plantarum TW1-1 alleviates diethylhexylphthalate-induced testicular damage in mice by modulating gut microbiota and decreasing inflammation. *Frontiers in Cellular and Infection Microbiology*, *9*, 221. doi:10.3389/fcimb.2019.00221 PMID:31297340

Tian, Y., Gui, W., Rimal, B., Koo, I., Smith, P. B., Nichols, R. G., Cai, J., Liu, Q., & Patterson, A. D. (2020). Metabolic impact of persistent organic pollutants on gut microbiota. *Gut Microbes*, *12*(1), 1–16. doi:10.1080/19490976.2020.1848209 PMID:33295235

Tilg, H., & Moschen, A. R. (2014). Microbiota and diabetes: An evolving relationship. *Gut*, *63*(9), 1513–1521. doi:10.1136/gutjnl-2014-306928 PMID:24833634

Van Cauwenberghe, L., & Janssen, C. R. (2014). Microplastics in bivalves cultured for human consumption. *Environmental Pollution*, *193*, 65–70. doi:10.1016/j.envpol.2014.06.010 PMID:25005888

Van de Wiele, T., Gallawa, C. M., Kubachk, K. M., Creed, J. T., Basta, N., Dayton, E. A., Whitacre, S., Du Laing, G., & Bradham, K. (2010). Arsenic metabolism by human gut microbiota upon in vitro digestion of contaminated soils. *Environmental Health Perspectives*, *118*(7), 1004–1009. doi:10.1289/ehp.0901794 PMID:20603239

Van Den Elsen, L. W. J., Poyntz, H. C., Weyrich, L. S., Young, W., & Forbes-Blom, E. E. (2017). Embracing the gut microbiota: The new frontier for inflammatory and infectious diseases. *Clinical & Translational Immunology*, *6*(1), e125. Advance online publication. doi:10.1038/cti.2016.91 PMID:28197336

Vardakou, M., Palop, C. N., Christakopoulos, P., Faulds, C. B., Gasson, M. A., & Narbad, A. (2008). Evaluation of the prebiotic properties of wheat arabinoxylan fractions and induction of hydrolase activity in gut microflora. *International Journal of Food Microbiology*, *123*(1–2), 166–170. doi:10.1016/j.ijfoodmicro.2007.11.007 PMID:18279987

Vested, A., Giwercman, A., Bonde, J. P., & Toft, G. (2014). Persistent organic pollutants and male reproductive health. *Asian Journal of Andrology*, *16*(1), 71. doi:10.4103/1008-682X.122345 PMID:24369135

Wilding, L. A., Bassis, C. M., Walacavage, K., Hashway, S., Leroueil, P. R., Morishita, M., Maynard, A. D., Philbert, M. A., & Bergin, I. L. (2016). Repeated dose (28-day) administration of silver nanoparticles of varied size and coating does not significantly alter the indigenous murine gut microbiome. *Nanotoxicology*, *10*(5), 513–520. doi:10.3109/17435390.2015.1078854 PMID:26525505

Williams, K., Milner, J., Boudreau, M. D., Gokulan, K., Cerniglia, C. E., & Khare, S. (2015). Effects of subchronic exposure of silver nanoparticles on intestinal microbiota and gut-associated immune responses in the ileum of Sprague-Dawley rats. *Nanotoxicology*, *9*(3), 279–289. doi:10.3109/17435390.2014.921346 PMID:24877679

Wright, S. L., & Kelly, F. J. (2017). Plastic and human health: A micro issue? *Environmental Science & Technology*, *51*(12), 6634–6647. doi:10.1021/acs.est.7b00423 PMID:28531345

Xu, C., Liu, Q., Huan, F., Qu, J., Liu, W., Gu, A., Wang, Y., & Jiang, Z. (2015). Changes in gut microbiota may be early signs of liver toxicity induced by epoxiconazole in rats. *Chemotherapy*, *60*(2), 135–142. doi:10.1159/000371837 PMID:25766299

Yang, D., Shi, H., Li, L., Li, J., Jabeen, K., & Kolandhasamy, P. (2015). Microplastic pollution in table salts from China. *Environmental Science & Technology*, *49*(22), 13622–13627. doi:10.1021/acs.est.5b03163 PMID:26486565

Yang, J., Feng, P., Ling, Z., Khan, A., Wang, X., Chen, Y., Ali, G., Fang, Y., Salama, E.-S., Wang, X., Liu, P., & Li, X. (2023). Nickel exposure induces gut microbiome disorder and serum uric acid elevation. *Environmental Pollution*, *324*, 121349. doi:10.1016/j.envpol.2023.121349 PMID:36870597

Ye, J., Zhao, M., Liu, J., & Liu, W. (2010). Enantioselectivity in environmental risk assessment of modern chiral pesticides. *Environmental Pollution*, *158*(7), 2371–2383. doi:10.1016/j.envpol.2010.03.014 PMID:20378218

Yilmaz, B., Terekeci, H., Sandal, S., & Kelestimur, F. (2020). Endocrine disrupting chemicals: Exposure, effects on human health, mechanism of action, models for testing and strategies for prevention. *Reviews in Endocrine & Metabolic Disorders*, *21*(1), 127–147. doi:10.100711154-019-09521-z PMID:31792807

Zarn, J. A., Brüschweiler, B. J., & Schlatter, J. R. (2003). Azole fungicides affect mammalian steroidogenesis by inhibiting sterol 14 alpha-demethylase and aromatase. *Environmental Health Perspectives*, *111*(3), 255–261. doi:10.1289/ehp.5785 PMID:12611652

Zhai, Q., Tian, F., Zhao, J., Zhang, H., Narbad, A., & Chen, W. (2016). Oral administration of probiotics inhibits absorption of the heavy metal cadmium by protecting the intestinal barrier. *Applied and Environmental Microbiology*, *82*(14), 4429–4440. doi:10.1128/AEM.00695-16 PMID:27208136

Zhang, F., Cui, B., He, X., Nie, Y., Wu, K., Fan, D., & Group, F. S. (2018). Microbiota transplantation: Concept, methodology and strategy for its modernization. *Protein & Cell*, *9*(5), 462–473. doi:10.100713238-018-0541-8 PMID:29691757

Zhang, F., Luo, W., Shi, Y., Fan, Z., & Ji, G. (2012). Should we standardize the 1,700-year-old fecal microbiota transplantation? [-author.]. *The American Journal of Gastroenterology*, *107*(11), 1755. doi:10.1038/ajg.2012.251 PMID:23160295

Zhang, L., Nichols, R. G., Correll, J., Murray, I. A., Tanaka, N., Smith, P. B., Hubbard, T. D., Sebastian, A., Albert, I., Hatzakis, E., Gonzalez, F. J., Perdew, G. H., & Patterson, A. D. (2015). Persistent organic pollutants modify gut microbiota–host metabolic homeostasis in mice through aryl hydrocarbon receptor activation. *Environmental Health Perspectives*, *123*(7), 679–688. doi:10.1289/ehp.1409055 PMID:25768209

Zhang, S., Jin, Y., Zeng, Z., Liu, Z., & Fu, Z. (2015). Subchronic exposure of mice to cadmium perturbs their hepatic energy metabolism and gut microbiome. *Chemical Research in Toxicology*, *28*(10), 2000–2009. doi:10.1021/acs.chemrestox.5b00237 PMID:26352046

Zhao, J., Ran, W., Teng, J., Liu, Y., Liu, H., Yin, X., Cao, R., & Wang, Q. (2018). Microplastic pollution in sediments from the Bohai Sea and the Yellow Sea, China. *The Science of the Total Environment*, *640*, 637–645. doi:10.1016/j.scitotenv.2018.05.346 PMID:29870939

Zhao, Y., Zhou, C., Wu, C., Guo, X., Hu, G., Wu, Q., Xu, Z., Li, G., Cao, H., Li, L., Latigo, V., Liu, P., Cheng, S., & Liu, P. (2020). Subchronic oral mercury caused intestinal injury and changed gut microbiota in mice. *The Science of the Total Environment*, *721*, 137639. doi:10.1016/j.scitotenv.2020.137639 PMID:32172103

Zhong, X., Li, J., Lu, F., Zhang, J., & Guo, L. (2022). Application of zebrafish in the study of the gut microbiome. *Animal Models and Experimental Medicine*, *5*(4), 323–336. doi:10.1002/ame2.12227 PMID:35415967

Chapter 9
Classification of Melanoma Skin Cancer Based on Transformer Deep Learning Model

Nagarjuna Telagam
https://orcid.org/0000-0002-6184-6283
GITAM University, India

Nehru Kandasamy
Madanapalle Institute of Science and Technology, India

ABSTRACT

An increasing number of genetic and metabolic anomalies have been determined to lead to cancer, which is generally fatal. Cancerous cells may spread to any body part, which can be life-threatening. Skin cancer is significant cancer, and its frequency is increasing worldwide. The main subtypes of skin cancer are squamous and basal cell carcinomas and melanoma. The deep learning methods were used to detect the two primary types of tumours, malignant and benign, by using the MELANOMA dataset. The proposed system utilizes a convolutional neural network (CNN), transformer, and InceptionV3 architecture to learn and extract meaningful features from skin lesion images. The CNN model was trained on a large dataset of dermoscopic images of melanoma and benign lesions. The transformer model in deep learning refers to a neural network architecture based on the transformer architecture specifically designed for image classification tasks. Inception is an image recognition model that has been shown to attain greater than 78.1% accuracy on the ImageNet dataset.

DOI: 10.4018/978-1-7998-9414-8.ch009

INTRODUCTION

When cells in the body develop slowly and divide uncontrollably, creating tumours, that condition is known as cancer. Skin cancer is a condition in which uncontrolled growth of aberrant skin cells might result in tumours. One type of skin cancer is called melanoma. The most dangerous kind, it can quickly spread to other bodily areas. Tumours come in two main categories: benign and malignant. Benign tumours don't spread to other body parts; they remain in their original place. They don't disperse to far-off areas of the body. Benign tumours have defined borders and a modest rate of growth. People rarely have problems with benign tumours. Tum tumours form as the body's cells divide and enlarge at their fastest speed. The body usually manages to keep cell division and development in check. Damaged cells are automatically replaced by new, healthy cells when they die.

Figure 1. Benign tumours

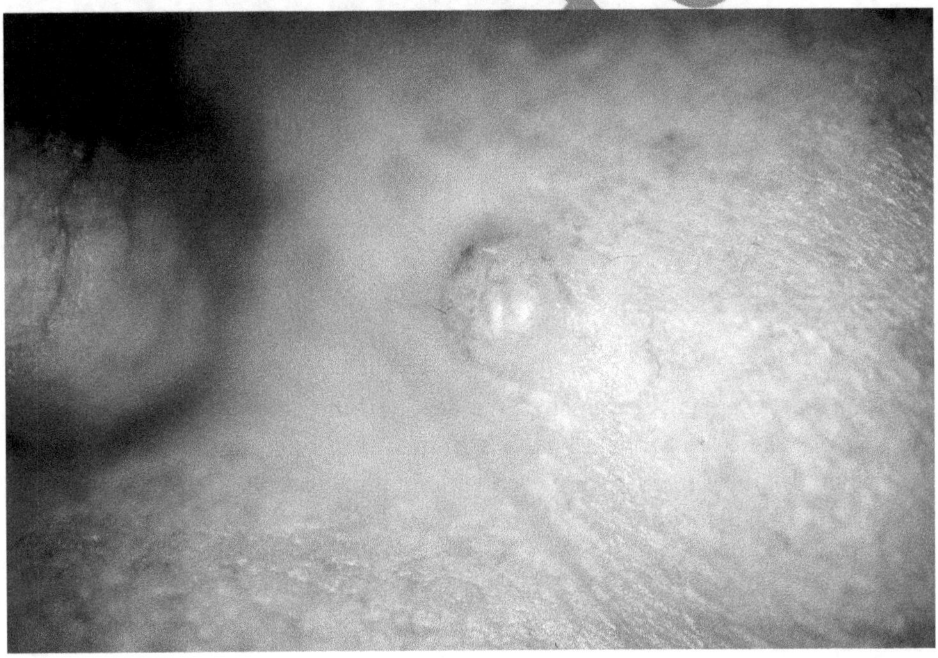

Basal and squamous cell carcinomas make up the majority of skin cancer cases. Despite being cancerous, if treated right away, they can readily spread to other body parts. If they aren't handled in time, they might sustain local injuries. Malignant melanomas are one kind of skin cancer that is uncommon but distinct. Melanoma has one of the most significant mortality rates among invasive skin cancers. Despite severe skin cancer, it is incredibly curable if detected early. The importance of early detection and prevention cannot be overstated, particularly if you have fair skin, red hair, and blue eyes.

Figure 2. Melanoma prognosis

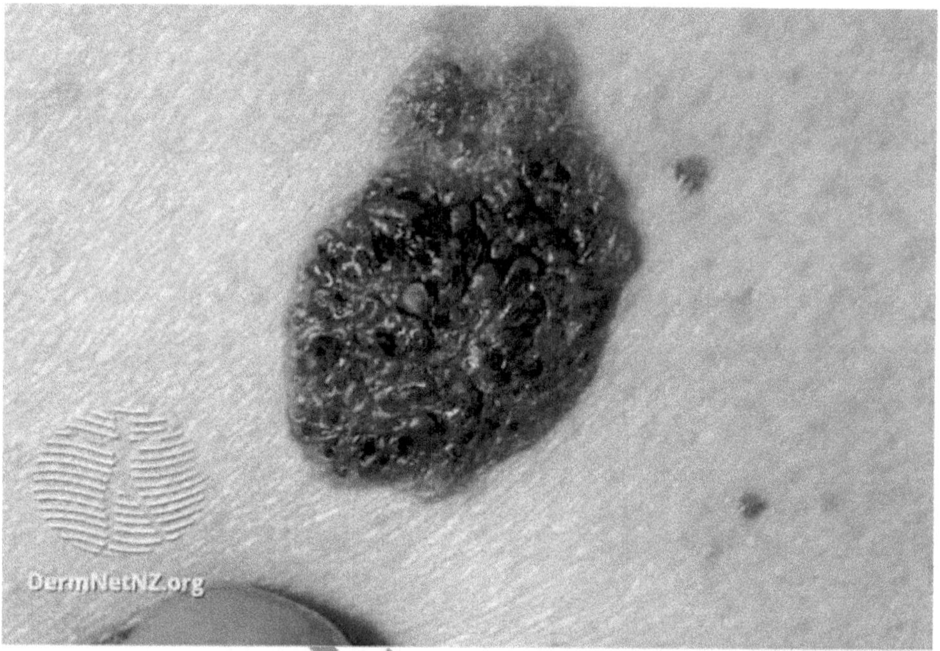

Early diagnosis and treatment of the deadly skin cancer melanoma can save many lives. There are various methods to treat this type of skin cancer. In this endeavour, deep learning is being used with multiple models. CNN, Transformer, and InceptionV3 algorithms are those we use.

LITERATURE SURVEY

The paper was proposed for Melanoma skin cancer 135 lesion images. The training algorithm ANN with Levenberg–Marquardt (LM), resilient backpropagation (RBP), and scaled conjugate gradient (GCG) learning algorithms. Combination of multiple classifiers to avoid misclassification. The proposed system achieved an overall accuracy score of (SCG:91.9, LM: 95.1, RBP:88.1) (Xie et al., 2017). The paper is Computer-Aided Melanoma Skin Cancer Detection Using Artificial Neural Network Classifier was proposed in 2016. It is a Melanoma skin cancer with 31 dermoscopic images. The ANN training algorithm with a backpropagation model classifies the input images. ABCD parameters for feature extraction, maximum entropy for thresholding, and grey level co-occurrence matrix for features extraction. The proposed system achieved an overall accuracy score of 96.9% (Kanimozhi & Murthi, 2016). The paper is Melanoma Detection by Analysis of Clinical Images Using Convolutional Neural Network and was proposed in 2016. It is a Melanoma/benign with 170 skin lesion images. The training algorithm CNN classifier. Two convolving layers in CNN. The proposed system achieved an overall accuracy score of Accuracy (81), sensitivity (81), specificity (80) (Nasr-Esfahani et al., 2016). The paper Vision-Based Classification of Skin Cancer Using Deep Learning proposed in 2016. It is a Benign/melanoma skin cancer with ISIC dataset. The training algorithm VGG-16 and CNN. Dataset was trained on three separate learning models. The proposed system achieved on overall accuracy score of Accuracy (78) (Kalouche, 2016). The paper Basal Cell Carcinoma Detection in Full-Field OCT Images Using Convolutional Neural Networks and the Authors was proposed in 2016.It is a Melanoma /nonmelanoma skin cancer with DermIS dataset and DermQuest data. The training algorithm is SVM-trained, with CNN, extracted features. A median filter for noise removal and CNN for feature extraction. The proposed system achieved on overall accuracy score of Accuracy (93.75) (Shoieb et al., 2016). The paper is Skin Cancer Detection Using Convolutional Neural Network and the Authors was proposed in 2019.It is a Benign/malignant skin cancer with ISIC database images .The training algorithm is CNN the input images . ABCD symptomatic checklist for feature extraction. The proposed system achieved on overall Accuracy (89.5) (Hasan et al., 2019). The paper is Skin Melanoma Segmentation Using Recurrent and Convolutional Neural Networks and the Authors was proposed in 2017.It is a Melanoma/non melanoma skin cancer with ISIC dataset. The training algorithm is Hybrid of fully CNN with auto encoder and decoder and RNN. Proposed model outperformed state-of-art SegNet, FCN, and ExB architecture. The proposed system achieved on overall Accuracy (98) Jaccard index (93), sensitivity (95), specificity (94) (Attia et al., 2017).

The paper is Skin Lesion Classification Using Convolutional Neural Network With Novel Regularizer and the Authors proposed in 2019.It is a Benign/malignant skin cancer with ISIC dataset images. The training algorithm 2-layer CNN with a novel regularizer input images. Proposed regularisation technique controlled complexity by adding a penalty on the dispersion value of classifier's weight matrix. The proposed system achieved on overall Accuracy (97.49) AUC (98), sensitivity (94.3), specificity (93.6) (Albahar, 2019). The paper is Analysing Skin Lesions in Dermoscopy Images Using Convolutional Neural Networks and the Authors proposed in 2018. It is a Benign/malignant skin cancer with ISIC dataset images. The training algorithm Deep CNN images. Data augmentation was performed for data balancing. The proposed system achieved on overall Accuracy (80.3), precision (81), AUC (69) (Singh & Nwogu, 2018). The paper is Convolutional Neural Networks for Classifying Melanoma Images and the proposed in 2020.It is a Melanoma/ non melanoma skin cancer with 3600 lesion images from the ISIC dataset. The training algorithm ResNet-50 with deep transfer learning. The proposed model showed better performance than o InceptionV3, Densenet169, Inception ResNetV2, and Mobilenet. The proposed system achieved on overall Accuracy (93.5), precision (94) recall (77), F1score (85) (Sagar & Dheeba, 2020).

Data Set Used and Contributions of Paper

The dataset used is melanoma skin cancer because melanoma is more dangerous than other skin cancers; the total number of images is 10605; 1000 images are for testing, and the remaining 9605 images are to train the model. With the help of this dataset, we can classify whether it is benign or malignant.

The deep learning model was developed using transformer model for accurate classification of melanoma skin cancer classification, the main objective of this paper was to identify the best deep learning algorithm for skin cancer classification, the performance metrics of various deep learning methods are calculated based on confusion matrix and number of epochs.

PROPOSED SYSTEM

Transformers are a subset of artificial neural networks used in deep learning applications to address the issue of transduction or the change of input sequences into output sequences. Two different kinds of transformer models exist. Vision and Swin are them. Convolutional Neural Networks (CNNs), the present state-of-the-art for various image recognition computer vision tasks, have recently been challenged by Vision Transformer (ViT), which has recently gained popularity as a viable substitute. Regarding accuracy and computational efficiency, ViT models outperform the present state-of-the-art (CNN) by almost a factor of four.

Figure 3. Basic vision transformer architecture

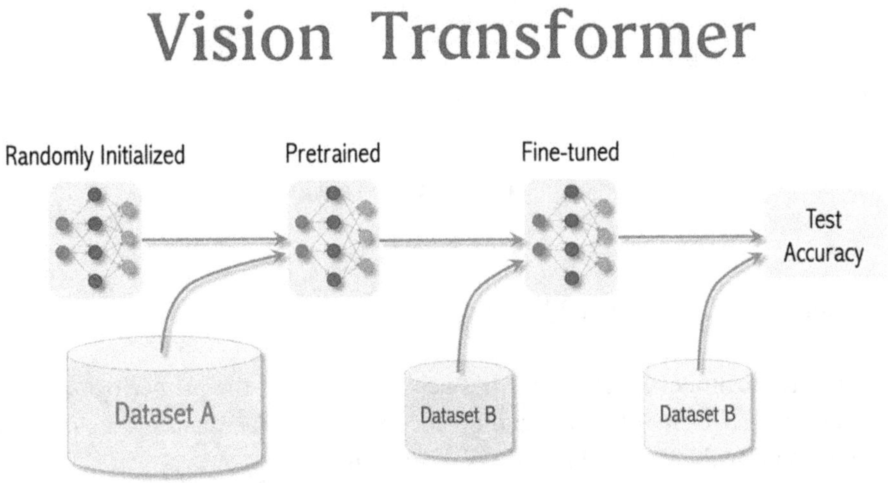

Advantages of Transformer Model:

1. Self-attention Mechanism: The self-attention mechanism in the Transformer model allows it to capture long-term dependencies in the input sequence, which is particularly useful in NLP tasks.
2. Parallelization: The Transformer model can be parallelized across multiple GPUs, allowing faster training and evaluation.
3. Scalability: The Transformer model can scale to handle large datasets and input sequences, making it suitable for real-world NLP applications.
4. State-of-the-art Performance: Transformer models have achieved state-of-the-art performance on various NLP tasks, including machine translation and text generation

The foundation of Inception V3 is the concept of using "inception modules," which enable the network to learn various filters and feature maps at multiple sizes. To capture features at varying levels of abstraction, these modules employ a variety of convolutional filters with varied sizes and strides. The network also uses batch normalisation, dropout, and supplementary classifiers to enhance training and lessen overfitting. Convolutional, pooling, fully connected, and auxiliary classifiers are some of the elements that make up Inception V3's architecture. The network is a potent and complicated model with over 23 million parameters.

Figure 4. Basic inception V3 architecture

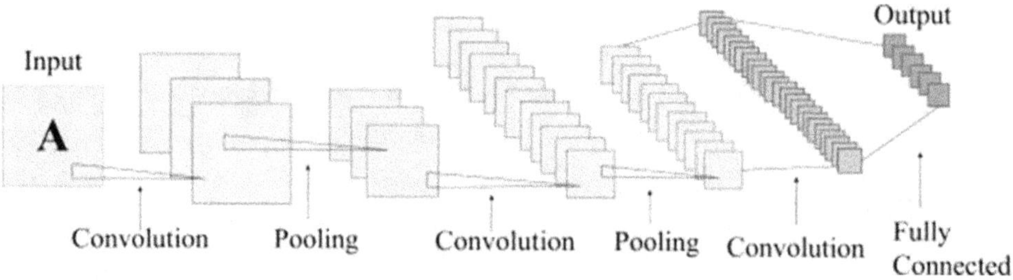

Advantages of Inception V3 Model:

1. Highly Accurate: InceptionV3 has achieved state-of-the-art performance on various image classification and object detection tasks, thanks to its multiple convolutional layers and the use of Inception modules.
2. Efficient: InceptionV3 uses fewer parameters than similar models, making it more computationally efficient and faster to train.
3. Generalizability: InceptionV3 has been pre-trained on large datasets such as ImageNet, which means it will be calibrated on smaller datasets for specific applications without requiring extensive training data.

SKIN CANCER DETECTION FLOW CHART

Figure 5. Skin cancer detection flowchart

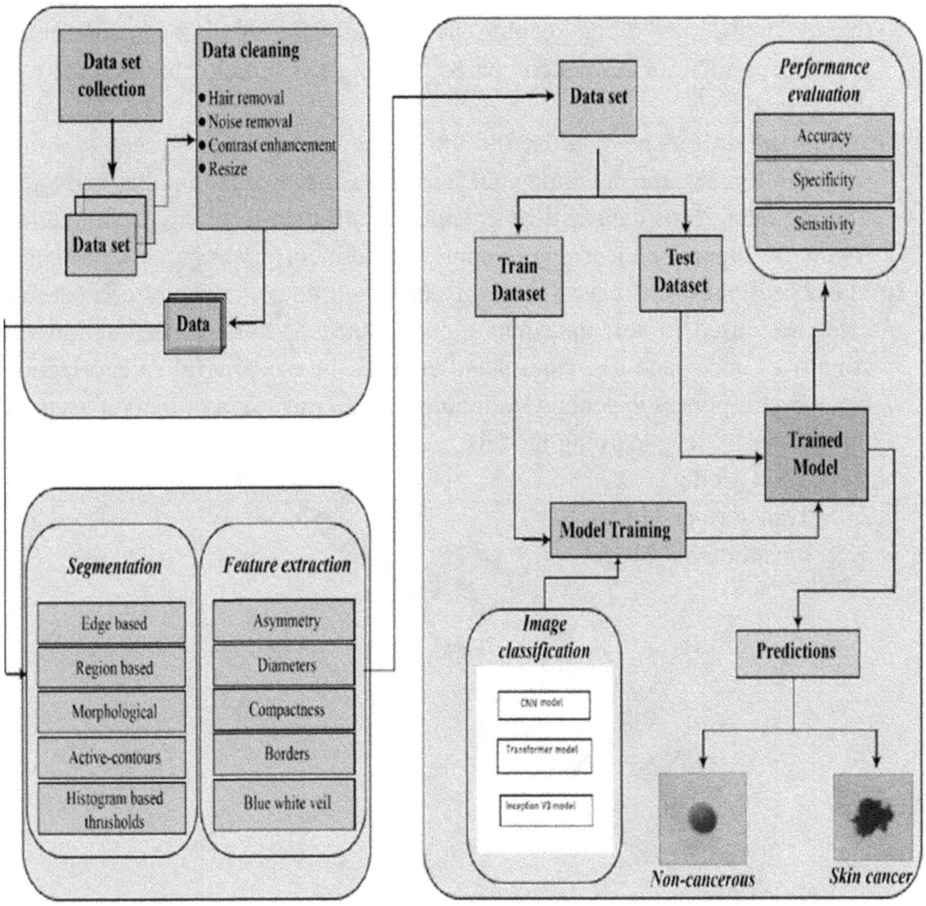

Ø **Data collection:** It is the process of gathering information pertinent to your AI project's aims and purposes. Ultimately, you create a dataset, your data collection, ready to be fed into an ML model for training.

Ø **Preprocessing:** Data preprocessing is a part of data preparation. It is the term used to define any processing done on raw data to prepare it for another data processing procedure. It has historically been a crucial first step in the data mining procedure.

Ø **Segmentation:** In an artificial vision, segmentation is a crucial stage. Segment-specific processing requires the segmentation of visual input by machines. In well-known industries like robotics, medical imaging, autonomous vehicles, and intelligent video analytics, image segmentation is thus used.

Ø **Feature extraction:** Feature extraction turns raw data into numerical features that can be handled while keeping the original data set's information intact. It produces superior results compared to using machine learning on the raw data directly.

Ø **Training the data:** The data used to introduce an algorithm or deep learning model to forecast the outcome your model was designed to predict known as training data. Your data will be enhanced with data labelling or annotation if you use supervised learning or some hybrid incorporating that technique.

Ø **Image classification:** A classification is an organised collection of comparable categories used to sort data into groups based on their commonalities. It comprises codes and descriptors and enables the meaningful categorization of survey responses to generate valuable data. In image classification we used three types of deep learning models.

1. CNN Model.
2. Transformer Model.
3. Inception V3 Model

ALGORITHM

CNN Model

Figure 6. Algorithm flowchart of the CNN-SVR

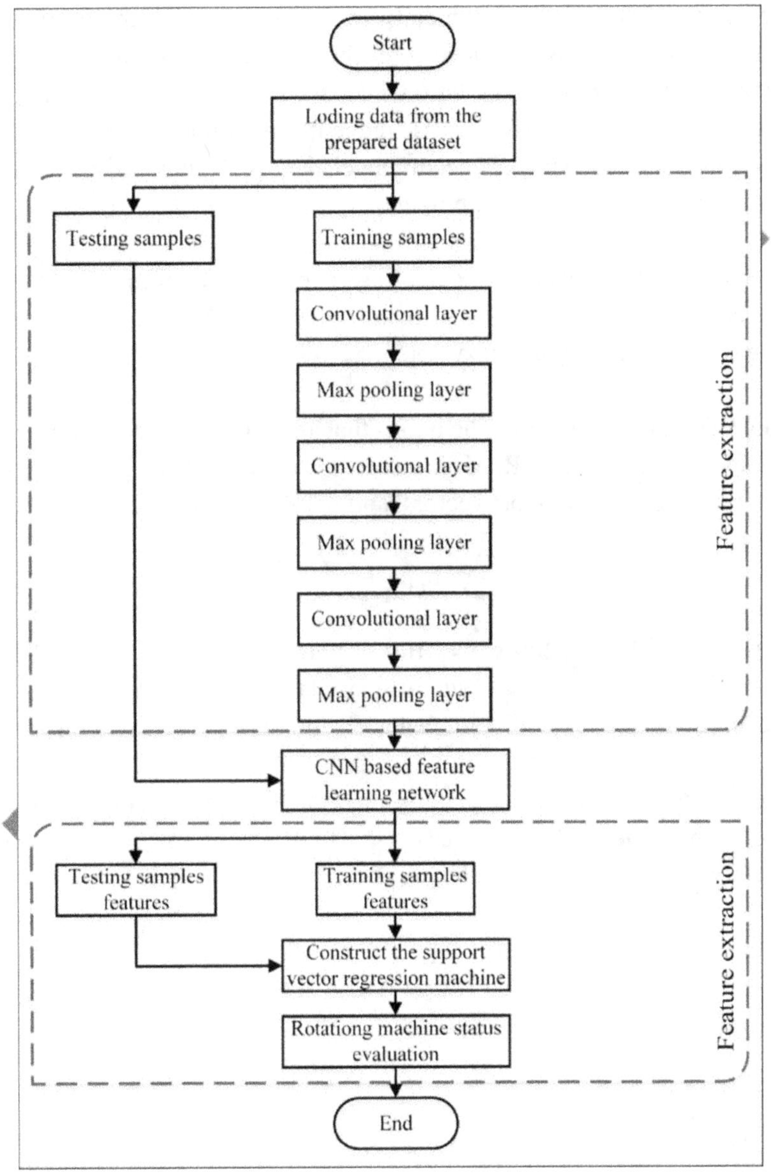

The CNN Algorithm starts by loading the data from a prepared dataset, and it trains the data according to the algorithm of CNN. Those data/samples go to the next step, which is the convolution layer.

- Ø **Convolution layer-**The fundamental component of a CNN is a convolutional layer. It has a selection of filters (or kernels), the parameters of which are trained into the system. The filters are typically smaller in size than the original image. The activation map is produced by each filter's convolution with the picture.
- Ø **Max pooling layer-**In Convolutional Neural Network (CNN) architectures, pooling is a characteristic that is frequently incorporated. The fundamental goal of a pooling layer is to "accumulate" features from maps produced by the convolution of a filter over an image.

The convolution layer and max pool layer repeat, and from that, trained data /samples are sent to the CNN-based feature learning network, and those testing sample features are sent to the next step.

- Ø **Support Vector Regression-**The points that are outside the -tube are considered to support vectors in SVR. More issues will be outside the line; consequently, there will be more support vectors, and the smaller the value of ε.

$f(x) = å\ n = 1\ N\ (\alpha\ n - \alpha\ n\ ^*)\ G\ (x\ n, x) + b$.

- Ø **Rotating machine status evaluation-**Rotating machinery is a defect mode recognition procedure that functions similarly to image recognition. Pattern recognition includes both fault categorization and picture recognition.

After this evaluation step, it gets to know all features of skin images, and it determines whether that skin has cancer; it also determines which skin has cancer and which types it is, and it ends the process.

Transformer Model

Figure 7. Algorithm flowchart of the transformer model

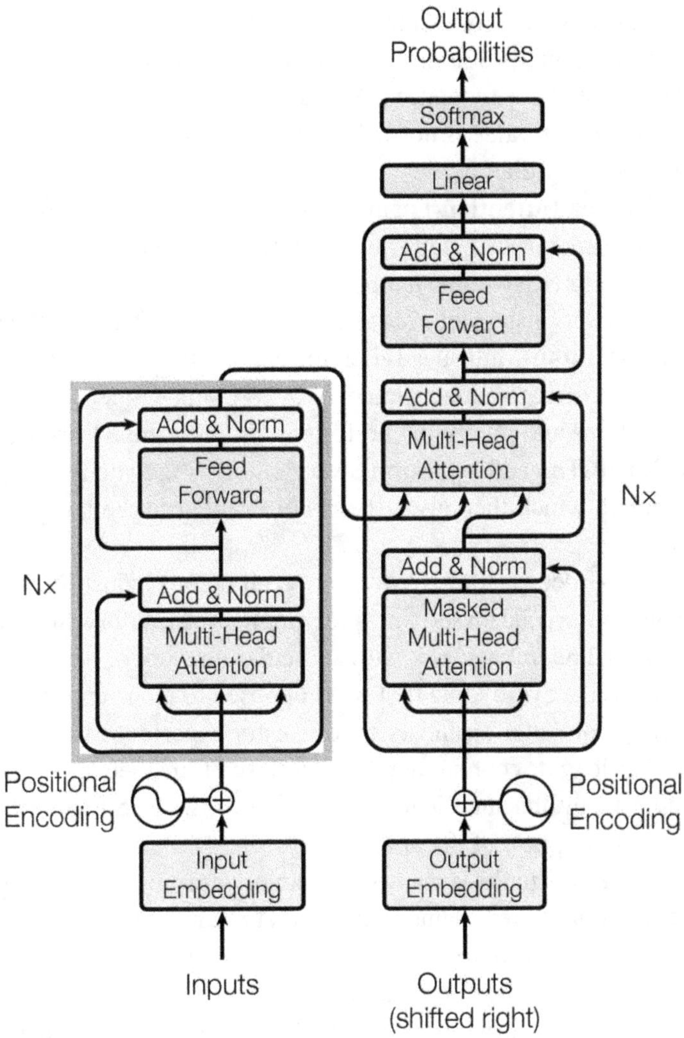

Ø **Input Embedding-**There are two embedding layers in the Transformer. The Input Embedding, the initial Embedding layer, receives the input sequence. After moving the target's one position to the right and inserting a Start token in the first position, the target sequence is sent to the second embedding layer.

- Ø **Output Embedding-**The ultimate output of a transformer model is the sum of the word and position embeddings. Thus, when setting them up, you must ensure that the output length is the same for both embedding layers.
- Ø **Positional encoding-**To give each position a distinct representation, positional encoding represents the location or position of an entity in a sequence. There are various reasons why a single integer, such as the index value, does not represent an item's status in transformer models. The indices can become very large over long sequences. Variable length sequences may experience issues if the index value is normalised to fall between 0 and 1, as they would be normalised differently.
- Ø **Masked multi-head attention-**For improving causal speech, a deep neural network (DNN) utilising masked multi-head attention (MHA) is suggested. Recurrent neural networks (RNNs) and temporal convolutional networks cannot simulate long-range relationships of noisy speech as effectively as MHA can (TCNs).
- Ø **Multi-head attention-**Multi-head attention enables the neural network to manage the blending of data across elements of an input sequence, creating richer representations and improving performance on machine learning tasks.
- Ø **Add and Norm-**The add and norm layer combines the attention layer's output and its input, which then gives the result as input to the layer normalisation algorithm.
- Ø **Feedforward-**As deep learning grows across numerous industries, it is becoming more and more crucial for software programmers. A feedforward neural network is an artificial neural network with a circular network of nodes. Feedforward neural networks cycle some routes, as opposed to recurrent neural networks.
- Ø **Linear-**Many computer scientists need to gain experience with linear algebra because it belongs to continuous rather than discrete mathematics. Understanding and using many machine learning algorithms, intense learning algorithms, requires a solid grasp of linear algebra.
- Ø **Softmax-**Numerous multi-layer neural networks terminate with a penultimate layer that produces real-valued ratings that are not easily scaled and may be challenging. The softmax is helpful in this situation since it transforms the scores into a normalised probability distribution that may be shown to the user or used as input to other systems. For this reason, a softmax function is frequently used as the neural network's last layer.
- Ø **Output Probabilities-**The model's confidence or certainty in its forecast can be shown as the output probability. Output probabilities are crucial in many deep learning applications, including speech recognition, natural language processing, and image classification. We can anticipate the results of the input data using output probabilities, and we can also calculate the degree of uncertainty or confidence in those predictions.

Inception V3-Model

Figure 8. Model inception V3's algorithm flowchart

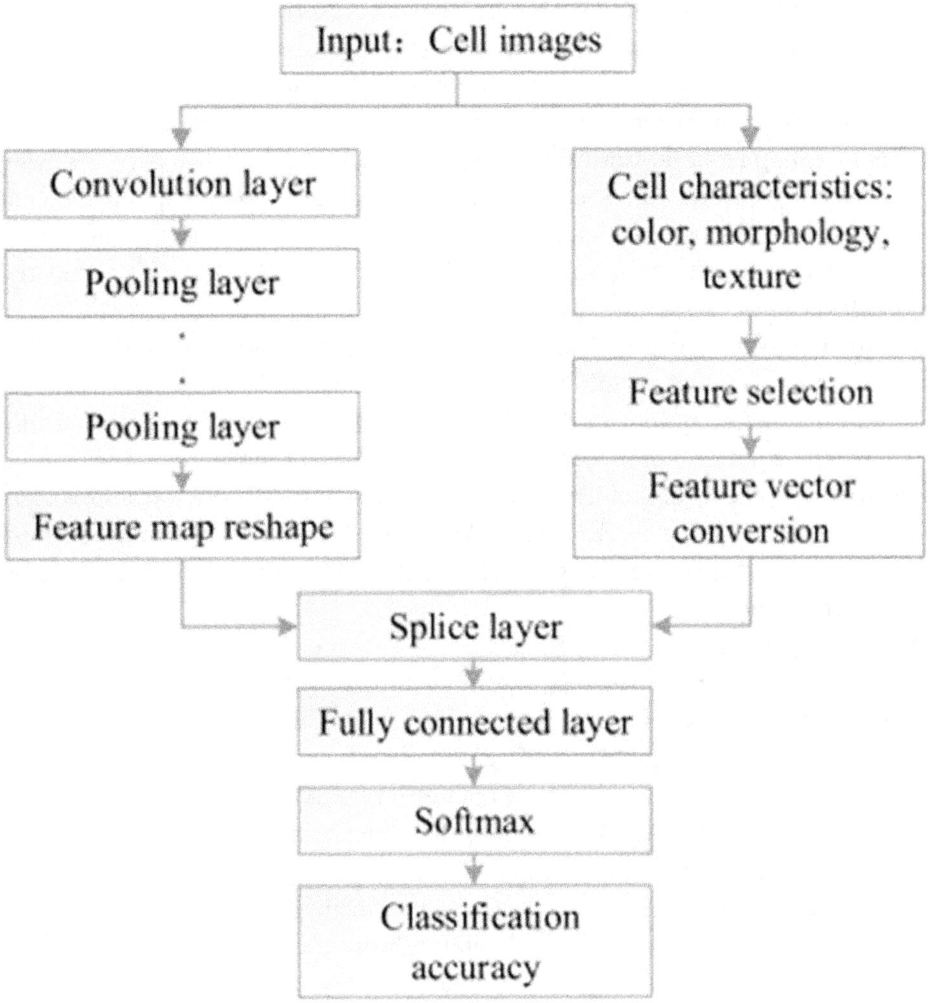

In the Inception V3 algorithm, we should first take input as cell images. A data cell for images displays an image. The default cell type for a bound column will be an image cell type if the data type for that bound column is a bit array. A cell's Value property can be set to an image object. The data model needs to contain the picture or serialised image object, and this data is sent in two different steps. Those are-

- Ø **Convolution layer-**The fundamental component of a CNN is a convolutional layer. It has a selection of filters (or kernels), the parameters of which are trained into the system. The filters are typically smaller in size than the original image. The activation map is produced by each filter's convolution with the picture.
- Ø **Pooling layer-**In Convolutional Neural Network (CNN) architectures, pooling is a characteristic that is frequently incorporated. The fundamental goal of a pooling layer is to "accumulate" features from maps produced by the convolution of a filter over an image.
- Ø **Feature map reshape-**Reshape the data into a series of input-output pairs using a loop. We extract the specified number of values from the input and output sequences for each team and append them to the x_train and y_train arrays, respectively. After that, we define the LSTM model using the Keras Sequential class.
- Ø **Feature Selection**-By using only pertinent data and eliminating data noise. Feature selection is a technique for decreasing the input variable to your model. It involves selecting your machine-learning model's characteristics automatically based on the kind of problem you're attempting to address.
- Ø **Feature vector conversion-**The broad method of creating numerical feature vectors from text documents. The Bag of Words or "Bag of n-grams" representation is the name given to this particular method of tokenization, counting, and normalising.
- Ø **Splice Layer-**A clip or image that appears as a separate layer can be added to your video as an overlay. The top track of the timeline contains all of the overlays.
- Ø **Fully connected layer-**A fully connected layer adds a bias vector after multiplying the input by a weight matrix. One or more fully combined layers come after the convolutional (and down-sampling) layers. As the name implies, every neuron in a layer that is fully linked has connections to every neuron in the layer above it.
- Ø **Softmax-**The softmax function is the activation function that predicts a multinomial probability distribution in the neural network models' output layer. When class membership is required on more than two class labels in multi-class classification problems, softmax is used as the activation function.
- Ø **Classification accuracy-**One parameter for assessing classification models is accuracy. The percentage of predictions that our model correctly predicted is known as accuracy. The Amount of accurate guesses equals accuracy amount of guesses overall.

RESULTS AND ANALYSIS

For running of 100 epochs accuracy is of 89%

Figure 9. CNN model epoch

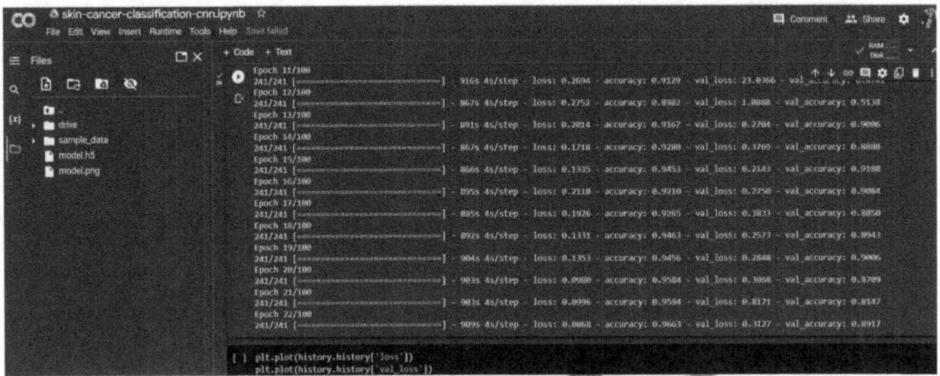

confusion metrix TP =16, FP= 91, FN=91, TN=802

Figure 10. CNN model confusion matrix

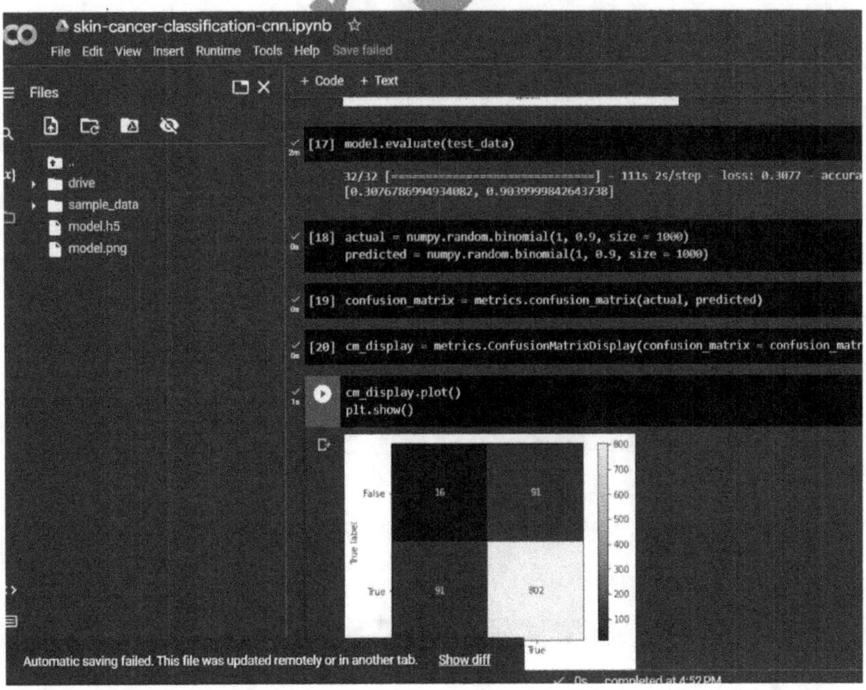

For running of 100 epochs accuracy is of 91%

Figure 11. Transformer model epoch

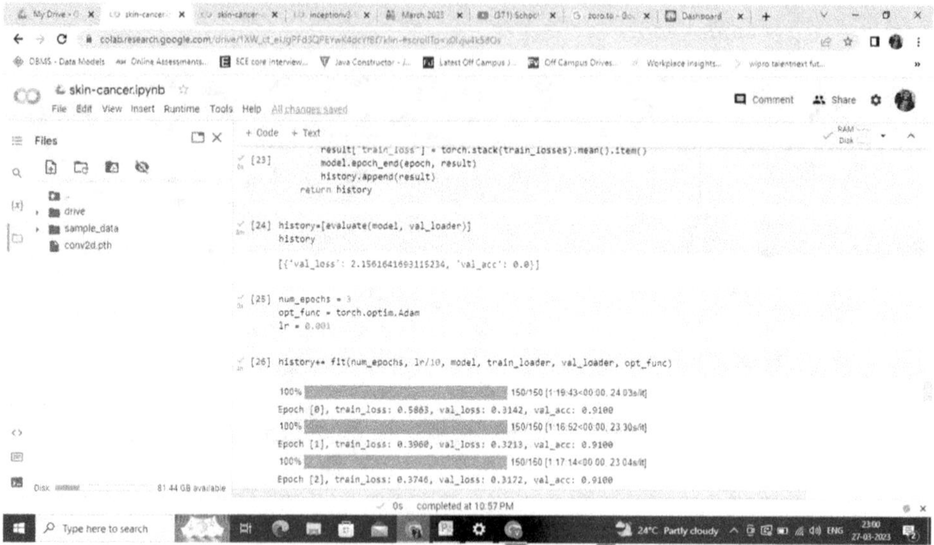

In confusion metrix TP =16, FP= 91, FN=91, TN=802

Figure 12. Transformer model confusion matrix

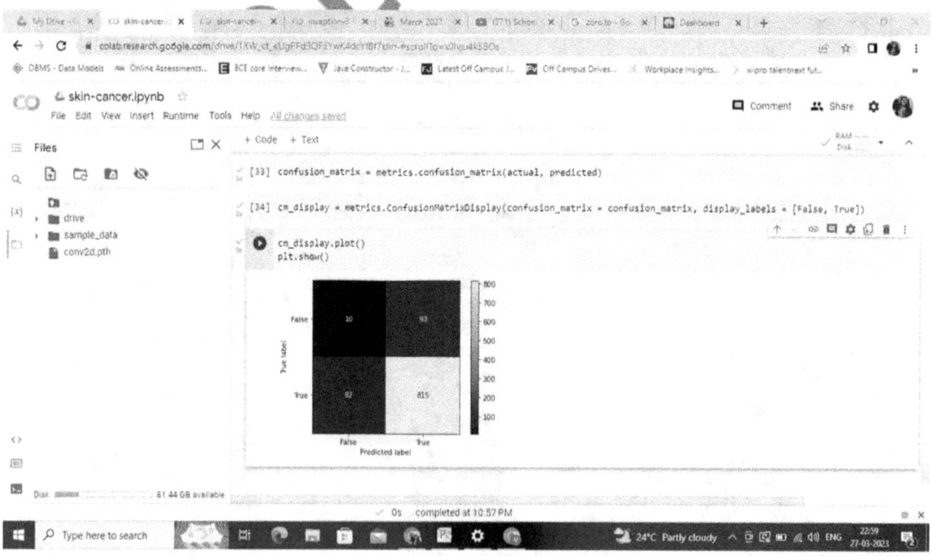

For running of 100 epochs accuracy is of 89

Figure 13. Inception V3 model epochs

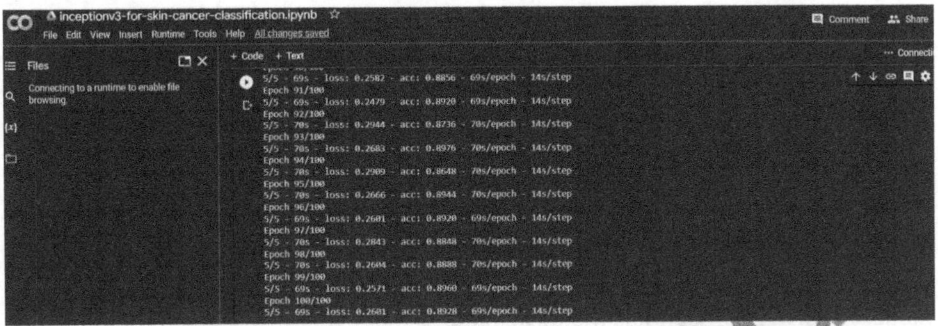

confusion metrix TP =8, FP= 88, FN=99, TN=805

Figure 14. Inception V3 model confusion matrix

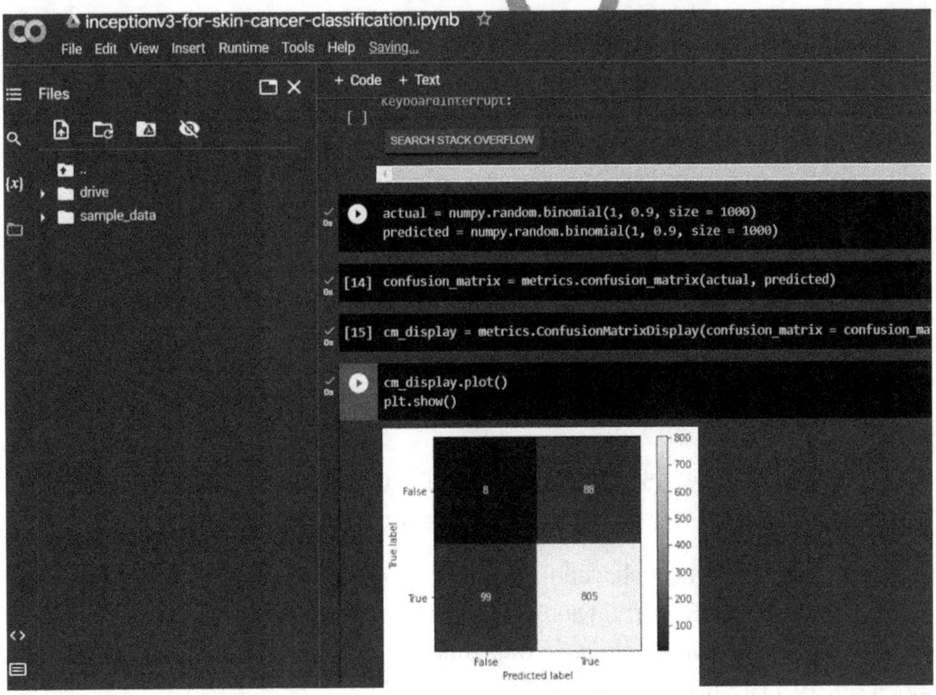

CONCLUSION

We concluded that the model is best suited for the classification of skin cancer. Because here, we used one of the most dangerous datasets, i.e., the melanoma skin cancer dataset. With this first, we have trained and tested for 1500+ images in google collabo using Vision transformer. Then, later on, we used more images, i.e., 10605 images; with this, we again trained and tested the dataset with different models, which gave the highest accuracy. Models used are CNN, InceptionV3, and Vision Transformer. The main disadvantage of the CNN model is it cannot encode the location and orientation of objects, and a large amount of training data is required for the CNN to function correctly. Because of operations like max pool, CNNs typically run substantially more slowly. The disadvantage of Inception V3 model is naive approach has the disadvantage that even the 55 convolutional layer is computationally demanding, time-consuming, and expensive regarding CPU power. So from these models. Transformer model got the highest accuracy, and it is the best classification of skin cancer.

Future Work

In the future, the classification of skin cancer can be determined by different models. Our project used three models to find the best value accuracy. Each model has various disadvantages, but when we compared three models (CNN, Transformer, and inception v3), the best and newest model is the transformer model. It allows parallel processing and is much faster than any other model with the same performance. This model is one of the most recent models using an encoder-decoder model that uses self-awareness to understand speech sequences compared to CNN and the Inception V3 model.

REFERENCES

Albahar, M.A. (2019). Skin Lesion Classification Using Convolutional Neural Network With Novel Regularizer. *IEEE Access, 7*, 38306–38313.

Attia, M., Hossny, M., Nahavandi, S., & Yazdabadi, A. (2017). Skin Melanoma Segmentation Using Recurrent and Convolutional Neural Networks. *Proceedings of the 2017 IEEE 14th International Symposium on Biomedical Imaging (ISBI 2017)*, 292–296. 10.1109/ISBI.2017.7950522

Hasan, M., Barman, S. D., Islam, S., & Reza, A. W. (2019). Skin Cancer Detection Using Convolutional Neural Network. In *Proceedings of the 2019 5th International Conference on Computing and Artificial Intelligence-ICCAI 19*. Association for Computing Machinery. 10.1145/3330482.3330525

Kalouche, S. (2016). *Vision-Based Classification of Skin Cancer Using Deep Learning*. Available online:https://www.semanticscholar.org/paper/Vision-Based-Classification-of-Skin-Cancer-using-Kalouche/b57ba909756462d812dc20fca157b3972bc1f533

Kanimozhi, T., & Murthi, D. A. (2016). Computer-Aided Melanoma Skin Cancer Detection Using Artificial Neural Network Classifier. *J. Sel. Areas Microelectron.*, *8*, 35–42.

Nasr-Esfahani, E., Samavi, S., Karimi, N., Soroushmehr, S. M. R., Jafari, M. H., Ward, K., & Najarian, K. (2016). Melanoma Detection by Analysis of Clinical Images Using Convolutional Neural Network. *Proceedings of the 2016 38th Annual International Conference of the IEEE Engineering in Medicine and Biology Society (EMBC)*, 1373. 10.1109/EMBC.2016.7590963

SagarA.DheebaJ. (2020). Convolutional Neural Networks for Classifying Melanoma Images. bioRxiv. doi:10.1101/2020.05.22.110973

Shoieb, D. A., Aly, W. M., & Youssef, S. M. (2016). Basal Cell Carcinoma Detection in Full-Field OCT Images Using Convolutional Neural Networks. *JOIG*, *4*, 122–129. doi:10.18178/joig.4.2.122-129

Singh, V., & Nwogu, I. (2018). Analysing Skin Lesions in Dermoscopy Images Using Convolutional Neural Networks. *Proceedings of the 2018 IEEE International Conference on Systems, Man, and Cybernetics (SMC)*, 4035–4040. 10.1109/SMC.2018.00684

Xie, F., Fan, H., Li, Y., Jiang, Z., Meng, R., & Bovik, A. (2017). Melanoma Classification on Dermoscopy Images Using a Neural Network Ensemble Model. *IEEE Transactions on Medical Imaging*, *36*(3), 849–858. doi:10.1109/TMI.2016.2633551 PMID:27913337

Chapter 10
Clinical Characteristics and Relationship of Acute Kidney Injury With the Severity of Disease and Death in COVID-19 Patients:
A Review

Donovan McGrowder
The University of the West Indies, Jamaica

Fabian Miller
The Mico University College, Jamaica & The University of the West Indies, Jamaica

Magdalene Nwokocha
The University of the West Indies, Jamaica

Melisa Anderson Cross
University of Technology, Jamaica

Cameil Wilson-Clarke
The University of the West Indies, Jamaica

Shelly McFarlane
The University of the West Indies, Jamaica

Lorenzo Gordon
Caribbean School of Medical Sciences, Jamaica

ABSTRACT

Globally, the coronavirus disease 2019 (COVID-19) pandemic has adversely affected healthcare with significant morbidity and mortality among patients due to the virulent effects of the severe acute respiratory syndrome coronavirus 2 (SARS-

DOI: 10.4018/978-1-7998-9414-8.ch010

CoV-2). The kidneys and other organs are infected and the development of renal disorders such as acute kidney injury (AKI) and subsequently chronic renal disease if there is no clinical intervention. There is increasing evidence of more COVID-19 patients developing AKI, and incidence of mild, moderate and severe stages of the disease. There is also evidence of significant morbidity and mortality, especially in patients with severe AKI admitted to the intensive care units (ICU). We reviewed the recent publications in PubMed, Google Scholar, Embase and Cochrane library relating to AKI in COVID-19 patients. This review examines (i) multifactorial causes of AKI in COVID-19 and the pathophysiology of this renal disorder, (ii) incidence of AKI in COVID-19 as well as the same in mild, moderate and severe disease, (iii) the association between the incidence of AKI and disease severity as well as mortality in patients with confirmed COVID-19, and (iv) incidence of renal replacement therapy among COVID-19 patients with AKI. The evidence points to the need for early therapeutic intervention and multidisciplinary supportive care for COVID-19 patients. This is critical for COVID-19 patients with comorbidities such as diabetes mellitus, hypertension and cardiovascular diseases as their renal function may be compromised.

INTRODUCTION

In December 2019, several patients in Wuhan, Hubei province, China, presented to hospitals with unknown causes of infectious viral pneumonia (Li et al., 2020). The causative agent of this viral illness with severe respiratory tract association was identified and named by the International Committee on Taxonomy of viruses as severe acute respiratory syndrome coronavirus 2 (SARS-CoV-2). SARS-CoV-2 is the pathogen responsible for the coronavirus disease 2019 (COVID-19) (Coronaviridae Study Group, 2020).

Globally, there has been a fast and significant spread of SARS-CoV-2 since its description in December 2019, and on March 11, 2020, the World Health Organization (WHO) declared COVID-19 an international public health emergency. The WHO encouraged all countries to take the necessary precautions to prevent and contain the spread of the SARS-CoV-2 viral infection (WHO, 2020). As of May 3, 2021, the WHO reported that SARS-CoV-2 has spread to 216 countries and worldwide, there have been 152,534,452 established cases of COVID-19. There were also 3198528 deaths, with the majority in the United States of America, Brazil, India, Mexico and the United Kingdom (WHO, 2021).

The clinical expressions of COVID-19 patients comprise symptoms of a viral infection such as fatigue, dyspnea, myalgia, fever, dry or unproductive cough, and the most common serious clinical manifestation of the disease is pneumonia with chest images showing bilateral infiltrates. Less common symptoms include diarrhoea, loss of smell and taste, sore throat, headache, and conjunctivitis. Besides, other severe symptoms include loss of movement or speech and chest pain or pressure (Yan et al., 2020). In addition to the wide range of symptoms is the highly variable nature of the disease going from asymptomatic infection to severe and critical cases of lung disease and respiratory failure requiring mechanical ventilation leading to dysfunction of multiple organs and subsequently death (Zhou et al., 2020). Pathological findings of COVID-19 patients with critical illness who present with features of acute respiratory distress syndrome (ARDS) include significant diffuse alveolar damage due to the direct effect of SARS-CoV-2 and generalized inflammation (Batah and Fabro, 2021).

There are published reports of severe or critically ill COVID-19 cases with 15.8% - 26.0%. There is a reported median time of 16.0 to 18.6 days of progression from severe early illness to death, with up to 60% of patients dying of the disease (Zhou et al., 2020; Guan et al., 2020; Chen et al., 2020). However, another report shows that approximately 5% of COVID-19 patients become severely ill present with ARDS, septic shock with life-threatening multiple-organ failure and eventually death if they do not recover from the illness (Cummings et al., 2020).

It is well documented that advanced age is an independent risk factor for mortality in COVID-19 patients (Yang et al., 2020). Older patients are more at risk of contracting the SARS-CoV-2 viral infection and are more likely to develop and present with severe illness and a cascade of complications (Chronopoulos et al., 2020). The SARS-CoV-2 infection may rapidly spread and progress to ARDS, which necessitates intensive care and possibly the demise of the elderly patients, especially if they have underlying medical comorbidities (Sanyaolu et al., 2020). The risk factors for developing severe COVID-19 among elderly patients include hypertension, diabetes mellitus, malignancy, obesity, cardiovascular disease, smoking, chronic renal disease and chronic obstructive pulmonary disease, which affect their clinical progression and adversely impact recovery resulting in poor clinical outcomes (Osibogun et al., 2021).

The clinical development and manifestations of COVID-19 are extremely variable, with initial asymptomatic infection and progression to multi-organ dysfunction, including acute kidney injury (AKI), respiratory failure and death. The pathophysiology of AKI in COVID-19 is unclear. However, multifactorial causes have been documented. There is growing data on the increasing incidence of AKI in COVID-19 patients with adverse clinical outcomes, including significant infection, higher respiratory morbidity and mortality. This review aims to assess AKI incidence in COVID-19 patients and adverse clinical outcomes associated with renal disease severity and prognosis.

METHOD

Literature Search Strategies

A literature search and review of recent publications in PubMed, Google Scholar, Embase and Cochrane library relating to AKI in COVID-19 was conducted. The search examined all relevant published studies from January 1, 2020, to August 31, 2021. The search strategy included keywords such as COVID-19, acute kidney injury, mortality, severity, incidence, renal replacement therapy, severe acute respiratory syndrome, coronavirus 2 and pathophysiology.

Study Eligibility Criteria: Inclusion and Exclusion

The studies included in this review were published in peer-reviewed journals during the period mentioned above, written in English, and reported medical and scientific findings. In addition, pertinent data extracted from published studies include authors, year of publication, study design and population, and laboratory-confirmed COVID-19 cases. Information from peer-review articles concerning AKI in patients with COVID-19 and multifactorial causes, incidence, disease severity and renal replacement therapy were reviewed and documented.

PATHOPHYSIOLOGY OF ACUTE KIDNEY INJURY IN COVID-19 PATIENTS

Interaction of SARS-Co-V-2 With ACE and Entry Into Renal Cells

The pathophysiology and mechanisms involved in AKI development in COVID-19 patients are multi-factorial and not clearly determined but may involve pre-renal and intrinsic renal insults with subsequent structural damage and reduced kidney function (Batah et al., 2012).

One of the probable mechanisms of kidney injury is the direct entry of SARS-CoV-2 into the kidney's nephrons. The primary entrance of SARS-CoV-2 into the renal parenchyma involves the binding of the viral spike protein to the angiotensin-converting enzyme 2 (ACE2), a SARS-CoV-2 receptor, as well as transmembrane protease, serine 2 (TMPRSS2) present on the apical brush borders of the proximal convoluted tubules and podocytes (Zhang et al., 2020). TMPRSS2 participates in the SARS-CoV-2 cell invasion process by cleaving the viral spike (S) proteins. There is the fusion of the SARS-CoV-2 envelope and the cellular membrane of the renal cells with the subsequent entry of the virus and its replication in the host cell (Hoffmann et al., 2020).

Recently, a transmembrane spike glycoprotein called CD147, which has a high expression on proximal tubular epithelial cells, has been implicated as a facilitator or target for SARS-Co-V-2 entry into the nephrons as the C-terminal domain of its receptor binds to the external sub-domain of the virus (Helal et al., 2020). The SARS-CoV-2 enters the glomerular capillaries of the nephrons with infection of the endothelial cells of the glomerulus and subsequent entry into the tubular fluid, thus, the binding of SARS-CoV-2 to the ACE2 receptors on the proximal convoluted tubules (Yin and Zhang, 2020).

Histological evidence, though limited, supports the direct infection of renal cells by SARS-CoV-2. Su et al., (2020) published histological data from the autopsies of 26 confirmed COVID-19 patients where kidney abnormalities were determined by light and electron microscopy and immunostaining. Approximately 35.0% showed renal injury with non-isometric vacuolar degeneration and varying degrees of diffuse and frank necrosis of the proximal tubule accompanied by loss of brush border on light microscopy. There was also evidence of masses of corona virus-like elements in the podocytes and proximal tubular epithelium on electron microscopy and SARS-CoV-2 nucleoprotein antibody in the latter on immunostaining. According to the researchers, this data points to the direct infection of renal tissue by the invasion of SARS-CoV-2 (Su et al., 2020).

A recent study provides conclusive proof that SARS-CoV-2 directly infects kidney tissue in COVID-19 patients as viral RNA from replicated SARS-CoV-2 was found in 60% of patients in a post-mortem series comprising 63 individuals who presented with ARDS (Braun et al., 2020). Other studies have reported the findings of kidney tropism of SARS-CoV-2 and viral RNA formed by the replication of SARS-CoV-2 taken from different compartments of the kidney during the autopsy and analysis performed by employing real-time polymerase chain reaction and fluorescence in situ hybridization (Puelles et al., 2020; Hanley et al., 2020; Daio et al., 2020).

There have been recent reports of collapsing glomerulopathy in COVID-19 patients with AKI. There is a severe tubule-interstitial disease in these patients and a breakdown of the glomerular tuft and capillaries accompanied by hyperplasia and hypertrophy of the superimposing podocytes (Nicholas Cossey et al., 2017). Recent studies described collapsing glomerulopathy in COVID-19 patients with severe AKI, hypoalbuminemia, and heavy proteinuria (Peleg et al., 2020; Larsen et al., 2020). Besides, Kissling et al. (2020) were the first to report a case of collapsing glomerulopathy in a 63-year old patient with confirmed COVID-19 who presented with massive proteinuria hypoalbuminemia, lymphopenia and elevated serum levels of D-Dimer and C-reactive protein. Electron microscopy revealed many spherical particles in vacuoles within the cytoplasm of the podocytes that resemble viral inclusion bodies of SARS-CoV-2 (Kissling et al., 2020). These reports indicate that collapsing glomerulopathy caused by SARS-CoV-2 may be present in COVID-19 patients. This renal manifestation results from the virus's direct effect on podocytes and/or cytokine injury of the podocytes induced by the virus (Nasr and Koop, 2020).

However, some studies do not agree with the histopathological changes in COVID-19 patients with AKI and do not support the hypothesis that SARS-CoV-2 directly targets and infects the kidneys, causing significant damage. In two reports from Italy and Germany, there was no evidence of SARS-CoV-2 in renal tissue taken on biopsy from a critically ill COVID-19 patient with AKI or noticeable morphological changes in renal tissues of 10 postmortem COVID-19 patients (Rossi et al., 2020; Shaller et al., 2020). That points to the need for more histopathological studies in this area using a significant number of COVID-19 patients in each cohort.

Overactivation of the Renin-Angiotensin-Aldosterone System (RAAS)

The binding of the spike protein of SARS-CoV-2 to the ACE2 receptor results in the down-regulation of its surface expression on renal cells and decreased ACE2 activity. This leads to the reduced breakdown of angiotensin II to angiotensin 1-7 (Mizuiri and Ohashi, 2015). The SARS-CoV-2 mediated elevation of angiotensin II may stimulate an overactive local renin-angiotensin-aldosterone system. This results in a reduced reno-protective effect with the development of renal fibrosis, vasoconstriction, sodium retention, inflammation, and subsequent AKI (Vaduganathan et al., 2020). There are inadequate mechanistic investigations in this area. An early Chinese report in December 2019 of 12 cases of hospitalized COVID-19 patients who developed pneumonia and ARDS showed markedly elevated angiotensin II levels, which were concomitant with viral load and lung injury. Hence, the study investigators suggested that angiotensin receptor-blocker drugs might be promising for repurposing therapy for COVID-19 patients (Liu et al., 2020).

ACE2 acts together with the angiotensin-II type 1 receptor (AT1R), forming a complex that averts the internalization and degradation of the enzyme. The reduction of ACE enzymatic activity and accumulated angiotensin-II levels related to the interaction of the enzyme with AT1R induces ubiquitination and the internalization and degradation of ACE in lysosomes via the processes of phagocytosis, autophagy and endocytosis (Deshotels et al., 2014). The increased angiotensin II levels promote, through its interaction with the AT1 receptor, a cascade of activities with ensuing vasoconstriction, inflammation, epithelial cell apoptosis, atherogenesis, thrombosis and cytokine-induced organ damage, including kidneys (Zhang and Baker et al., 2017).

Dysregulated Immune Response and Cytokine Storm Mediated Acute Kidney Disease

The binding of SARS-CoV-2 to ACE2 receptors on the surface of renal cells initiates its entry into the host cells. The virus enters macrophages in renal cells and offers its antigen to the T-helper 1 (CD4+ and Th1) cells resulting in the activation of the latter, which in turn stimulates B lymphocytes to produce T-killer cells (CD8+ and Tk) (Rabi et al., 2020). CD8+ T cells are cytotoxic and exert their effect against SARS-CoV-2 in a direct manner or by the synthesis of pro-inflammatory cytokines such as interleukin-6 (IL-6), IL-8, tumor necrosis factor (TNF)-α, IL-1β via the stimulation of the NF-κB signalling pathway (Li et al., 2020). The activated macrophage expresses vital inflammatory mediators such as chemokines namely CCL2, CCL3, CCL4 and CCL5, which can also be induced by TNF-α and IL-1 (Chi et al., 2020). The elevated levels of pro-inflammatory cytokines and the activated systemic inflammatory response is regarded as a 'cytokine storm', which is concomitant with decreased CD4+ and CD8+ T-cell counts (Prompetchara et al., 2020). Besides, the excessive IL-6 produced during the cytokine storm may influence endothelial cells of the kidney to release other cytokines and chemokines and compromise endothelial and vascular permeability. This results in dysfunction of the microcirculation system in the kidney (Desai et al., 2002).

The immunopathology in COVID-19 patients with AKI also involves elevated levels of pro-inflammatory cytokines triggering hyper-inflammation and inducing capillary leak syndrome. There is the development of thrombosis, which may lead to disseminated intravascular coagulation. In addition to the development of lymphopenia and thrombocytopenia, these immune-pathological processes contribute to renal cell death and tissue damage akin to AKI (Henderson et al., 2020; Ahmadian et al., 2020).

The cytokine storm causes hemo-phagocytosis, which contributes to the dysregulated immune response (Lima et al., 2020). The entry of SARS-CoV-2 into renal cells results in the recruitment and infiltration of CD56+ natural killer cells, CD68+ macrophages and CD4+T cells into the tubular interstitium (Diao et al., 2020). The excessive activity of these immune cells, and significantly elevated cytokine levels, induce microvascular alterations, cell apoptosis, and interstitial fibrosis in renal cells (Meng et al., 2014; Pober et al., 2014).

There is over-activation and dysregulation of the complement system during the cytokine storm. The dysregulation of the complement system is initiated by its activation by SARS-CoV-2 via the alternative and lectin pathways (Java et al., 2020). The resulting C5 convertase commences the membrane attack system and the virus controls the assembly and deposition of complement C5b to C9 in the apical brush border of the tubules and in the tubular lumen with resulting tubular interstitial injury (Shagdarsuren et al. 2005). Laurence and colleagues demonstrated that the anti-C5 monoclonal antibody eculizumab effectively managed three patients with severe COVID-19, as there were significant reductions in neutrophil count, serum D-dimer levels as well creatinine. It was noted that one of the patients with AKI regained kidney function (Laurence et al., 2020).

The role of the lectin and complement pathways and proteins in AKI pathogenesis is a key area that warrants further investigation, mainly using drugs that block the complement pathway, particularly in severe and critically ill COVID-19 patients (Laurence et al., 2020). Besides, the activated lectin and alternative complement pathways can stimulate the pro-coagulation pathway with subsequent enhanced coagulation activity, depletion of coagulation factors and thus, microvascular thrombosis (Terpos et al., 2020). Complement-mediated thrombotic micro-angiopathy can cause expensive glomerular micro-thrombi, diffuse coagulative necrosis of the cortex and endothelial injury in renal cells leading to AKI (Jhaveri et al., 2020).

Organ Crosstalk: Lung-Kidney and Heart-Kidney

The pathogenesis of AKI may involve the lung-kidney axis, which is affected by cytokines, particularly high levels of IL-6 in COVID-19 patients with ARDS (Panitchote et al., 2019). IL-6 interacts with the endothelial cells and, with inflammatory/immune reaction, damages the nephrons (Husain-Syed et al., 2020). Heart and kidney crosstalk may be also a contributing factor to AKI in COVID-19 patients. The cardio-renal syndrome in these patients is characterized by bi-directional damage to both the organs and cytokine-induced cardiomyopathy and acute viral myocarditis that causes diminished cardiac output and compromises kidney perfusion. Lower blood pressure, hypovolemia, and renal vein compression result in decreased glomerular filtration rate (GFR) and less functioning kidneys (Ronco et al., 2020). Furthermore, there may be a right ventricular failure in COVID-19 patients present with pneumonia that, in conjunction with left ventricular dysfunction, can result in arterial under-filling and reduced cardiac output and consequent attenuation of GFR (Ronco et al., 2020). This review provides the pathological rationales for developing AKI in COVID-19 patients. There need to be more standardized and guided treatment options, and the development of such therapies should be informed by the clinical and laboratory data gathered from these patients.

INCIDENCE OF ACUTE KIDNEY DISEASE (AKI)

Ever since the occurrence of the COVID-19 pandemic, numerous studies have provided information on the incidence, severity, morbidity and prognostic outcomes, including mortality of the disease along with AKI. One of the earliest studies that examined the effect of SARS-CoV-2 infection on the kidney of 116 hospitalized COVID-19 patients in Wuhan, China, found no case of AKI in 111 of these patients with no prior case of acute kidney disease (Wang et al., 2020). In another retrospective study that involved extracting clinical data from 1,099 COVID-19 hospitalized patients from 30 provinces in China, 0.5% had AKI (Guan et al., 2020). Moreover, an incidence of 5.1% was reported by Cheng et al. of 701 hospitalized adult COVID-19 patients in Wuhan, China (Cheng et al., 2020). A higher AKI incidence of 7.3% was reported in a retrospective study comprising 492 hospitalized COVID-19 in Wuhan, China (Table 1; Figure 2). The researchers indicated that independent risk factors of AKI in these patients were gender (male), elevated leukocytes and diuretic use (Dai et al., 2020).

Table 1. Incidence of Acute Kidney Injury (AKI) in COVID-19 patients

Reference	Study Design	Study Region	Enrollment Period	Number of Participants	Incidence of AKI (%)
Wang et al. 2020	Prospective, single center	Wuhan, China	January 14, 2020 - February 13, 2020	116	0.0
Guan et al., 2020	Retrospective, multi-center	China	December 1, 2019 – January 29, 2020	1,099	0.5
Cheng et al., 2020	Prospective cohort, single center	Wuhan, China	Not given	701	5.1
Dai et al., 2021	Retrospective, multi-center	Wuhan, China	January 29, 2020 - February 25, 2020	492	7.1
Zhou et al., 2020	Retrospective, multi-center	Wuhan, China	December 29, 2019 – January 31, 2020	119	15.0
Yan et al., 2021	Retrospective, multi-center	Wuhan, China	January 27, 2020 - February 17, 2020	1,764	13.0
Hirsch et al., 2020	Retrospective, multi-center	New York, United States	March 1, 2020 - April 5, 2020	5,449	36.6
Chan et al., 2021	Retrospective, multi-center	New York, United States	February 27, 2020 - May 30, 2020	3,993	46.0
Lim et al., 2020	Retrospective, multi-center	South Korea	February 17, 2020 – March 22, 2020	164	18.3
Diebold et al., 2021	Retrospective, single center	Switzerland	February 1, 2020 - June 30, 2020	188	22.0
Hardenberg et al., 2021	Prospective cohort, single center	Germany	March 1, 2020 – June 3, 2020	223	31.0
Gupta et al., 2021	Retrospective, multi-center	United States	March 4, 2020 - April 11, 2020	3,099	20.6

Later studies of COVID-19 patients in China have demonstrated a higher incidence of AKI. In a retrospective, multi-centre cohort study of 191 COVID-19 patients, the incidence of AKI (diagnosed according to KDIGO clinical practice guidelines) was 15.0%, and renal disorder developed at a median of 15 days (Zhou et al., 2020). In a larger retrospective, an observational cohort study on 1,764 hospitalized COVID-19 patients, of which 882 were elderly, 13.0% developed AKI (Yan et al., 2021) (Table 1; Figure 2). Similarly, in another retrospective observational study of 287 COVID-19 patients, 14.3% were with stage-1 AKI, while 4.9% was the incidence of stage-2 or stage-3. It was noted that the patients who developed AKI were older, with hypoxia, and more likely to have underlying co-morbidity such as cerebrovascular diseases and hypertension (Xiao et al., 2020). The disparity in the incidence of AKI in the cohorts of Chinese patients may partly be explicated by the age and gender of the participants, various definitions of the disorder, including baseline or reference, ranges for serum creatinine, the degree of COVID-19 severity and the proportions of severely ill individuals (Diebold et al., 2021).

In contrast, reports of incidences of AKI in COVID-19 patients in the United States, Europe and other parts of the world are higher. A study by Hirsch et al. (2020) conducted in New York on 5,449 COVID-19 patients indicated that AKI (defined according to KDIGO criteria) developed in 36.6% of individuals, out of which 46.5% had stage-1, 22.4% stage-2 and 31.1% stage-3 (Hirsch et al., 2020). Another study on 3,993 COVID-19 patients admitted to the Mount Sinai Health System indicated that 46.0% developed AKI; of this 30.0% of survivors regained complete renal function at the time of hospital discharge (Chan et al., 2021) (Table 1; Figure 2). In the Mount Sinai Health System cohort, more patients had comorbidities such as diabetes mellitus and hypertension compared to those covered in the studies from Italy and China. However, no relationship between diabetes mellitus and AKI was found (Chan et al., 2021).

INCIDENCE OF SEVERE AKI AND ASSOCIATED OUTCOMES

AKI can be staged using "The Kidney Disease: Improving Global Outcomes (KDIGO) criteria". As per the criteria, an elevation of serum creatinine greater than 0.3 mg/dL from the baseline within seven days is deemed stage-1; an elevation greater than 2.0 to 2.9 times is stage-2 and stage-3 is where the elevation is greater than 3.0 times with the initiation of renal replacement therapy (RRT) (Kdigo, 2012). A meta-analysis covering 13 studies on 3,027 COVID-19 patients found that serum creatinine levels higher than 133 μmol/L were associated with severe disease and a higher risk of death in 979 of the patients (Zheng et al., 2020). Some researchers believe that the risk of AKI in COVID-19 patients is relatively low, and the development of this renal disorder is generally a sign of disease severity as well as multi-organ dysfunction (Ng et al., 2020; Makris and Spanou, 2016). In a study conducted in South Korea by Lim and colleagues, the clinical outcomes of 164 hospitalized patients with COVID-19 and AKI were investigated. The authors reported an incidence of AKI of 18.3% (30 patients), and of these, 46.7%, 13.3% and 40.0% had stage-1, 2 and 3 diseases, respectively (Lim et al., 2020) (Figure 3). COVID19 patients with severe AKI and having serum creatinine levels greater than 300% above baseline values had significantly higher mortality than either non-AKI (Hazard Ratio HR = 3.62, 95% CI = 1.75-7.48, p= 0.001) or stage-1 AKI (HR = 15.65, 95% CI = 2.43-100.64, p= 0.004) (Lim et al., 2020).

Some studies have stated the incidence of overall AKI (using the KDIGO definition) in COVID-19 patients admitted to ICU ranges from 50.0% to 81.0% (Chan et al., 2021; Hirsch et al., 2020; Doher et al., 2020; Fominskiy et al., 2021) (Figure 3). These reports have also posited incidence of severe AKI in COVID-19 patients admitted to ICU ranging from 40.6% to 57.4%. The inconsistencies in the incidence of severe AKI in COVID-19 patients could be due to several factors including different ways of defining severity (and applying the KDIGO definition) and methods used to determine baseline serum creatinine that could result in incongruence in incidences up to 15% (Wiersema et al., 2020). Notably, in early studies of COVID-19 patients admitted to ICU in China the incidence of AKI ranged from 20.0 to 30.0% and this renal disorder was viewed as a marker of disease severity (Yang et al., 2020; Yu et al., 2020).

There is increasing attention worldwide regarding the development of acute renal impairment in COVID-19 patients. In a retrospective study of 102 COVID-19 patients admitted to the ICU in Brazil, the incidence of AKI (defined according to KDIGO criteria) was 55.9%, and the majority of these patients (66.7%) presented with stage-3 disease (severe) and most required RR via hemodialysis (Costa et al., 2021). Chinese researchers Chen and colleagues, conducting a comparative perspective of the incidence of renal impairment in COVID-19 patients, found 0.5%, 0.1% and 2.9% AKI in patients with mild, moderate and severe indications of COVID-19, respectively (Chen et al., 2020).

Studies have demonstrated a positive relationship between AKI and poor outcomes in patients with COVID-19 (Wang et al., 2020; Huang et al., 2020; Hirsch et al., 2020; Cheng et al., 2020). Conducting a meta-analysis of 14415 COVID-19 patients Menon et al. (2021) found a pooled prevalence of 11.0% for AKI. The study reported a significant relationship between AKI and severe COVID-19 with a pooled odds ratio (OR) of 8.45 as well as considerably higher mortality with an OR of 13.52 ($P < 0.0001$). Similarly, a systematic review and meta-analysis of six studies by Ali et al. (2020) found an association between AKI and elevated mortality with a relative risk of 3.08 (95% CI: 1.54-6.19). Furthermore, in a retrospective study of 188 COVID-19 patients with an incidence of AKI of 22.0%, the stages of the renal disorder were 39.0% had stage-1, 24.0% stage-2 and 37.0% stage-3. AKI was related to high in-hospital death as just over one-quarter (27.0%) of hospitalized COVID-19 patients with AKI (Diebold et al., 2021). These studies all suggested that AKI in patients with COVID-19 is a significant marker of disease severity and has an adverse impact on patient morbidity and mortality (Menon et al., 2021; Ali et al., 2020; Diebold et al., 2021; Zhou et al., 2020).

The pathophysiology of COVID-19 and the severity of AKI, particularly in hospitalized patients have contributed to poor outcomes. There are several risk factors for severe AKI. Hardenberg and colleagues conducted a prospective study of 223 consecutive COVID-19 patients and found that 31.0% developed AKI, of which 95.7% required RRT (Hardenberg et al., 2021). The authors indicated that risk factors of severe AKI in COVID-19 patients were male, older particularly over 65 years and had comorbidities such as diabetes mellitus, cardio-cerebrovascular disease and hypertension. It was observed that severe AKI occurred entirely in those admitted to ICU, and 97.3% of these patients developed respiratory failure (Hardenberg et al., 2021). Moreover, independent risk factors of severe AKI include elevated procalcitonin (PCT) levels and leukocyte count, vasopressor treatment and use of mechanical ventilation (Figure 1) (Hardenberg et al., 2021). In another study, Lowe et al., 2021 performed a cohort study involving 81 COVID-19 patients, of whom 44.0% had stage-3 AKI and required RRT, found that these individuals needed

Figure 1. Risk factors for the development of Acute Kidney Injury (AKI) during COVID-19

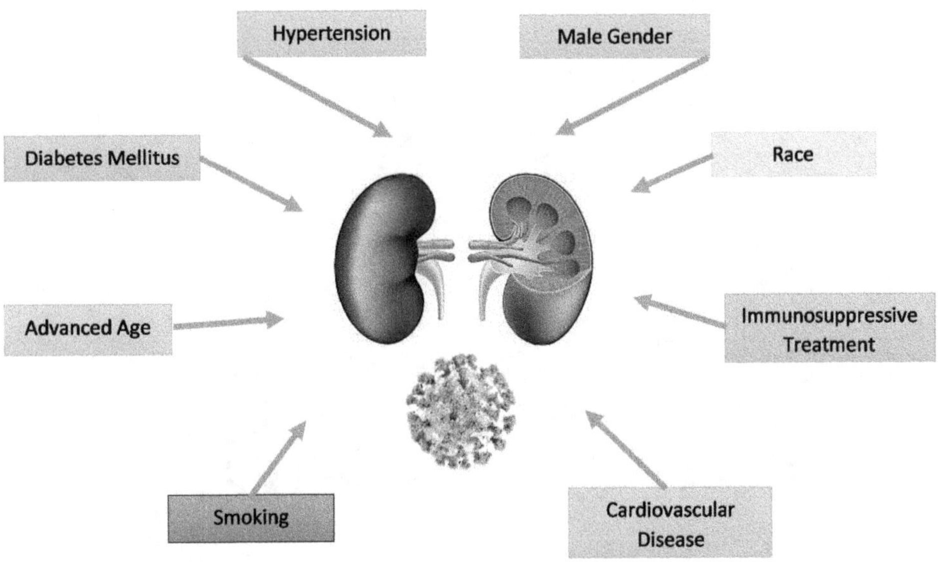

Figure 2. Incidence of acute kidney injury in COVID-19 patients

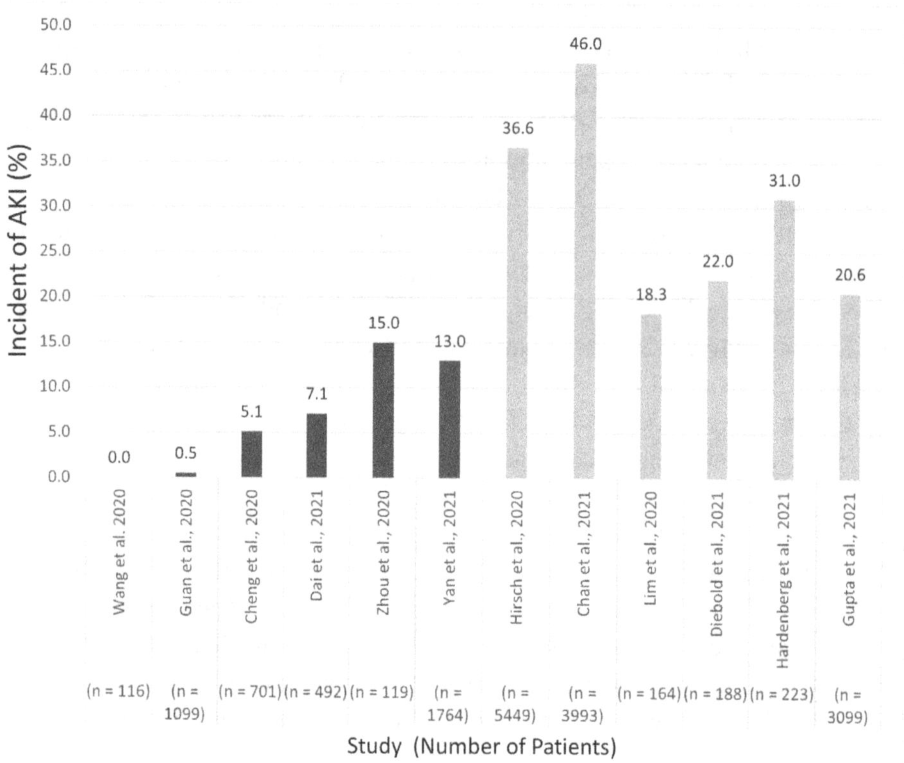

Figure 3. Relationship between disease severity and acute kidney injury

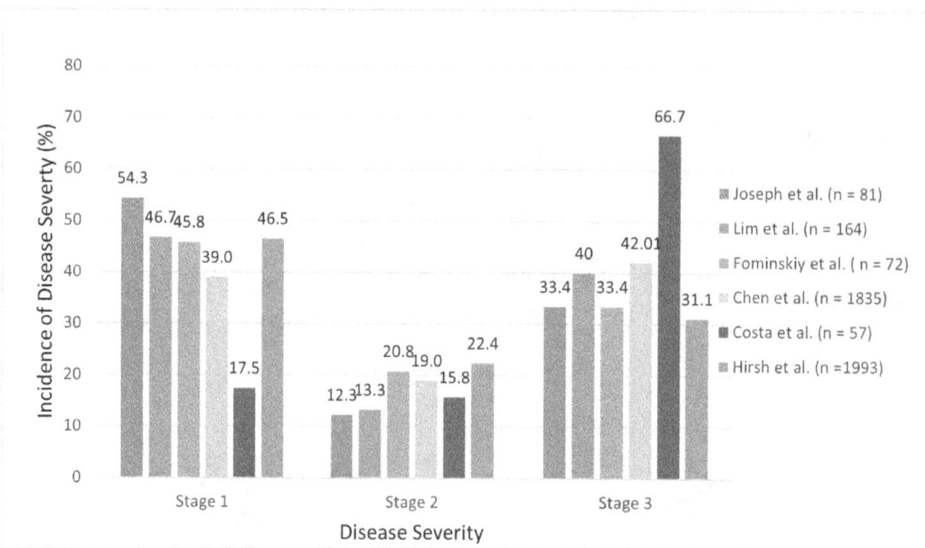

invasive mechanical ventilation, vasopressor therapy as well as they presented with lymphopenia and had greater illness severity scores.

In a recent study, Ghosn and colleagues retrospectively examined the outcomes of 110 critically ill COVID-19 patients having severe AKI and respiratory failure and were admitted to ICU. The incidence of severe AKI was 45.4%, and investigation showed that disease severity was independently related to age, serum creatinine levels on admission, longer hospital stay, extracorporeal membrane oxygenation and hospital mortality. However, there was no association of severe AKI with thrombotic markers (fibrinogen, D-Dimer) or inflammatory markers (ferritin, interleukin-6, C-reactive protein) (Ghosn et al., 2021). In an earlier study by Paek and colleagues, who conducted a retrospective study of 704 Korean hospitalized, the incidence of COVID-19 patients who developed AKI was 4.0%. There were 35.7% of patients with AKI stage-3 disease and elevated creatine kinase, advanced age, and high neutrophil-to-lymphocyte ratio were independent risk factors associated with the in-hospital deaths (Paek et al., 2020). Moreover, in another retrospective observational study comprising 211 COVID-19 patients admitted to ICU, there was the incidence of severe AKI in 26.0% (KDIGO stage-3) and 28-day mortality of 56.0% among these patients (Jian et al., 2019).

INCIDENCE OF RENAL REPLACEMENT THERAPY (RRT)

In the past few months, various studies have reported RRT incidence among COVID-19 patients with AKI. In China, Guan et al., (2019) conducted a retrospective study involving 1,099 laboratory-confirmed COVID-19 patients and found that 5.0% were admitted to ICU, 0.5% had AKI, and 0.8% of these persons received continuous RRT (CRRT). In a later retrospective single-center case series consisting of 138 hospitalized COVID-19 patients conducted in Wuhan, China 36 (26.0%) were admitted to ICU and 2 of these individuals received RRT (Wang et al., 2020). Likewise, in a retrospective, multi-centre cohort study of 191 adult COVID-19 patients, there were 137 survivors and 54 non-survivors. Of the non-survivors, 10 (5.0%) received RRT (Zhou et al., 2020). Likewise, Huang et al. (2020) reported 3 (7.0%) hospitalized COVID-19 patients requiring CRRT.

In contrast to China, the incidence of RRT is higher in reports from researchers in the United States and Europe. In a retrospective, observational study of 3,993 hospitalized COVID-19 patients admitted to the Mount Sinai Care Health System, AKI was diagnosed in 1,835 (46.0%) individuals, of which 347 (19.0%) received RRT (Chan et al., 2021) (Figure 4). Also, in the United States Gupta and colleagues directed a multi-center cohort study comprising 3,099 critically ill COVID-19 adult patients in 67 hospitals and observed that 637 (20.6%) developed AKI and required

Figure 4. Incidence of acute renal injury requiring renal replacement therapy

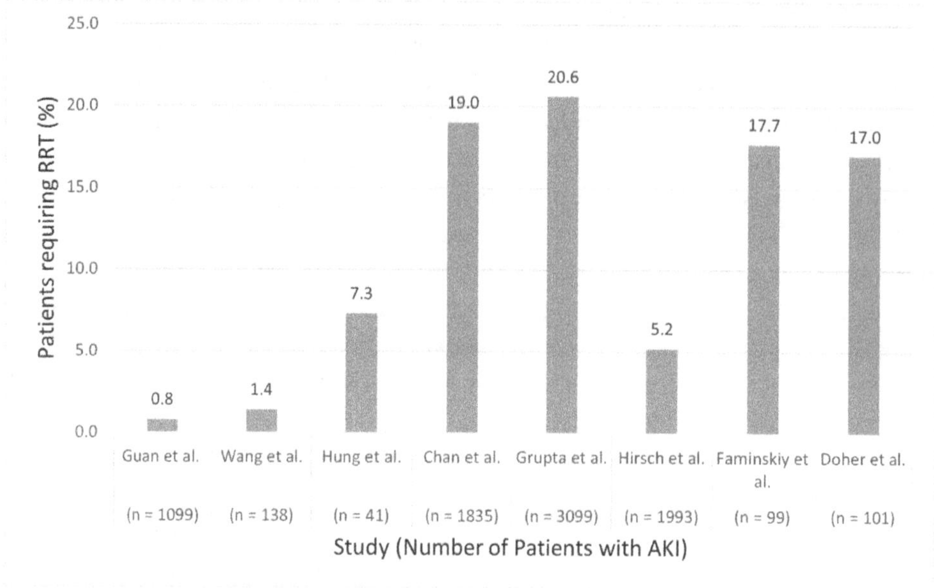

RRT. The researchers noted that the risk factors for AKI-RTT were elevated D-dimer, co-morbidities such as diabetes mellitus and hypertension, higher body mass index (BMI), male, non-Caucasian and greater severity of arterial hypoxia on admission to ICU (Gupta et al., 2021). Similarly, in New York, researchers reviewed the health records of 5,449 hospitalized COVID-19 patients and reported that 1,993 (36.6%) developed AKI. They also noted that 285 of the cases required RRT, which is 14.3% of patients with AKI (Hirsch et al., 2021). Other studies have reported incidence rates of RRT in COVID-19 patients diagnosed with AKI ranging from 13.0% to 24.5% (Fominskiy et al., 2021; Joseph et al., 2020; Doher et al. 2020; Ghosn et al., 2020).

COVID-19 patients with severe AKI are expected to have significantly greater requirements for dialysis (Nugent et al., 2021) and may be at higher risk for developing chronic kidney disease (Coca et al., 2012). Early commencement of CRRT in patients with COVID-19 -AKI where there is the removal of low to high molecular weight inflammatory markers using polystyrene-based hemo-absorber can improve clinical outcomes (Morgera et al., 2003). CRRT has been found to be effective in the treatment of AKI and in reducing cytokine storms in COVID-19 patients admitted to ICU (Forest et al., 2021) (Figure 4). However, there is the risk of CRRT circuit clotting due to the hypercoagulable state brought about by COVID-19, and therefore an anticoagulation strategy using regional citrate should be adopted (Ronco et al., 2020).

AKI AND MORTALITY

In COVID-19 patients, especially those admitted to ICU, the presence of AKI has been shown to herald worse survival. Fu et al. (2020) conducted a comprehensive systematic review and meta-analysis of 142 studies involving 49,048 hospitalized COVID-19 patients with AKI. They found an incidence of 28.6% for individuals from Europe and the USA and 5.5% among those from China. Overall, AKI was associated with a greater risk of death, and the pooled risk ratio was 4.6 (Fu et al., 2020). In another study published at the same time, a systematic review and meta-analysis of 15 studies involving 5832 COVID-19 patients, found that the occurrence of AKI was related to increased risk of COVID-19 mortality (OR = 23.9) (Cheruiyot et al., 2020). Likewise, a systematic review and meta-analysis by Yang and colleagues of 24 studies comprising 4,963 confirmed COVID-19 patients found an overall AKI incidence of 4.5%. The incidence of AKI in survivors was 0.7% while it was 52.9% in non-survivors (Yang et al., 2020). Furthermore, the pooled incidence of AKI in a meta-analysis of 26 studies comprising 5,497 COVID-19 patients was 8.4%, and this was found associated with elevated baseline serum creatinine, hypertension, diabetes mellitus and age. AKI occurrence was associated with an elevated pooled estimated risk for death from AKI of 13.33% (OR = 13.33%; 95% CI, 4.05 – 43.91%) (Hansrivijit et al., 2020). These systematic and meta-analysis demonstrates that the overall pooled mortality risk of patients with COVID-19 and AKI ranged from 5.5% to 52.9% and demonstrates a higher death rate in these persons.

Studies have reported that elderly hospitalized COVID-19 patients have a higher incidence of AKI (Xu et al., 2021; Zahid et al., 2020). In another recent observational study comprising 316 elderly hospitalized COVID-19 patients and 876 non-COVID-19 individuals, the risk of in-hospital mortality was found uppermost in persons with AKI and COVID-19 (OR = 80.3) versus lower risks in COVID-19 without AKI (OR = 16.3) and non-COVID-19 without AKI (OR = 10.2) (Xu et al., 2021). A single-centered, retrospective cohort study of 469 elderly hospitalized COVID-19 patients (median age of 66.0 years) conducted in the United States showed 27.3% incidence of AKI which increased to 39.1% in individuals with low estimated GFR (eGFR). The mortality rate in this group of patients was 71.1%. The Cox proportional hazard model revealed that risk factors for in-hospital mortality were low eGFR, elevated serum urea on admission, and AKI stage-3 (Zahid et al., 2020). Adverse outcomes such as mortality and pulmonary edema in elderly AKI-COVID-19 patients could be due to abnormal volume status and uremic state, secondary effects of hypoxia, inflammatory cytokine or shock, and the direct effects of SARS-CoV-2 (Kang et al., 2020). Histological studies of kidney tissues of dead AKI and COVID-19 patients have shown diffuse acute injury of tubules, glomerular ischemia, SARS-CoV-2 virus-like particles in the podocytes and cytoplasm of proximal tubular epithelial cells, with the latter expressing SARS-CoV-2 nucleoprotein and clathrin-coated structures (Su et al., 2020; Goldsmith et al., 2020; Santoriello et al., 2020).

A number of studies have investigated the incidence of mortality in AKI and COVID-19 patients and its associated risk. A recent retrospective study by Dai and colleagues in Wuhan, China, of 492 COVID-19 patients found a prevalence for AKI of 7.3% and a mortality of 63.0%, significantly different from those without renal disease (11.4%) (Dai et al., 2021). There were similar findings in a retrospective, observational cohort study of 1,764 COVID-19 in-hospital patients comprising 882 elderly persons. The Chinese researchers found 13.0% AKI incidence among the older patients, of which 14.5% died. Notably, patients with AKI had a mortality of 59.1% versus 7.8% in those without AKI. Cox regression analysis suggested that elevated serum interleukin-6 levels on admission, the presence of AKI and myocardial injury increased the risk of in-hospital mortality (Yan et al., 2021).

In terms of risk profile, a recent meta-analysis by Wang and colleagues of 42 studies and 8,832 COVID-19 patients showed a significant association between AKI development and increased mortality (OR = 30.46) (Wang et al., 2021). Turkish researchers, in a descriptive study of a sample population of 55 COVID-19 patients, reported that based on regression analysis, stage-3 AKI (OR = 10.30) along with low albumin (OR = 12.79) were significant independent risk factors of mortality in persons who had AKI (Tocoglu et al., 2021). Likewise, researchers in Mexico found that AKI (58.6%) was common in 99 COVID-19 patients with severe pneumonia, and AKI (OR = 8.6%), as well as obesity (OR = 5.57), were significant risk factors for in-hospital mortality (Casas-Aparicio et al., 2021). These studies suggest that taking appropriate therapeutic interventions and early diagnosis of AKI are essential to reduce mortality. Notably, early AKI (adjusted HR = 2.46) and late AKI (adjusted HR = 3.09) are associated with an elevated risk of in-hospital death (Peng et al., 2020). On the contrary, Fominskiy and colleagues reported no association between AKI development in COVID-19 patients on mechanical ventilation and the risk of sudden death (adjusted HR = 1.32) (Fominskiy et al., 2021).

There are biochemical indices such as serum urea and creatinine that are elevated in AKI and related to death in patients with AKI and COVID-19. Cheng and colleagues in a prospective cohort study of 701 COVID-19 patients found that AKI was present in 5.1% of persons and overall in-hospital mortality of 16.1%, and more deaths in patients with significantly elevated baseline serum creatinine (33.7% vs 13.2%). Based on Cox regression analysis, higher baseline serum creatinine, urea, and AKI stages were found independent risk factors for in-hospital mortality (Cheng et al., 2020). Spanish researchers reported that 11.4% of a cohort of 1,603 hospitalized COVID-19 patients had AKI, with an overall mortality of 12.3%. Mortality was higher among patients with elevated serum creatinine (32.4%) at admission and AKI (15.9%), and there was a positive relationship between the latter and death (Portolés et al., 2020). Elevated serum creatinine and urea levels in pre-renal and intrinsic AKI have been significantly related to higher mortality in COVID-19 patients (HR = 3.2 for pre-renal and HR = 7.7 for intrinsic) (Hansrivijit et al., 2021). The varied nature of AKI and clinical outcomes including death in COVID-19 patients warrants its characterisation, particularly given its high incidence in these individuals.

DISCUSSION

AKI is damage to the kidney resulting in sudden or declining function over a short period. It is characterized by elevated proteins, including albumin and globulins in the urine, hematuria, and increased serum urea and creatinine.

Since the COVID-19 pandemic, numerous studies have published findings on the incidence, severity, prognosis and mortality of COVID-19 patients who developed AKI. The incidence of AKI in COVID-19 varies widely in reported studies, with low values in patients from China and higher numbers in those from the United States of America and European countries, particularly in those admitted to ICU. Further, early reported studies presented data involving cohort COVID-19 patients at a single hospital or multi-centered in several public or academic clinical institutions. The differences in incidence could be due to varying COVID-19 patient population characteristics, racial and ethnic composition, possible dissimilar criteria for diagnosing AKI, and maybe many other factors.

This review attests to the growing evidence in the literature that demonstrates that AKI occurs in a significant number of patients with COVID-19. The renal injury is considerably associated with negative clinical outcomes in these individuals. Most of the studies in this review showed a significant positive association between AKI in patients with severe COVID-19 and a similar trend of increased morbidity and mortality. Clinically, this points to the fact that early detection of AKI using the KDIGO criteria for diagnosis with the appropriate determination of serum urea and creatinine, and applying appropriate therapeutic interventions are important to reduce morbidity and mortality in these patients. Recognizing renal participation in early-stage COVID-19 and implementing supportive measures irrespective of comorbidities such as hypertension and diabetes mellitus would be protective and could halt further progressive decline to moderate or severe stages. Clinical trials and randomized controlled studies should be conducted to investigate the possible mechanisms associated with AKI in COVID-19 patients, and the primary cause of morbidity and mortality is necessary.

Studies that we examined also documented a number of risk factors for AKI in patients, which vary among studies and include advanced age, male gender, race, immunosuppressive treatment, immunosuppression due to haematological cancers, smoking and chronic disorders such as hypertension, diabetes mellitus and cardiovascular diseases. On the other hand, there are independent risk factors of severe AKI in COVID-19 patients that are time-varying and include mechanical ventilation, vasopressor management, ICU admission, higher levels of serum procalcitonin and leucocyte count and persistently high D-dimer levels (Figure 1).

There is no precise anti-SARS-CoV-2 therapy for COVID-19 patients with AKI, particularly those admitted to ICU. However, patients may benefit from sequential extracorporeal treatments and supportive care from a multidisciplinary staff. Findings from more research will provide a more in-depth understanding of the pathophysiology of AKI secondary to COVID-19, which may result in improved clinical outcomes. It is important to identify early COVID-19 patients at risk of developing AKI and most likely to benefit from emerging treatment approaches, particularly those from clinical trials. This will aid physicians in decreasing the incidence of AKI and, in so doing, reduce morbidity and mortality of COVID-19 patients.

The pathophysiology of AKI associated with COVID-19 is not yet completely elucidated and it appears to be multifactorial. Numerous mechanisms have been suggested, and there is evidence of the involvement of both pre-renal and intrinsic renal injuries. The COVID-19 specific mechanisms entail significant damage to tubular cells as a result of the binding of SARS-CoV-2 to ACE2 receptors and dysregulated activation of RAAS leading to the elevation of angiotensin II. Multifactorial direct and indirect causes include cytokine storm (elevated IL-6 and procalcitonin), hypoxia, hypo-perfusion, microvascular thrombosis, hypovolemia, and ARDS. The ongoing research in this area and new data may provide better insight into the precise etiology of AKI in COVID-19.

One of the limitations of this review is that most of the studies were either prospective or retrospective, and there was no inclusion of any randomised clinical trials. The findings from most of these studies could also bring in positive publication bias, and it is unlikely that negative results would be published.

There are significant increases in the number of emerging data and few longitudinal studies or those with significant follow-up periods, which limits detailed information on COVID-19 and AKI. Thus, in this age of rapid emergent data and the negative impact of AKI on COVID-19, researchers could afford to examine more wide-ranging data on the degree and severity of AKI, thus facilitating a better upstanding of the disease pathogenesis and prognosis, for adopting strategies, more imaginative, during the pandemics.

CONCLUSION

Reports of the high incidence of AKI in COVID-19 patients continue to increase, and this renal disorder is significantly related to negative outcomes for these persons. In the early stage of COVID-19, renal function should be closely monitored and supportive care adopted to protect kidney function and prevent the development of AKI or halt progression to more severe stages. Large randomized controlled studies and clinical trials are needed to investigate the precise mechanism of the relationship of AKI with COVID-19 and the primary causes of morbidity and mortality.

REFERENCES

Ahmadian, E., Hosseiniyan Khatibi, S.M., Razi, Soofiyani, S., Abediazar, S., Shoja, M. M., Ardalan, M., & Zununi Vahed, S. (2020). Covid-19 and kidney injury: Pathophysiology and molecular mechanisms. *Reviews in Medical Virology*, e2176. doi:/ doi:10.1002/rmv.2176

Ali, H., Daoud, A., Mohamed, M. M., Salim, S. A., Yessayan, L., Baharani, J., Murtaza, A., Rao, V., & Soliman, K. M. (2020). Survival rate in acute kidney injury superimposed COVID-19 patients: A systematic review and meta-analysis. *Renal Failure*, *42*(1), 393–397.

Basile, D. P., Anderson, M. D., & Sutton, T. A. (2012). Pathophysiology of acute kidney injury. *Comprehensive Physiology*, *2*(2), 1303–1353. doi:10.1002/cphy.c110041

Batah, S. S., & Fabro, A. T. (2021). Pulmonary pathology of ARDS in COVID-19: A pathological review for clinicians. *Respiratory Medicine*, 176, 106239. doi:.rmed.2020.106239 doi:10.1016/j

Braun, F., Lütgehetmann, M., Pfefferle, S., Wong, M. N., Carsten, A., Lindenmeyer, M. T., Nörz, D., Heinrich, F., Meißner, K., Wichmann, D., Kluge, S., & .... (2020). SARS-CoV-2 renal tropism associates with acute kidney injury. *Lancet*, *396*(10251), 597–598. doi:10.1016/S0140-6736(20)31759-1

Casas-Aparicio, G. A., León-Rodríguez, I., Alvarado-de la Barrera, C., González-Navarro, M., Peralta-Prado, A. B., Luna-Villalobos, Y., Velasco-Morales, A., Calderón-Dávila, N., Ormsby, C. E., & Ávila-Ríos, S. (2021). Acute kidney injury in patients with severe COVID-19 in Mexico. *PLoS One*, *16*(2), e0246595. doi:10.1371/journal.pone.0246595

Chan, L., Chaudhary, K., Saha, A., Chauhan, K., Vaid, A., Zhao, S., & Nadkarni, G. N. (2021). AKI in hospitalized patients with COVID-19. *Journal of the American Society of Nephrology*, *32*(1), 151–160.

Chan, L., Chaudhary, K., Saha, A., Chauhan, K., Vaid, A., Zhao, S., Paranjpe, I., Somani, S., Richter, F., Miotto, R., Lala, A., Kia, A., Timsina, P., Li, L., Freeman, R., Chen, R., Narula, J., Just, A. C., Horowitz, C., ... Nadkarni, G. N.Mount Sinai COVID Informatics Center (MSCIC). (2021). AKI in hospitalized patients with COVID-19. *Journal of the American Society of Nephrology*, *32*(1), 151–160. doi:10.1681/ASN.2020050615

Chen, Lu., & Caixia, G. (2020). Focus on kidney disease among the coronavirus disease 2019 patients: A comparative perspective between China, Italy and the United States. *International Journal of Clinical Practice*, *74*(9), e13561. doi:10.1111/ijcp.13561

Chen, T., Wu, D., Chen, H., Yan, W., Yang, D., Chen, G., Ma, K., Xu, D., Yu, H., Wang, H., et al. (2020). Clinical characteristics of 113 deceased patients with coronavirus disease 2019: Retrospective study. *British Medical Journal*, 368, m1091, 368. 10.1136/bmj.m1091

Cheng, Y., Luo, R., Wang, K., Zhang, M., Wang, Z., Dong, L., Li, J., Yao, Y., Ge, S., & Xu, G. (2020). Kidney disease is associated with in-hospital death of patients with COVID-19. *Kidney International*, *97*(5), 829–838.

Cheruiyot, I., Henry, B., Lippi, G., Kipkorir, V., Ngure, B., Munguti, J., & Misiani, M. (2020). Acute kidney injury is associated with worse prognosis in COVID-19 patients: A systematic review and meta-analysis. *Acta Biomedica*, *91*(3), e2020029. doi:10.23750/abm.v91i3.10222

Chi, Y., Ge, Y., Wu, B., Zhang, W., Wu, T., Wen, T., Liu, J., Guo, X., Huang, C., Jiao, Y., Zhu, F., & Cui, L. (2020). Serum cytokine and chemokine profile in relation to the severity of coronavirus disease 2019 in China. *The Journal of Infectious Diseases*, *222*(5), 746–754.

Chronopoulos, A., Cruz, D.N., & Ronco, C. (2010). Hospital-acquired acute kidney injury in the elderly. *Nature Reviews Nephrology*, 6(3), 141-149. doi: eph.2009.234 doi:10.1038/nrn

Coca, S. G., Singanamala, S., & Parikh, C. R. (2012). Chronic kidney disease after acute kidney injury: A systematic review and meta-analysis. *Kidney International*, *81*(5), 442–448.

Coronaviridae Study Group of the International Committee on Taxonomy of Viruses. (2020). The species Severe acute respiratory syndrome-related coronavirus: classifying 2019-nCoV and naming it SARS-CoV-2. *Nature Microbiology*, 5, 536-544.doi:. 10 38 /s41564-020-0695-z. doi:10

CostaR.SóriaT. C.SallesE. F.GerechtA. V.CorvisierM. F.MenezesM.ÁvilaC.SilvaE. PereiraS.SimvoulidisL. (2021). Acute kidney injury in patients with Covid-19 in a Brazilian ICU: incidence, predictors and in-hospital mortality. *Journal brasileiro de nefrologia: 'orgao oficial de Sociedades Brasileira e Latino-Americana de Nefrologia*, 43(3), 349-358. doi.org/ doi:10.1590/2175-8239-JBN-2020-0144

Cummings, M. J., Baldwin, M. R., Abrams, D., Jacobson, S. D., Meyer, B. J., Balough, E. M., Aaron, J. G., Claassen, J., Rabbani, L. E., Hastie, J., Hochman, B. R., Salazar-Schicchi, J., Yip, N. H., Brodie, D., & O'Donnell, M. R. (2020). Epidemiology, clinical course, and outcomes of critically ill adults with COVID-19 in New York City: A prospective cohort study. *Lancet*, *395*(10239), 1763–1770. doi:10.1016/S0140-6736(20)31189-2

Dai, Y., Liu, Z., Du, X., Wei, H., Wu, Y., Li, H., Tian, M., Li, C., Song, X., Wu, W., Cai, Y., Yu, Y., Hu, S., & Dong, J. (2021). Acute kidney injury in hospitalized patients infected with COVID-19 from Wuhan, China: A retrospective study. *BioMed Research International*, *6655185*. Advance online publication. doi:10.1155/2021/6655185

Desai, T. R., Leeper, N. J., Hynes, K. L., & Gewertz, B. L. (2002). Interleukin-6 causes endothelial barrier dysfunction via the protein kinase C pathway. *The Journal of Surgical Research*, *104*, 118–123. doi:10.1006/jsre.2002.6415

Deshotels, M. R., Xia, H., Sriramula, S., Lazartigues, E., & Filipeanu, C. M. (2014). Angiotensin II mediates angiotensin converting enzyme type 2 internalization and degradation through an angiotensin II type I receptor-dependent mechanism. *Hypertension*, *64*(6), 1368–1375. doi:10.1161/HYPERTENSIONAHA.114.03743

Diao, B., Wang, C., & Wang, R. (2020). Human kidney is a target for novel severe acute respiratory syndrome coronavirus 2 (SARS-CoV-2) infection. MedRxiv, 2020.03. 04.20031120, 1010.1101/2020.03.04.20031120

Diebold, M., Schaub, S., Landmann, E., Steiger, J., & Dickenmann, M. (2021). Acute kidney injury in patients with COVID-19: A retrospective cohort study from Switzerland. *Swiss Medical Weekly*, *151*, w20482. doi:10.4414mw.2021.20482

Doher, M. P., De Carvalho, F. R. T., Scherer, P. F., Matsui, T. N., Ammirati, A. L., Da Silva, B. C., Barbeiro, B. G., Carneiro, F. D., Corrêa, T. D., Ferraz, L. J. R., & … . (2020). Acute kidney injury and renal replacement therapy in critically Ill COVID-19 patients: Risk factors and outcomes: A single-center experience in Brazil. *Blood Purification*, *18*, 1–11. doi:10.1159/000513425

Fominskiy, E. V., Scandroglio, A. M., Monti, G., Calabrò, M. G., Landoni, G., Dell'Acqua, A., Beretta, L., Moizo, E., Ravizza, A., Monaco, F., & … . (2021). Prevalence, characteristics, risk factors, and outcomes of invasively ventilated COVID-19 patients with acute kidney injury and renal replacement therapy. *Blood Purification*, *50*, 102–109.

Fu, E. L., Janse, R. J., de Jong, Y., van der Endt, V. H. W., Milders, J., van der Willik, E. M., de Rooij, E. N. M., Dekkers, O. M., Rotmans, J. I., & van Diepen, M. (2020). Acute kidney injury and kidney replacement therapy in COVID-19: A systematic review and meta-analysis. *Clinical Kidney Journal*, *13*(4), 550–563. doi:10.1093/ckjfaa160

Ghosn, M., Attallah, N., Badr, M., Abdallah, K., De Oliveira, B., Nadeem, A., Varghese, Y., Munde, D., Salam, S., Abduljawad, B., Saleh, K., Elkambergy, H., Wahla, A., Taha, A., Dibu, J., Bayrlee, A., Hamed, F., Rahman, N., & Mallat, J. (2021). Severe acute kidney injury in critically ill patients with COVID-19 admitted to ICU: Incidence, risk factors, and outcomes. *Journal of Clinical Medicine*, *10*(6), 1217. doi:10.3390/jcm10061217

Goldsmith, C. S., Miller, S. E., Martines, R. B., Bullock, H. A., & Zaki, S. R. (2020). Electron microscopy of SARS-CoV-2: A challenging task. *Lancet*, *395*, e99.

Guan, W. J. (2019). ni ZY, hu Y, liang Wh, ou CQ, he JX, et al. (2021). Clinical characteristics of coronavirus disease. *Int Rep*, *6*(4), 905–915. doi:10.1016/j.ekir.2021.01.011

Guan, W.J., Ni, Z.Y., Hu, Y., Liang, W.H., Ou, C.Q., He, J.X., Liu, L., Shan, H., Lei, C.L., Hui, D.S.C., Du, B., Li, L.J., Zeng, G., Yuen, K.Y., Chen, R.C., Tang, C.L., Wang, T., Chen, P.Y., Xiang, J., Li, S.Y., Wang, J.L., Liang, Z.J., Peng, Y.X., Wei, L., Liu, Y., Hu, Y.H., Peng, P., Wang, J.M., Liu, J.Y., Chen, Z., Li, G., Zheng, Z.J., Qiu, S.Q., Luo, J., Ye, C.J., Zhu, S.Y., & Zhong, N.S.; China Medical Treatment Expert Group for Covid-19. (2020). Clinical characteristics of coronavirus disease 2019 in China. *The New England Journal of Medicine*, *382*(18), 1708–1720. doi:10.1056/NEJMoa2002032

Gupta, S., Coca, S. G., Chan, L., Melamed, M. L., Brenner, S. K., Hayek, S. S., Sutherland, A., Puri, S., Srivastava, A., Leonberg-Yoo, A., Shehata, A. M., Flythe, J. E., Rashidi, A., Schenck, E. J., Goyal, N., Hedayati, S. S., Dy, R., Bansal, A., Athavale, A., ... Leaf, D. E. (2021). STOP-COVID Investigators. (2021). AKI treated with renal replacement therapy in critically ill patients with COVID-19. *Journal of the American Society of Nephrology*, *32*(1), 161–176. doi:10.1681/ASN.2020060897

Hanley, B., Naresh, K. N., Roufosse, C., Nicholson, A. G., Weir, J., Cooke, G. S., Thursz, M., Manousou, P., Corbett, R., Goldin, R., Al-Sarraj, S., Abdolrasouli, A., Swann, O. C., Baillon, L., Penn, R., Barclay, W. S., Viola, P., & Osborn, M. (2020). Histopathological findings and viral tropism in UK patients with severe fatal COVID-19: A post-mortem study. *The Lancet Microbe*, *1*(6), e245–e253. doi:10.1016/S2666-5247(20)30115-4

Hansrivijit, P., Gadhiya, K. P., Gangireddy, M., & Goldman, J. D. (2021). Risk factors, clinical characteristics, and prognosis of acute kidney injury in hospitalized COVID-19 patients: A retrospective cohort study. *Medicines (Basel, Switzerland)*, *8*(1), 4. doi:10.3390/medicines8010004

Hardenberg, J. B., Stockmann, H., Aigner, A., Gotthardt, I., Enghard, P., Hinze, C., Balzer, F., Schmidt, D., Zickler, D., Kruse, J., Körner, R., Stegemann, M., Schneider, T., Schumann, M., Müller-Redetzky, H., Angermair, S., Budde, K., Weber-Carstens, S., Witzenrath, M., ... Schmidt-Ott, K. M. (2021). Critical illness and systemic inflammation are key risk factors of severe acute kidney injury in patients with COVID-19. *Kidney International Reports*, (4), 905–915. doi:10.1016/j.ekir.2021.01.011

Helal, M. A., Shouman, S., Abdelwaly, A., Elmehrath, A. O., Essawy, M., Sayed, S. M., Saleh, A. H., & El-Badri, N. (2020). Molecular basis of the potential interaction of SARS-CoV-2 spike protein to CD147 in COVID-19 associated-lymphopenia. *Journal of Biomolecular Structure & Dynamics*, *16*, 1–11. doi:10.1080/07391102.2020.1822208

Henderson, L. A., Canna, S. W., Schulert, G. S., Volpi, S., Lee, P. Y., Kernan, K. F., Caricchio, R., Mahmud, S., Hazen, M. M., Halyabar, O., Hoyt, K. J., Han, J., Grom, A. A., Gattorno, M., Ravelli, A., De Benedetti, F., Behrens, E. M., Cron, R. Q., & Nigrovic, P. A. (2020). On the alert for cytokine storm: Immunopathology in COVID-19. *Arthritis & Rheumatology (Hoboken, N.J.)*, *72*(7), 1059–1063. doi:10.1002/art.41285

Hirsch, J. S., Ng, J. H., Ross, D. W., Sharma, P., Shah, H. H., Barnett, R. L., Hazzan, A. D., Fishbane, S., Jhaveri, K. D., Abate, M., & .... (2020). Acute kidney injury in patients hospitalized with COVID-19. *Kidney International*, *98*, 209–218. doi:10.1016/j.kint.2020.05.006

Hoffmann, M., Kleine-Weber, H., Schroeder, S., Krüger, N., Herrler, T., Erichsen, S., Schiergens, T. S., Herrler, G., Wu, N. H., Nitsche, A., Müller, M. A., Drosten, C., & Pöhlmann, S. (2020). SARS-CoV-2 cell entry depends on ACE2 and TMPRSS2 and is blocked by a clinically proven protease inhibitor. *Cell*, *181*(2), 271–280.

Huang, C., Wang, Y., Li, X., Ren, L., Zhao, J., Hu, Y., & .... (2020). Clinical features of patients infected with 2019 novel coronavirus in Wuhan, China. *Lancet*, *395*(10223), 497–506.

Husain-Syed, F., Slutsky, A. S., & Ronco, C. (2016). Lung-kidney cross-talk in the critically ill patient. *American Journal of Respiratory and Critical Care Medicine*, *194*, 402–414. doi:10.1164/rccm.201602-0420CP

Java, A., Apicelli, A. J., Liszewski, M. K., Coler-Reilly, A., Atkinson, J. P., Kim, A. H., & Kulkarni, H. S. (2020). The complement system in COVID-19: Friend and foe? *JCI Insight*, *5*(15), e140711. doi:10.1172/jci.insight.140711

Jhaveri, K. D., Meir, L. R., Flores Chang, B. S., Parikh, R., Wanchoo, R., Barilla-LaBarca, M. L., Bijol, V., & Hajizadeh, N. (2020). Thrombotic microangiopathy in a patient with COVID-19. *Kidney International*, *98*(2), 509–512. doi:10.1016/j.kint.2020.05.025

Jiang, L., Zhu, Y., Luo, X., Wen, Y., Du, B., Wang, M., & .... (2019). Beijing Acute Kidney Injury Trial (BAKIT) workgroup. Epidemiology of acute kidney injury in intensive care units in Beijing: The multi-center BAKIT study. *BMC Nephrology*, *20*, 468.

Joseph, A., Zafrani, L., Mabrouki, A., Azoulay, E., & Darmon, M. (2020). Acute kidney injury in patients with SARS-CoV-2 infection. *Annals of Intensive Care*, *10*, 117. doi:10.118613613-020-00734-z

Kang, S. H., Kim, S. W., Kim, A. Y., Cho, K. H., Park, J. W., & Do, J. Y. (2020). Association between chronic kidney disease or acute kidney injury and clinical outcomes in COVID-19 patients. *Journal of Korean Medical Science*, *35*(50), e434. doi:10.3346/jkms.2020.35.e434

Kdigo, A. (2012). Work Group. KDIGO clinical practice guideline for acute kidney injury. *Kidney International. Supplement*, *2*(1), 1–138.

Kissling, S., Rotman, S., Gerber, C., Halfon, M., Lamoth, F., Comte, D., Lhopitallier, L., Sadallah, S., & Fakhouri, F. (2020). Collapsing glomerulopathy in a COVID-19 patient. *Kidney International*, *98*(1), 228–231. doi:10.1016/j.kint.2020.04.006

Larsen, C. P., Bourne, T. D., Wilson, J. D., Saqqa, O., & Sharshir, M. A. (2020). Collapsing glomerulopathy in a patient with COVID-19. *Kidney International Reports*, *5*(6), 935–939. doi:10.1016/j.ekir.2020.04.002

Laurence, J., Mulvey, J. J., Seshadri, M., Racanelli, A., Harp, J., Schenck, E. J., & .... (2020). Anti-complement C5 therapy with eculizumab in three cases of critical COVID-19. *Clinical Immunology (Orlando, Fla.)*, *219*, 108555.

Li, Q., Guan, X., Wu, P., Wang, X., Zhou, L., Tong, Y., Ren, R., Leung, K. S. M., Lau, E. H. Y., Wong, J. Y., & .... (2020). Early transmission dynamics in Wuhan, China, of novel coronavirus-infected pneumonia. *The New England Journal of Medicine*, *382*(13), 1199–1207. doi:10.1056/NEJMoa2001316

Li, X., Geng, M., Peng, Y., Meng, L., & Lu, S. (2020). Molecular immune pathogenesis and diagnosis of COVID-19. *Journal of Pharmaceutical Analysis*, *10*(2), 102–108. doi:10.1016/j.jpha.2020.03.001

Lim, J. H., Park, S. H., Jeon, Y., Cho, J. H., Jung, H. Y., Choi, J. Y., Kim, C. D., Lee, Y. H., Seo, H., Lee, J., Kwon, K. T., Kim, S. W., Chang, H. H., & Kim, Y. L. (2020). Fatal outcomes of COVID-19 in patients with severe acute kidney injury. *Journal of Clinical Medicine*, *9*(6), 1718. doi:10.3390/jcm9061718

Lowe, R., Ferrari, M., Nasim-Mohi, M., Jackson, A., Beecham, R., Veighey, K., Cusack, R., Richardson, D., Grocott, M., Levett, D., & Dushianthan, A.University Hospital Southampton Critical Care Team and the REACT COVID investigators. (2021). Clinical characteristics and outcome of critically ill COVID-19 patients with acute kidney injury: A single centre cohort study. *BMC Nephrology*, *22*(1), 92. doi:10.118612882-021-02296-z

Makris, K., & Spanou, L. (2016). Acute kidney injury: Definition, pathophysiology and clinical phenotypes. *The Clinical Biochemist. Reviews / Australian Association of Clinical Biochemists*, *37*(2), 85–98.

Mallhi, T. H., Khan, Y. H., & Adnan, A. S. (2020). Stratification of Acute Kidney Injury in COVID-19. *The American Journal of Tropical Medicine and Hygiene*, *103*(6), 2164–2167. Retrieved December 15, 2021, from https://www.ajtmh.org/view/journals/tpmd/103/6/article-p2164.xml

Meng, X. M., Nikolic-Paterson, D. J., & Lan, H. Y. (2014). Inflammatory processes in renal fibrosis. *Nature Reviews. Nephrology*, *10*, 493–503. doi:10.1038/nrneph.2014.114

Menon, T., Sharma, R., Kataria, S., Sardar, S., Adhikari, R., Tousif, S., Khan, H., Rathore, S. S., Singh, R., & Ahmed, Z. (2021). The association of acute kidney injury with disease severity and mortality in COVID-19: A systematic review and meta-analysis. *Cureus, 13*(3), e13894. doi:10.7759/cureus.13894

Mizuiri, S., & Ohashi, Y. (2015). ACE and ACE2 in kidney disease. *World Journal of Nephrology*, *4*(1), 74–82. doi:10.5527/wjn.v4.i1.74

Morgera, S., Rocktäschel, J., Haase, M., & Lehmann, C., von, Heymann, C., Ziemer, S., et al. (2003). Intermittent high permeability hemofiltration in septic patients with acute renal failure. *Intensive Care Medicine*, *29*(11), 1989–1995.

Nasr, S.H., & Kopp, J.B. (2020). COVID-19-associated collapsing glomerulopathy: An emerging entity. *Kidney International Reports*, 5(6), 759-761. doi: ir.2020 .04.030 doi:10.1016/j.ek

Ng, J. J., Luo, Y., Phua, K., & Choong, A. (2020). Acute kidney injury in hospitalized patients with coronavirus disease 2019 (COVID-19): A meta-analysis. *The Journal of Infection*, *81*(4), 647–679. doi:10.1016/j.jinf.2020.05.009

Nicholas Cossey, L., Larsen, C. P., & Liapis, H. (2017). Collapsing glomerulopathy: A 30-year perspective and single, large center experience. *Clinical Kidney Journal*, *10*(4), 443–449. doi:10.1093/ckjfx029

Nugent, J., Aklilu, A., Yamamoto, Y., & … . (2021). Assessment of acute kidney injury and longitudinal kidney function after hospital discharge among patients with and without COVID-19. *JAMA Network Open*, *4*(3), e211095. doi:10.1001/ jamanetworkopen.2021.1095

Osibogun, A., Balogun, M., Abayomi, A., Idris, J., Kuyinu, Y., Odukoya, O., Wright, O., Adeseun, R., Mitiu, B., & Saka, B. (2021). Outcomes of COVID-19 patients with comorbidities in southwest Nigeria. *PLoS One*, *16*(3), e0248281. doi:10.1371/ journal. pone.0248281

Paek, J. H., Kim, Y., Park, W. Y., Jin, K., Hyun, M., Lee, J. Y., Kim, H. A., Kwon, Y. S., Park, J. S., & Han, S. (2020). Severe acute kidney injury in COVID-19 patients is associated with in-hospital mortality. *PLoS One*, *15*(12), e0243528.

Panitchote, A., Mehkri, O., Hastings, A., Hanane, T., Demirjian, S., Torbic, H., Mireles-Cabodevila, E., Krishnan, S., & Duggal, A. (2019). Factors associated with acute kidney injury in acute respiratory distress syndrome. *Annals of Intensive Care*, *9*(1), 74. doi:10.118613613-019-0552-5

Peleg, Y., Kudose, S., D'Agati, V., Siddall, E., Ahmad, S., Nickolas, T., Kisselev, S., Gharavi, A., & Canetta, P. (2020). Acute kidney injury due to collapsing glomerulopathy following COVID-19 infection. *Kidney International Reports*, *5*(6), 940–945. doi:10.1016/j.ekir.2020.04.017

Peng, S., Wang, H. Y., Sun, X., Li, P., Ye, Z., Li, Q., Wang, J., Shi, X., Liu, L., Yao, Y., Zeng, R., He, F., Li, J., Ge, S., Ke, X., Zhou, Z., Dong, E., Wang, H., Xu, G., ... Zhao, M. H. (2020). Early versus late acute kidney injury among patients with COVID-19-a multicenter study from Wuhan, China. *Nephrology, Dialysis, Transplantation*, *35*(12), 2095–2102. doi:10.1093/ndt/gfaa288

Pober, J. S., & Sessa, W. C. (2015). Inflammation and the blood microvascular system. *Cold Spring Harbor Perspectives in Biology*, *7*(1), a016345.

Portolés, J., Marques, M., Lopez-Sanchez, P., Valdenebro, M., Muñez, E., & Serrano, M. (2020). Chronic kidney disease and acute kidney injury in the COVID-19 Spanish Outbreak. *Nephrology, Dialysis, Transplantation*, *35*, 1353–1361.

Prompetchara, E., Ketloy, C., & Palaga, T. (2020). Immune responses in COVID-19 and potential vaccines: Lessons learned from SARS and MERS epidemic. *Asian Pacific Journal of Allergy and Immunology*, *38*, 1–9. doi:10.12932/ap-200220-0772

Puelles, V. G., Lütgehetmann, M., Lindenmeyer, M. T., Sperhake, J. P., Wong, M. N., Allweiss, L., Chilla, S., Heinemann, A., Wanner, N., Liu, S., & ... . (2020). Multiorgan and renal tropism of SARS-CoV-2. *The New England Journal of Medicine*, *383*(6), 590–592. doi:10.1056/NEJMc2011400

Rabi, F. A., Al Zoubi, M. S., Kasasbeh, G. A., Salameh, D. M., & Al-Nasser, A. D. (2020). SARS-CoV-2 and coronavirus disease 2019: What we know so far. *Pathogens (Basel, Switzerland)*, *9*(3), 231. doi:10.3390/pathogens9030231

Ronco, C., Reis, T., & Husain-Syed, F. (2020). Management of acute kidney injury in patients with COVID-19. *The Lancet. Respiratory Medicine*, *8*(7), 738–742. doi:10.1016/S2213-2600(20)30229-0

Rossi, G. M., Delsante, M., Pilato, F. P., Gnetti, L., Gabrielli, L., Rossini, G., Re, M. C., Cenacchi, G., Affanni, P., Colucci, M. E., Picetti, E., Rossi, S., Parenti, E., Maccari, C., Greco, P., Di Mario, F., Maggiore, U., Regolisti, G., & Fiaccadori, E. (2020). Kidney Biopsy findings in a critically ill COVID-19 patient with dialysis-dependent acute kidney injury: A case against "SARS-CoV-2 nephropathy". *Kidney International Reports*, *5*(7), 1100–1105. doi:10.1016/j.ekir.2020.05.005

Santoriello, D., Khairallah, P., Bomback, A. S., Xu, K., Kudose, S., Batal, I., Barasch, J., Radhakrishnan, J., D'Agati, V., & Markowitz, G. (2020). Postmortem kidney pathology findings in patients with COVID-19. *Journal of the American Society of Nephrology*, *31*, 2158–2167.

Sanyaolu, A., Okorie, C., Marinkovic, A., Patidar, R., Younis, K., Desai, P., Hosein, Z., Padda, I., Mangat, J., & Altaf, M. (2020). Comorbidity and its impact on patients with COVID-19. *SN Comprehensive Clinical Medicine*, *25*, 1–8. doi:10.100742399-020-00363-4

Schaller, T., Hirschbühl, K., Burkhardt, K., Braun, G., Trepel, M., Märkl, B., & Claus, R. (2020). Postmortem examination of patients with COVID-19. *Journal of the American Medical Association*, *323*(24), 2518–2520. doi:10.1001/jama.2020.8907

Shagdarsuren, E., Wellner, M., Braesen, J. H., Park, J. K., Fiebeler, A., Henke, N., & … . (2005). Complement activation in angiotensin II-induced organ damage. *Circulation Research*, *97*(7), 716–724.

Su, H., Yang, M., Wan, C., Yi, L. X., Tang, F., Zhu, H. Y., Yi, F., Yang, H. C., Fogo, A. B., Nie, X., & Zhang, C. (2020). Renal histopathological analysis of 26 postmortem findings of patients with COVID-19 in China. *Kidney International*, *98*(1), 219–227. doi:10. 1016 / j.kint.2020.04.003

Terpos, E., Ntanasis-Stathopoulos, I., Elalamy, I., Kastritis, E., Sergentanis, T. N., Politou, M., & … . (2020). Hematological findings and complications of COVID-19. *American Journal of Hematology*, *95*(7), 834–847.

Tocoglu, A., Dheir, H., Bektas, M., Acikgoz, S. B., Karabay, O., & Sipahi, S. (2021). Predictors of mortality in patients with COVID-19 infection-associated acute kidney injury. *Journal of the College of Physicians and Surgeons—Pakistan*, *31*(1), S60–S65. doi:10.29271/jcpsp.2021.Supp1.S60

Vaduganathan, M., Vardeny, O., Michel, T., McMurray, J., Pfeffer, M., & Solomon, S. (2020). Renin-angiotensin aldosterone system inhibitors in patients with COVID-19. *The New England Journal of Medicine*, *382*, 1653–1659. doi:10.1056/NEJMsr2005760

Wang, B., Luo, Q., Zhang, W., Yu, S., Cheng, X., Wang, L., Chen, X., & Chen, Y. (2021). The involvement of chronic kidney disease and acute kidney injury in disease severity and mortality in patients with COVID-19: A meta-analysis. *Kidney & Blood Pressure Research*, *46*(1), 17–30. doi:10.1159/000512211

Wang, D., Hu, B., Hu, C., Zhu, F., Liu, X., Zhang, J., Wang, B., Xiang, H., Cheng, Z., Xiong, Y., Zhao, Y., Li, Y., Wang, X., & Peng, Z. (2020). Clinical characteristics of 138 hospitalized patients with 2019 novel coronavirus-infected pneumonia in Wuhan, China. *Journal of the American Medical Association*, *323*(11), 1061–1069. doi:10.1001/jama.2020.1585

Wang, L., Li, X., Chen, H., Yan, S., Li, D., Li, Y., & Gong, Z. (2020). Coronavirus disease 19 infection does not result in acute kidney injury: An analysis of 116 hospitalized patients from Wuhan, China. *American Journal of Nephrology*, *51*(5), 343–348.

Wiersema, R., Jukarainen Eck, R. J., Kaufmann, T., Koeze, J., Keus, F., Pettilä, V., Van Der Horst, I. C. C., & Vaara, S. T. (2020). Different applications of the KDIGO criteria for AKI lead to different incidences in critically ill patients: A post hoc analysis from the prospective observational SICS-II study. *Critical Care (London, England)*, *24*, 1–8.

World Health Organization. (2020). Coronavirus disease (COVID-19) outbreak. Available at: https://www.who.int

World Health Organization. (2021). Coronavirus disease. Available at: https://covid19.who.int/

Xiao, G., Hu, H., Wu, F., Sha, T., Zeng, Z., Huang, Q., & Cai, S. (2021). Acute kidney injury in patients hospitalized with COVID-19 in Wuhan, China: A single-center retrospective observational study. *Journal of Southern Medical University*, *41*(2), 157.

Xu, H., Garcia-Ptacek, S., Annetorp, M., Bruchfeld, A., Cederholm, T., Johnson, P., Kivipelto, M., Metzner, C., Religa, D., & Eriksdotter, M. (2021). Acute kidney injury and mortality risk in older adults with COVID-19. *Journal of Nephrology*, *34*(2), 295–304. doi:10.100740620-021-01022-0

Yan, Q., Zuo, P., Cheng, L., Li, Y., Song, K., Chen, Y., Dai, Y., Yang, Y., Zhou, L., Yu, W., Li, Y., Xie, M., Zhang, C., & Gao, H. (2021). Acute kidney injury is associated with in-hospital mortality in older patients with COVID-19. *The Journals of Gerontology: Series A*, *76*(3), 456–462. doi:10.1093/gerona/glaa181

Yan, X., Han, X., Peng, D., Fan, Y., Fang, Z., Long, D., Xie, Y., Zhu, S., Chen, F., Lin, W., & Zhu, Y. (2020). Clinical characteristics and prognosis of 218 patients with COVID-19: A retrospective study based on clinical classification. *Frontiers in Medicine*, *7*, 485. doi:10.3389/fmed.2020.00485

Yang, X., Jin, Y., Li, R., Zhang, Z., Sun, R., & Chen, D. (2020). Prevalence and impact of acute renal impairment on COVID-19: A systematic review and meta-analysis. *Critical Care (London, England)*, *24*, 1–8.

Yang, X., Yu, Y., Xu, J., Shu, H., Xia, J., Liu, H., Wu, Y., Zhang, L., Yu, Z., Fang, M., Yu, T., Wang, Y., Pan, S., Zou, X., Yuan, S., & Shang, Y. (2020). Clinical course and outcomes of critically ill patients with SARS-CoV-2 pneumonia in Wuhan, China: A single-centered, retrospective, observational study. *The Lancet. Respiratory Medicine*, *8*(5), 475–481. doi:10.1016/S2213-2600(20)30079-5

Yin, W., & Zhang P.L. (2020). Infectious pathways of SARS-CoV-2 in renal tissue. *Journal of Nephropathology*, 9, 4, e37. 10.34172/jnp.2020.37

Yu, Y., Xu, D., Fu, S., Zhang, J., Yang, X., Xu, L., & ... . (2020). Patients with COVID-19 in 19 ICUs in Wuhan, China: A cross-sectional study. *Critical Care (London, England)*, *24*, 1–10. doi:10.118613054-019-2683-3

Zahid, U., Ramachandran, P., Spitalewitz, S., Alasadi, L., Chakraborti, A., Azhar, M., Mikhalina, G., Sherazi, A., & Narh, J.T., Khattar, & P., Bedi, P. (2020). Acute kidney injury in COVID-19 patients: An inner city hospital experience and policy implications. *American Journal of Nephrology*, *51*(10), 786–796. doi:10.1159/000511160

Zhang, H., & Baker, A. (2017). Recombinant human ACE2: Acing out angiotensin II in ARDS therapy. *Critical Care*, *21*, 305. doi:10.118613054-017-1882-z

Zhang, H., Penninger, J. M., Li, Y., Zhong, N., & Slutsky, A. S. (2020). Angiotensin-converting enzyme 2 (ACE2) as a SARS-CoV-2 receptor: Molecular mechanisms and potential therapeutic target. *Intensive Care Medicine*, *46*(4), 586–590. doi:10.100700134-020-05985-9

Zheng, Z., Peng, F., Xu, B., Zhao, J., Liu, H., Peng, J., Li, Q., Jiang, C., Zhou, Y., Liu, S., Ye, C., Zhang, P., Xing, Y., Guo, H., & Tang, W. (2020). Risk factors of critical & mortal COVID-19 cases: A systematic literature review and meta-analysis. *The Journal of Infection*, *81*(2), e16–e25. doi:10.1016/j.jinf.2020.04.021

Zhou, F., Yu, T., Du, R., Fan, G., Liu, Y., Liu, Z., Xiang, J., Wang, Y., Song, B., Gu, X., Guan, L., Wei, Y., Li, H., Wu, X., Xu, J., Tu, S., Zhang, Y., Chen, H., & Cao, B. (2020). Clinical course and risk factors for mortality of adult inpatients with COVID-19 in Wuhan, China: A retrospective cohort study. *Lancet*, *395*(10229), 1054–1062. doi:10.1016/S0140-6736(20)30566-3

Compilation of References

Abagli, A. Z., Alavo, T. B. C., Perez-Pacheco, R., & Platzer, E. G. (2019). Efficacy of the mermithid nematode, *Romanomermis iyengari*, for the biocontrol of *Anopheles gambiae*, the major malaria vector in Sub-Saharan Africa. *Parasites & Vectors*, *12*(1), 253. doi:10.118613071-019-3508-6 PMID:31118105

Abrahamse, H., & Hamblin, M. (2016). New photosensitisers for photodynamic therapy. *The Biochemical Journal*, *473*(4), 347–364. doi:10.1042/BJ20150942 PMID:26862179

Achee, N. L., Grieco, J. P., Vatandoost, H., Seixas, G., Pinto, J., Ching-Ng, L., Martins, A. J., Juntarajumnong, W., Corbel, V., Gouagna, C., David, J. P., Logan, J. G., Orsborne, J., Marois, E., Devine, G. J., & Vontas, J. (2019). Alternative strategies for mosquito-borne arbovirus control. *PLoS Neglected Tropical Diseases*, *13*(1), e0006822. doi:10.1371/journal.pntd.0006822 PMID:30605475

Acosta, A. L., Pinter, A., Faria, N., Sabino, E., & Mureb Sallum, M. A. (2021). Landscape connectivity for yellow fever: A proxy approach to detect displacement and circulation through environmental corridors and interfaces. *The Lancet. Planetary Health*, *5*, S22. doi:10.1016/S2542-5196(21)00106-6

Ade, A. D., Nagaraj, K., & Vallepalii, C. (2020). School environment and sanitation: A comparative study. *National Journal of Community Medicine*, *11*(4), 6–12.

Adsul, B. B., Laad, P. S., Howal, P. V., & Chaturvedi, R. M. (2011). Health problems among migrant construction workers: A unique public-private partnership project. *Indian Journal of Occupational and Environmental Medicine*, *15*(1), 29. doi:10.4103/0019-5278.83001 PMID:21808498

Afify, A., & Potter, C. J. (2020). Insect repellents mediate species-specific olfactory behaviours in mosquitoes. *Malaria Journal*, *19*(1), 127. doi:10.118612936-020-03206-8 PMID:32228701

Afroz, S., Debsarma, S., Dutta, S., Rhaman, M. M., & Mohsena, M. (2019). Prevalence of helminthic infestations among Bangladeshi rural children and its trend since mid-seventies. *IMC J Med Sci*, *13*(1), 1–8. doi:10.3329/imcjms.v13i1.42038

Ahuja, S. C., Ahuja, S., & Ahuja, U. (2015). Nirgundi (*Vitex negundo*) – Nature's Gift to Mankind. *Asian Agri-History*, *19*(1), 5–32.

Aitbali, Y., Ba-M'hamed, S., Elhidar, N., Nafis, A., Soraa, N., & Bennis, M. (2018). Glyphosate based-herbicide exposure affects gut microbiota, anxiety and depression-like behaviors in mice. *Neurotoxicology and Teratology, 67*, 44–49. doi:10.1016/j.ntt.2018.04.002 PMID:29635013

Akther, N., Begum, M., Islam, F., Khan, S. S., Doha, M., & Chowdhury, S., S. (2021). Practices of hygiene among primary school children in a selected semi urban area of Bangladesh. *Journal of Z H Sikder Women's Medical College, 3*(2), 16-20. doi:10.47648/zhswmcj.2021.v0302.04

Albahar, M.A. (2019). Skin Lesion Classification Using Convolutional Neural Network With Novel Regularizer. *IEEE Access, 7*, 38306–38313.

Alberti, M. (2005). The Effects of Urban Patterns on Ecosystem Function. *International Regional Science Review, 28*(2), 168–192. doi:10.1177/0160017605275160

Alemayehu, B., & Alemayehu, M. (2017). Leishmaniasis: A Review on Parasite, Vector and Reservoir Host. *Health Science Journal, 11*(4), 519. doi:10.21767/1791-809X.1000519

Alkhayat, F. A., Ahmad, A. H., Rahim, J., Dieng, H., Ismail, B. A., Imran, M., Sheikh, U. A. A., Shahzad, M. S., Abid, A. D., & Munawar, K. (2020). Characterization of mosquito larval habitats in Qatar. *Saudi Journal of Biological Sciences, 27*(9), 2358–2365. doi:10.1016/j.sjbs.2020.07.006 PMID:32884417

Allison, A. C. (1965). Population genetics of abnormal haemoglobins and glucose-6-phosphate dehydrogenase deficiency. International Journal for HP Jonxis.

Amini, M., Hanafi-Bojd, A. A., Aghapour, A. A., & Chavshin, A. R. (2020). Larval habitats and species diversity of mosquitoes (Diptera: Culicidae) in West Azerbaijan Province, Northwestern Iran. *BMC Ecology, 20*(1), 60. doi:10.118612898-020-00328-0 PMID:33213441

Aminov, R. (2010). A Brief History of the Antibiotic Era: Lessons Learned and Challenges for the Future. *Frontiers in Microbiology, 1*. Advance online publication. doi:10.3389/fmicb.2010.00134 PMID:21687759

Anderson, S. E., & Meade, B. J. (2014). Potential health effects associated with dermal exposure to occupational chemicals. *Environmental Health Insights, 8*.

Andrews, J. M., & Tan, M. (2012). Probiotics in luminal gastroenterology: The current state of play. *Internal Medicine Journal, 42*(12), 1287–1291. doi:10.1111/imj.12015 PMID:23252997

Anthonj, C., Nkongolo, O. T., Schmitz, P., Hango, J. N., & Kistemann, T. (2015). The impact of flooding on people living with HIV: A case study from the Ohangwena Region, Namibia. *Global Health Action, 8*(1), 26441. Advance online publication. doi:10.3402/gha.v8.26441 PMID:25813771

Antonio-Nkondjio, C., Sandjo, N. N., Awono-Ambene, P., & Wondji, C. S. (2018). Implementing a larviciding efficacy or effectiveness control intervention against malaria vectors: Key parameters for success. *Parasites & Vectors, 11*(1), 57. doi:10.118613071-018-2627-9 PMID:29368633

Anyamba, A., Chretien, J.-P., Britch, S. C., Soebiyanto, R. P., Small, J. L., Jepsen, R., Forshey, B. M., Sanchez, J. L., Smith, R. D., Harris, R., Tucker, C. J., Karesh, W. B., & Linthicum, K. J. (2019). Global Disease Outbreaks Associated with the 2015–2016 El Niño Event. *Scientific Reports, 9*(1), 1–14. doi:10.1038/s41598-018-38034-z

APA. (2020). *New APA Poll Reveals That Americans are Increasingly Anxious About Climate Change's Impact on Planet, Mental Health.* https://www.psychiatry.org/newsroom/news-releases/climate-poll-2020

Aprihatin, Y., Barlian, S., Fatimah, S., Yanti, E., & Armaita, A. (2020). Impact of environmental sanitation and infection disease as a determining stunting factor for children. *Sumatra Journal of Disaster. Geography and Geography Education*, *4*(2), 209–211. doi:10.24036jdgge.v4i2.336

Araújo, H. R., Carvalho, D. O., Ioshino, R. S., Costa-da-Silva, A. L., & Capurro, M. L. (2015). *Aedes aegypti* control strategies in Brazil: Incorporation of new technologies to overcome the persistence of dengue epidemics. *Insects*, *6*(2), 576–594. doi:10.3390/insects6020576 PMID:26463204

Aresti Sanz, J., & El Aidy, S. (2019). Microbiota and gut neuropeptides: A dual action of antimicrobial activity and neuroimmune response. *Psychopharmacology*, *236*(5), 1597–1609. doi:10.100700213-019-05224-0 PMID:30997526

Arman, N. Z., Salmiati, S., Aris, A., Salim, M. R., Nazifa, T. H., Muhamad, M. S., & Marpongahtun, M. (2021). A review on emerging pollutants in the water environment: Existences, health effects and treatment processes. *Water (Switzerland)*, *13*(22), 1–31. doi:10.3390/w13223258

Arnbjerg, C. J., Vestad, B., Hov, J. R., Pedersen, K. K., Jespersen, S., Johannesen, H. H., Holm, K., Halvorsen, B., Fallentin, E., Hansen, A. E., Lange, T., Kjær, A., Trøseid, M., Fischer, B. M., & Nielsen, S. D. (2018). Effect of Lactobacillus rhamnosus GG supplementation on intestinal inflammation assessed by PET/MRI scans and gut microbiota composition in HIV-infected individuals. *JAIDS Journal of Acquired Immune Deficiency Syndromes*, *78*(4), 450–457. doi:10.1097/QAI.0000000000001693 PMID:29874201

Artificial intelligence yields new antibiotics. (2022). MIT News, Massachusetts Institute of Technology. Retrieved on 3 June 2022, from https://news.mit.edu/2020/artificial-intelligence-identifies-new-antibiotic-0220

Asad, H., & Carpenter, D. O. (2018). Effects of climate change on the spread of zika virus: A public health threat. *Reviews on Environmental Health*, *33*(1), 31–42. doi:10.1515/reveh-2017-0042 PMID:29500926

Asadullah, M. N., & Chaudhury, N. (2011). Poisoning the mind: Arsenic contamination of drinking water wells and children's educational achievement in rural Bangladesh. *Economics of Education Review*, *30*(5), 873–888. doi:10.1016/j.econedurev.2011.05.001

Ashford, R. W. (2000). The leishmaniases as emerging and reemerging zoonoses. *International Journal for Parasitology*, *30*(12-13), 1269–1281. doi:10.1016/S0020-7519(00)00136-3 PMID:11113254

Attia, M., Hossny, M., Nahavandi, S., & Yazdabadi, A. (2017). Skin Melanoma Segmentation Using Recurrent and Convolutional Neural Networks. *Proceedings of the 2017 IEEE 14th International Symposium on Biomedical Imaging (ISBI 2017)*, 292–296. 10.1109/ISBI.2017.7950522

Aturupane, H., Glewwe, P., & Wisniewski, S. (2013). The impact of school quality, socioeconomic factors, and child health on students' academic performance: Evidence from Sri Lankan primary schools. *Education Economics*, *21*(1), 2–37. doi:10.1080/09645292.2010.511852

Audrain, B., Farag, M. A., Ryu, C. M., & Ghigo, J. M. (2015). Role of bacterial volatile compounds in bacterial biology. *FEMS Microbiology Reviews*, *39*(2), 222–233. doi:10.1093/femsre/fuu013 PMID:25725014

Augustina, I., Jabal, A. R., Permana, G. I., & Ratnasari, A. (2021). Distribution and ecology of mosquito larvae in Pahandut subdistrict, Palangkaraya city. *Journal of Physics: Conference Series*, *1918*(5), 052018. doi:10.1088/1742-6596/1918/5/052018

Ayinde, A. A., Morakinyo, O. M., & Sridhar, M. K. C. (2020). Repellency and larvicidal activities of *Azadirachta indica* seed oil on *Anopheles gambiae* in Nigeria. *Heliyon*, *6*(5), e03920. doi:10.1016/j.heliyon.2020.e03920 PMID:32420486

Bai, L., Woodward, A., & Liu, Q. (2014). Temperature and mortality on the roof of the world: A time-series analysis in three Tibetan counties. *The Science of the Total Environment*, *485*, 41–48. doi:10.1016/j.scitotenv.2014.02.094 PMID:24704955

Bakken, J. S. (2009). Fecal bacteriotherapy for recurrent Clostridium difficile infection. *Anaerobe*, *15*(6), 285–289. doi:10.1016/j.anaerobe.2009.09.007 PMID:19778623

Balakrishnan, M., & Floch, M. H. (2012). Prebiotics, probiotics and digestive health. *Current Opinion in Clinical Nutrition and Metabolic Care*, *15*(6), 580–585. doi:10.1097/MCO.0b013e328359684f PMID:23037903

Bangladesh Bureau of Statistics. (2018). *Bangladesh demographic profile*. Bangladesh Bureau of Statistics.

Banu, S., Rahman, M. T., Uddin, M. K., Khatun, R., Ahmed, T., Rahman, M. M., Husain, M. A., & Leth, F. V. (2013). Epidemiology of Tuberculosis in an urban slum of Dhaka City, Bangladesh. *PLoS One*, *8*(10), 1–8. doi:10.1371/journal.pone.0077721 PMID:24204933

Barlow, G. M., Yu, A., & Mathur, R. (2015). Role of the gut microbiome in obesity and diabetes mellitus. *Nutrition in Clinical Practice*, *30*(6), 787–797. doi:10.1177/0884533615609896 PMID:26452391

Barnes, D. K. A., Galgani, F., Thompson, R. C., & Barlaz, M. (2009). Accumulation and fragmentation of plastic debris in global environments. *Philosophical Transactions of the Royal Society of London. Series B, Biological Sciences*, *364*(1526), 1985–1998. doi:10.1098/rstb.2008.0205 PMID:19528051

Baronia, A., & Ahmed, A. (2014). Current concepts in combination antibiotic therapy for critically ill patients. *Indian Journal of Critical Care Medicine : Peer-Reviewed, Official Publication of Indian Society of Critical Care Medicine, 18*(5), 310–314. doi:10.4103/0972-5229.132495 PMID:24914260

Barreto, M. L., Genser, B., Strina, A., Teixeira, M. G., Assis, A. M., Rego, R. F., Teles, C. A., Prado, M. S., Matos, S. M., Santos, D. N., dos Santos, L. A., & Cairncross, S. (2007). Effect of city-wide sanitation program on reduction in the rate of childhood diarrhea in northeast Brazil: Assessment by two cohort studies. *Lancet, 370*(9599), 1622–1628. doi:10.1016/S0140-6736(07)61638-9 PMID:17993362

Bassat, Q., & Alonso, P. L. (2011). Defying malaria: Fathoming severe *Plasmodium vivax* disease. *Nature Medicine, 17*(1), 48–49. doi:10.1038/nm0111-48 PMID:21217683

Batterman, S., Eisenberg, J., Hardin, R., Kruk, M. E., Lemos, M. C., Michalak, A. M., Mukherjee, B., Renne, E., Stein, H., Watkins, C., & Wilson, M. L. (2009). Sustainable control of water-related infectious diseases: A review and proposal for interdisciplinary health-based systems research. *Environmental Health Perspectives, 117*(7), 1023–1032. doi:10.1289/ehp.0800423 PMID:19654908

Becker, D. J., Streicker, D. G., & Altizer, S. (2015). Linking anthropogenic resources to wildlife-pathogen dynamics: Are view and meta-analysis. *Ecology Letters, 18*(5), 483–495. doi:10.1111/ele.12428 PMID:25808224

Beet, E. A. (1946). Sickle cell disease in the Balovale district of Northern Rhodesia. *East African Medical Journal, 23*, 75–86. PMID:21027890

Begon, M., Telfer, S., Smith, M. J., Burthe, S., Paterson, S., & Lambin, X. (2009). Seasonal host dynamics drive the timing of recurrent epidemics in a wildlife population. *Proceedings of the Royal Society B, 276*(1662), 1603-1610. 10.1098/rspb.2008.1732

Begon, M., Bennett, M., Bowers, R. G., French, N. P., Hazel, S. M., & Turner, J. (2002). A clarification of transmission terms in host-microparasite models: Numbers, densities, and areas. *Epidemiology and Infection, 129*(1), 147–153. doi:10.1017/S0950268802007148 PMID:12211582

Beker, B. M., Cervellera, C., De Vito, A., & Musso, C. G. (2018). Human physiology in extreme heat and cold. *International Archieves of Clinical Physiology, 1*(1). Advance online publication. doi:10.23937/iacph-2017/1710001

Bennett, L. M., & Gadlin, H. (2012). Collaboration and team science: From theory to practice. *Journal of Investigative Medicine, 60*(5), 768–775. doi:10.2310/JIM.0b013e318250871d PMID:22525233

Berglund, N., Piggot, T., Jefferies, D., Sessions, R., Bond, P., & Khalid, S. (2015). Interaction of the Antimicrobial Peptide Polymyxin B1 with Both Membranes of E. coli: A Molecular Dynamics Study. *PLoS Computational Biology, 11*(4), e1004180. doi:10.1371/journal.pcbi.1004180 PMID:25885324

Berhanu, A., Abera, A., Nega, D., Mekasha, S., Fentaw, S., Assefa, A., Gebrewolde, G., Wuletaw, Y., Assefa, A., Dugassa, S., Tekie, H., & Tasew, G. (2019). Isolation and identification of microflora from the midgut and salivary glands of *Anopheles* species in malaria endemic areas of Ethiopia. *BMC Microbiology, 19*(1), 85. doi:10.118612866-019-1456-0 PMID:31035931

Bernier, U. R., Kline, D. L., Barnard, D. R., Schreck, C. E., & Yost, R. A. (2000). Analysis of human skin emanations by gas chromatography/mass spectrometry. 2. Identification of volatile compounds that are candidate attractants for the yellow fever mosquito (*Aedes aegypti*). *Analytical Chemistry, 72*(4), 747–756. doi:10.1021/ac990963k PMID:10701259

Bethony, J., Brooker, S., Albonico, M., Geiger, S. M., Loukas, A., Diemert, D., & Hotez, P. J. (2006). Soil-transmitted helminth infections: Ascariasis, trichuriasis, and hookworm. *Lancet, 367*(9521), 1521–1532. doi:10.1016/S0140-6736(06)68653-4 PMID:16679166

Bhatt, S., Gething, P. W., Brady, O. J., Messina, J. P., Farlow, A. W., Moyes, C. L., Drake, J. M., Brownstein, J. S., Hoen, A. G., Sankoh, O., Myers, M. F., George, D. B., Jaenisch, T., Wint, G. R., Simmons, C. P., Scott, T. W., Farrar, J. J., & Hay, S. I. (2013). The global distribution and burden of Dengue. *Nature, 496*(7446), 504–507. doi:10.1038/nature12060 PMID:23563266

Biran, A., Schmidt, W.-P., Varadharajan, K. S., Rajaraman, D., Kumar, R., Greenland, K., Gopalan, B., Aunger, R., & Curtis, V. (2014). Effect of a behavior-change intervention on handwashing with soap in India (SuperAmma): A cluster-randomized trial. *The Lancet. Global Health, 2*(3), e145–e154. doi:10.1016/S2214-109X(13)70160-8 PMID:25102847

Bist, P., & Choudhary, S. (2022). Impact of Heavy Metal Toxicity on the Gut Microbiota and Its Relationship with Metabolites and Future Probiotics Strategy: A Review. *Biological Trace Element Research, 200*(12), 5328–5350. doi:10.100712011-021-03092-4 PMID:34994948

Biswal, D. (2017). Mosquito bio-control by indigestible green algae: Securing humans in a non-traditional way. In M. Gupta (Ed.), Non-traditional security: Problems and Prospects (pp. 184-193). Avenel Press.

Biswal, D. (2018). Impact of climate change on parasitic diseases: Do we need to bother? In M. Bose, S. A. Bandyopadhyay, & S. Poddar (Eds.), *Contemporary Health Issues and Environmental Impact* (pp. 32–40). Lincoln University College. doi:10.31674/books.2018.chiei.ch05

Biswas, D., Ahmed, M., Roguski, K., Ghosh, P. K., Parveen, S., Nizame, F. A., Rahman, M. Z., Chowdhury, F., Rahman, M., Luby, S. P., Sturm-Ramirez, K., & Iuliano, A. D. (2019). Effectiveness of a behavior change intervention with hand sanitizer use and respiratory hygiene in reducing laboratory-confirmed influenza among school children in Bangladesh: A cluster randomized controlled trial. *The American Journal of Tropical Medicine and Hygiene, 00*(0), 1–10. doi:10.4269/ajtmh.19-0376 PMID:31701861

Bogovic, P., & Strle, F. (2015). Tick-borne encephalitis: A review of epidemiology, clinical characteristics, and management. *World Journal of Clinical Cases, 3*(5), 430. doi:10.12998/wjcc.v3.i5.430 PMID:25984517

Bohbot, J. D., Fu, L., Le, T. C., Chauhan, K. R., Cantrell, C. L., & Dickens, J. C. (2011). Multiple activities of insect repellents on odorant receptors in mosquitoes. *Medical and Veterinary Entomology*, *25*(4), 436–444. doi:10.1111/j.1365-2915.2011.00949.x PMID:21395633

Borel, B. (2022). *iChip: The Future of Antibiotic Discovery*. Popular Science. Retrieved on 3 June 2022, from https://www.popsci.com/ichip-new-way-find-antibiotics-and-other-key-drugs/

Borody, T. J., & Khoruts, A. (2012). Fecal microbiota transplantation and emerging applications. *Nature Reviews. Gastroenterology & Hepatology*, *9*(2), 88–96. doi:10.1038/nrgastro.2011.244 PMID:22183182

Bouzid, D., Zanella, M., Kerneis, S., Visseaux, B., May, L., Schrenzel, J., & Cattoir, V. (2021). Rapid diagnostic tests for infectious diseases in the emergency department. *Clinical Microbiology and Infection*, *27*(2), 182–191. doi:10.1016/j.cmi.2020.02.024 PMID:32120036

Boyce, M. R., Osman, R., & Katz, R. (2019). A research agenda for urban pandemic preparedness: a call to action. Washington: Center for Global Health Science & Security, Georgetown University.

Bradley, C. A., & Altizer, S. (2007). Urbanization and the ecology of wildlife diseases. *Trends in Ecology & Evolution*, *22*(2), 95–102. doi:10.1016/j.tree.2006.11.001 PMID:17113678

Breidenbaugh, M. S., & de Szalay, F. A. (2010). Effects of aerial applications of naled on nontarget insects at Parris Island, South Carolina. *Environmental Entomology*, *39*(2), 591–599. doi:10.1603/EN09087 PMID:20388292

Breton, J., Massart, S., Vandamme, P., De Brandt, E., Pot, B., & Foligné, B. (2013). Ecotoxicology inside the gut: Impact of heavy metals on the mouse microbiome. *BMC Pharmacology & Toxicology*, *14*(1), 1–11. doi:10.1186/2050-6511-14-62 PMID:24325943

Bronfenbrenner, U. (1995). *Developmental ecology through space and time: A future perspective*. Academic Press.

Bronfenbrenner, U. (1974). Developmental research, public policy, and the ecology of childhood. *Child Development*, *45*(1), 1–5. doi:10.2307/1127743

Bronfenbrenner, U. (1977). Toward an experimental ecology of human development. *The American Psychologist*, *32*(7), 513–531. doi:10.1037/0003-066X.32.7.513

Bronfenbrenner, U., & Ceci, S. J. (1994). Nature-nurture reconceptualized: A bio-ecological model. *Psychological Review*, *10*(4), 568–586. doi:10.1037/0033-295X.101.4.568 PMID:7984707

Bronfenbrenner, U., & Evans, G. W. (2000). Developmental science in the 21st century: Emerging questions, theoretical models, research designs and empirical findings. *Social Development*, *9*(1), 115–125. doi:10.1111/1467-9507.00114

Bronzan, R. N., Taylor, T. E., Mwenechanya, J., Tembo, M., Kayira, K., Bwanaisa, L., Njobvu, A., Kondowe, W., Chalira, C., Walsh, A. L., Phiri, A., Wilson, L. K., Molyneux, M. E., & Graham, S. M. (2007). Bacteremia in Malawian children with severe malaria: Prevalence, etiology, HIV coinfection, and outcome. *The Journal of Infectious Diseases*, *195*(6), 895–904. doi:10.1086/511437 PMID:17299721

Brouwers, L., Cakici, B., Camitz, M., Tegnell, A., & Boman, M. (2009). Economic consequences to society of H1N1 influenza 2009-prelimianry results for Sweden. *Eurosurveillance*, *15*, 5–11.

Brown, H. E., Duik-Wasser, M., Andreadis, T., & Fish, D. (2008). Remotely-sensed vegetation indices identify mosquito clusters of West Nile virus vectors in an urban landscape in the northeastern United States. *Vector Borne and Zoonotic Diseases (Larchmont, N.Y.)*, *8*(2), 197–206. doi:10.1089/vbz.2007.0154 PMID:18452400

Buchenauer, L., Junge, K. M., Haange, S. B., Simon, J. C., von Bergen, M., Hoh, A. L., Aust, G., Zenclussen, A. C., Stangl, G. I., & Polte, T. (2022). Glyphosate differentially affects the allergic immune response across generations in mice. *The Science of the Total Environment*, *850*(April), 157973. doi:10.1016/j.scitotenv.2022.157973 PMID:35963408

Bueno, J. (2016). Antimicrobial Adjuvants Drug Discovery, the Challenge of Avoid the Resistance and Recover the Susceptibility of Multidrug-Resistant Strains. *Journal of Microbial & Biochemical Technology*, *8*(3). Advance online publication. doi:10.4172/1948-5948.1000281

Buffie, C. G., Bucci, V., Stein, R. R., McKenney, P. T., Ling, L., Gobourne, A., No, D., Liu, H., Kinnebrew, M., Viale, A., Littmann, E., van den Brink, M. R. M., Jenq, R. R., Taur, Y., Sander, C., Cross, J. R., Toussaint, N. C., Xavier, J. B., & Pamer, E. G. (2015). Precision microbiome reconstitution restores bile acid mediated resistance to *Clostridium difficile*. *Nature*, *517*(7533), 205–208. doi:10.1038/nature13828 PMID:25337874

Burzynski, J., & Schlunger, N. W. (2008). The epidemiology of Tuberculosis in the United States. *Seminars in Respiratory and Critical Care Medicine*, *29*(5), 492–498. doi:10.1055-0028-1085701 PMID:18810683

Butler, M., & Buss, A. (2006). Natural products - The future scaffolds for novel antibiotics. *Biochemical Pharmacology*, *71*(7), 919–929. doi:10.1016/j.bcp.2005.10.012 PMID:16289393

Byrne, B. B. (2016). *Structural equation modeling with Amos: Basic concepts, applications, and programming* (3rd ed.). Rutledge. doi:10.4324/9781315757421

Candon, S., Perez-Arroyo, A., Marquet, C., Valette, F., Foray, A.-P., Pelletier, B., Milani, C., Ventura, M., Bach, J.-F., & Chatenoud, L. (2015). Antibiotics in early life alter the gut microbiome and increase disease incidence in a spontaneous mouse model of autoimmune insulin-dependent diabetes. *PLoS One*, *10*(5), e0125448. doi:10.1371/journal.pone.0125448 PMID:25970503

Carding, S., Verbeke, K., Vipond, D. T., Corfe, B. M., & Owen, L. J. (2015). Dysbiosis of the gut microbiota in disease. *Microbial Ecology in Health and Disease*, *26*(1), 26191. doi:10.3402/mehd.v26.26191 PMID:25651997

Carlucci, C., Petrof, E. O., & Allen-Vercoe, E. (2016). Fecal microbiota-based therapeutics for recurrent Clostridium difficile infection, ulcerative colitis and obesity. *EBioMedicine*, *13*, 37–45. doi:10.1016/j.ebiom.2016.09.029 PMID:27720396

Carme, B., Bouquety, J. C., & Plassart, H. (1993). Mortality and sequelae due to cerebral malaria in African children in Brazzaville. *The American Journal of Tropical Medicine and Hygiene*, *48*(2), 216–221. doi:10.4269/ajtmh.1993.48.216 PMID:8447526

Caron, A., Cappelle, J., Cumming, G. S., Garine-Wichatitsky, M. D., & Gaidet, N. (2015). Bridge hosts, a missing link for disease ecology in multi-host systems. *Veterinary Research*, *46*(1), 1–11. doi:10.118613567-015-0217-9 PMID:26198845

Caruso, B. A., Freeman, M. C., Garn, J. V., Dreibelbis, R., Saboori, S., Muga, R., & Rheingans, R. (2014). Assessing the impact of a school-based latrine cleaning and handwashing program on pupil absence in Nyanza Province, Kenya: A cluster-randomized Trial. *Tropical Medicine & International Health*, *19*(10), 1185–1197. doi:10.1111/tmi.12360 PMID:25055716

Cavalli-Sforza, L. L., Menozzi, P., & Piazza, A. (1994). *The history and geography of human genes*. Princeton University Press.

CDC. (1999). Needs assessment following Hurricane Georges — Dominican Republic, 1998. *Morbidity and Mortality Weekly Report*, *48*, 93–95. PMID:10072265

CDC. (2010). *Waterborne diseases. Climate and Health Program*. Centers for Disease Control and Prevention.

Ceddia, M. G., Bardsley, N. O., Goodwin, R., Holloway, G. J., Nocella, G., & Stasi, A. (2013). A complex system perspective on the emergence and spread of infectious diseases: Integrating economic and ecological aspects. *Ecological Economics*, *90*, 124–131. doi:10.1016/j.ecolecon.2013.03.013

Cénit, M. C., Matzaraki, V., Tigchelaar, E. F., & Zhernakova, A. (2014). Rapidly expanding knowledge on the role of the gut microbiome in health and disease. *Biochimica et Biophysica Acta (BBA)-. Molecular Basis of Disease*, *1842*(10), 1981–1992. doi:10.1016/j.bbadis.2014.05.023 PMID:24882755

Centers for Disease Control and Prevention (U.S.) & National Center for Emerging Zoonotic and Infectious Diseases (U.S.). Division of Healthcare Quality Promotion. Antibiotic Resistance Coordination and Strategy Unit. (Eds.). (2019). *Antibiotic resistance threats in the United States, 2019* (CDC:82532). https://stacks.cdc.gov/view/cdc/82532

Centers for Disease Control and Prevention. (2021). *Antibiotic-resistant Germs: New Threats*. Retrieved on 9 October 2021, from https://www.cdc.gov/drugresistance/biggest-threats.html

Chadsuthi, S., Chalvet-Monfray, K., Wiratsudakul, A., & Modchang, C. (2021). The effects of flooding and weather conditions on leptospirosis transmission in Thailand. *Scientific Reports*, *11*(1), 1486. doi:10.103841598-020-79546-x PMID:33452273

Chae, Y., & An, Y.-J. (2017). Effects of micro-and nanoplastics on aquatic ecosystems: Current research trends and perspectives. *Marine Pollution Bulletin, 124*(2), 624–632. doi:10.1016/j.marpolbul.2017.01.070 PMID:28222864

Chae, Y., & An, Y.-J. (2018). Current research trends on plastic pollution and ecological impacts on the soil ecosystem: A review. *Environmental Pollution, 240*, 387–395. doi:10.1016/j.envpol.2018.05.008 PMID:29753246

Chakraborty, A., & Chatterjee, S. (2015). Studies on the fitness components and comparative oviposition preferences of *Aedes albopictus* in various larval microhabitats of Burdwan, West Bengal. *International Journal of Mosquito Research, 2*(3), 156–160.

Chan, B., Abedon, S., & Loc-Carrillo, C. (2013). Phage cocktails and the future of phage therapy. *Future Microbiology, 8*(6), 769–783. doi:10.2217/fmb.13.47 PMID:23701332

Chang, R., Chen, K., Wang, J., Wallin, M., Britton, W., Morales, S., Kutter, E., Li, J., & Chan, H.-K. (2018). Proof-of-Principle Study in a Murine Lung Infection Model of Antipseudomonal Activity of Phage PEV20 in a Dry-Powder Formulation. *Antimicrobial Agents and Chemotherapy, 62*(2), e01714-17. Advance online publication. doi:10.1128/AAC.01714-17 PMID:29158280

Chard, A. N., Garn, J. V., Chang, H. H., Clasen, T., & Freeman, M. C. (2019). Impact of a school-based water, sanitation, and hygiene intervention on school absence, diarrhea, respiratory function, and soil-transmitted helminths: Results from the WASH HELPS cluster-randomized trial. *Journal of Global Health, 9*(2), 1–14. doi:10.7189/jogh.09.020402 PMID:31360445

Charrahy, Z., Yaghoobi-Ershadi, M. R., Shirzadi, M. R., Akhavan, A. A., Rassi, Y., Hosseini, S. Z., Webb, N. J., Haque, U., Omid, F. B., & Hanafi-Bojd, A. A. (2021). Climate change and its effect on the vulnerability to zoonotic cutaneous leishmaniasis in Iran. *Transboundary and Emerging Diseases*. Advance online publication. doi:10.1111/tbed.14115 PMID:33876891

Chassaing, B., Koren, O., Goodrich, J. K., Poole, A. C., Srinivasan, S., Ley, R. E., & Gewirtz, A. T. (2015). Dietary emulsifiers impact the mouse gut microbiota promoting colitis and metabolic syndrome. *Nature, 519*(7541), 92–96. doi:10.1038/nature14232 PMID:25731162

Chen, G., Guo, Y., Yue, X., Tong, S., Gasparrini, A., Bell, M. L., Armstrong, B., Schwartz, J., Jaakkola, J. J. K., Zanobetti, A., Lavigne, E., Nascimento Saldiva, P. H., Kan, H., Royé, D., Milojevic, A., Overcenco, A., Urban, A., Schneider, A., Entezari, A., ... Li, S. (2021). Mortality risk attributable to wildfire-related PM2·5 pollution: A global time series study in 749 locations. *The Lancet. Planetary Health, 5*(9), e579–e587. doi:10.1016/S2542-5196(21)00200-X PMID:34508679

Chen, L. (2021). Gut microbiota manipulation to mitigate the detrimental effects of environmental pollutants. *Toxics, 9*(6), 2–3. doi:10.3390/toxics9060127 PMID:34205878

Chen, N. T., Chen, M. J., Da Wu, C., & Guo, Y. L. (2021). Emergency room visits for childhood atopic dermatitis are associated with floods? *The Science of the Total Environment, 773*, 145435. doi:10.1016/j.scitotenv.2021.145435 PMID:33940726

Chhetri, B. K., Takaro, T. K., Balshaw, R., Otterstatter, M., Mak, S., Lem, M., Zubel, M., Lysyshyn, M., Clarkson, L., Edwards, J., Fleury, M. D., Henderson, S. B., & Galanis, E. (2017). Associations between extreme precipitation and acute gastro-intestinal illness due to cryptosporidiosis and giardiasis in an urban Canadian drinking water system (1997–2009). *Journal of Water and Health*, *15*(6), 898–907. doi:10.2166/wh.2017.100 PMID:29215354

Chitnis, C. E., & Miller, L. H. (1994). Identification of the erythrocyte binding domains of *Plasmodium vivax* and *Plasmodium knowlesi* proteins involved in erythrocyte invasion. *The Journal of Experimental Medicine*, *180*(2), 497–506. doi:10.1084/jem.180.2.497 PMID:8046329

Chi, Y., Lin, Y., Lu, Y., Huang, Q., Ye, G., & Dong, S. (2019). Gut microbiota dysbiosis correlates with a low-dose PCB126-induced dyslipidemia and non-alcoholic fatty liver disease. *The Science of the Total Environment*, *653*, 274–282. doi:10.1016/j.scitotenv.2018.10.387 PMID:30412872

Choi, H. H., & Cho, Y.-S. (2016). Fecal microbiota transplantation: Current applications, effectiveness, and future perspectives. *Clinical Endoscopy*, *49*(3), 257–265. doi:10.5946/ce.2015.117 PMID:26956193

Choi, J. J., Eum, S. Y., Rampersaud, E., Daunert, S., Abreu, M. T., & Toborek, M. (2013). Exercise attenuates PCB-induced changes in the mouse gut microbiome. *Environmental Health Perspectives*, *121*(6), 725–730. doi:10.1289/ehp.1306534 PMID:23632211

Chown, S. L., Hodgins, K. A., Griffin, P. C., Oakeshott, J. G., Byrne, M., & Hoffmann, A. (2015). Biological invasions, climate change and genomics. *Evolutionary Applications*, *8*(1), 23–46. doi:10.1111/eva.12234 PMID:25667601

Chroboczek, T., Boisset, S., Rasigade, J. P., Meugnier, H., Akpaka, P. E., Nicholson, A., Nicolas, M., Olive, C., Bes, M., Vandenesch, F., Laurent, F., Etienne, J., & Tristan, A. (2013). Major West Indies MRSA clones in human beings: Do they travel with their hosts? *Journal of Travel Medicine*, *20*(5), 283–288. doi:10.1111/jtm.12047 PMID:23992570

Claus, S. P., Guillou, H., & Ellero-Simatos, S. (2016a). The gut microbiota: A major player in the toxicity of environmental pollutants? *NPJ Biofilms and Microbiomes*, *2*(1), 1–11. doi:10.1038/npjbiofilms.2016.3 PMID:28721242

Coatney G. R. & National Institute of Allergy and Infectious Diseases (U.S.). (1971). The primate malarias. U.S. Department of Health Education and Welfare National Institutes of Health National Institute of Allergy and Infectious Diseases.

Coban, C., Ishii, K. J., Kawai, T., Hemmi, H., Sato, S., Uematsu, S., Yamamoto, M., Takeuchi, O., Itagaki, S., Kumar, N., Horii, T., & Akira, S. (2005). Toll-like receptor 9 mediates innate immune activation by the malaria pigment hemozoin. *The Journal of Experimental Medicine*, *201*(1), 19–25. doi:10.1084/jem.20041836 PMID:15630134

Cohen, B. (2015). Urbanization, City growth, and the New United Nations development agenda. *Cornerstone*, *3*(2), 4–7.

Cole, M., Lindeque, P., Halsband, C., & Galloway, T. S. (2011). Microplastics as contaminants in the marine environment: A review. *Marine Pollution Bulletin, 62*(12), 2588–2597. doi:10.1016/j.marpolbul.2011.09.025 PMID:22001295

Collins, W. E. (2012). *Plasmodium knowlesi*: A malaria parasite of monkeys and humans. *Annual Review of Entomology, 57*(1), 107–121. doi:10.1146/annurev-ento-121510-133540 PMID:22149265

Colón-González, F. J., Harris, I., Osborn, T. J., Steiner São Bernardo, C., Peres, C. A., Hunter, P. R., & Lake, I. R. (2018). Limiting global-mean temperature increase to 1.5-2 °C could reduce the incidence and spatial spread of dengue fever in Latin America. *Proceedings of the National Academy of Sciences of the United States of America, 115*(24), 6243–6248. doi:10.1073/pnas.1718945115 PMID:29844166

Colón-González, F. J., Sewe, M. O., Tompkins, A. M., Sjödin, H., Casallas, A., Rocklöv, J., Caminade, C., & Lowe, R. (2021). Projecting the risk of mosquito-borne diseases in a warmer and more populated world: A multi-model, multi-scenario intercomparison modelling study. *The Lancet. Planetary Health, 5*(7), e404–e414. doi:10.1016/S2542-5196(21)00132-7 PMID:34245711

Condette, C. J., Bach, V., Mayeur, C., Gay-Quéheillard, J., & Khorsi-Cauet, H. (2015). Chlorpyrifos exposure during perinatal period affects intestinal microbiota associated with delay of maturation of digestive tract in rats. *Journal of Pediatric Gastroenterology and Nutrition, 61*(1), 30–40. doi:10.1097/MPG.0000000000000734 PMID:25643018

Confalonieri, U., Menne, B., Abeku, T., Alam, M., Beggs, P., Clot, B., Furgal, C., Curto, S., McMichael, A., Parry, M., Canziani, O., Palutikof, J., van der Linden, P., & Hanson, C. (2007). *Impacts, Adaptation and Vulnerability. Contribution of Working Group II to the Fourth Assessment Report of the Intergovernmental Panel on Climate Change*. Academic Press.

Conlen, M. (2021). *How Much Carbon Dioxide Are We Emitting?* NASA's Vital Signs of the Planet. https://climate.nasa.gov/news/3020/how-much-carbon-dioxide-are-we-emitting/

Conti, B., Flamini, G., Cioni, P. L., Ceccarini, L., Macchia, M., & Benelli, G. (2014). Mosquitocidal essential oils: Are they safe against non-target aquatic organisms? *Parasitology Research, 113*(1), 251–259. doi:10.100700436-013-3651-5 PMID:24146210

Control European Centre for Disease Prevention and Control. (2021). *Chikungunya worldwide overview*. https://www.ecdc.europa.eu/en/chikungunya-monthly

Cork, A., & Park, K. C. (1996). Identification of electrophysiologically-active compounds for the malaria mosquito, *Anopheles gambiae*, in human sweat extracts. *Medical and Veterinary Entomology, 10*(3), 269–276. doi:10.1111/j.1365-2915.1996.tb00742.x PMID:8887339

Costello, A., Abbas, M., Allen, A., Ball, S., Bell, S., Bellamy, R., Friel, S., Groce, N., Johnson, A., Kett, M., Lee, M., Maslin, M., McCoy, D., McGuire, B., Montgomery, H., Napier, D., Pagel, C., Patel, J., de Oliveira, J. A. P., ... Patterson, C. (2009). Managing the health effects of climate change. *Lancet, 373*(9676), 1773–1964. doi:10.1016/S0140-6736(09)60935-1 PMID:19447250

Crozier, A., Clifford, M. N., & Ashihara, H. (Eds.). (2008). *Plant secondary metabolites: occurrence, structure and role in the human diet*. John Wiley & Sons.

Cui, B., Li, P., Xu, L., Peng, Z., Xiang, J., He, Z., Zhang, T., Ji, G., Nie, Y., Wu, K., Fan, D., & Zhang, F. (2016). Step-up fecal microbiota transplantation (FMT) strategy. *Gut Microbes*, *7*(4), 323–328. doi:10.1080/19490976.2016.1151608 PMID:26939622

Cummings, D. A., Irizarry, R. A., Huang, N. E., Endy, T. P., Nisalak, A., Ungchusak, K., & Burke, D. S. (2004). Traveling waves in the occurrence of Dengue hemorrhagic fever in Thailand. *Nature*, *427*(6972), 344–347. doi:10.1038/nature02225 PMID:14737166

Cunningham, M. M., Tran, L., McKee, C. G., Polo, R. O., Newman, T., Lansing, L., Griffiths, J. S., Bilodeau, G. J., Rott, M., & Guarna, M. M. (2022). Honey bees as biomonitors of environmental contaminants, pathogens, and climate change. *Ecological Indicators*, *134*, 108457. doi:10.1016/j.ecolind.2021.108457

Curtin, P. D. (1968). Epidemiology and the Slave Trade. *Political Science Quarterly*, *83*(2), 190–216. doi:10.2307/2147089 PMID:11632739

Curtis, V., Schmidt, W., Luby, S., Florez, R., Touré, O., & Biran, A. (2011). Hygiene: New hopes, new horizons. *The Lancet. Infectious Diseases*, *11*(4), 312–321. doi:10.1016/S1473-3099(10)70224-3 PMID:21453872

D'Amato, M., Molino, A., Calabrese, G., Cecchi, L., Annesi-Maesano, I., & D'Amato, G. (2018). The impact of cold on the respiratory tract and its consequences to respiratory health. *Clinical and Translational Allergy*, *8*(1), 1–8. doi:10.1186/s13601-018-0208-9

D'Amato, G., Chong-Neto, H. J., Ortega, O. P. M., Vitale, C., Ansotegui, I., Rosario, N., Haahtela, T., Galan, C., Pawankar, R., Murrieta-Aguttes, M., Cecchi, L., Bergmann, C., Ridolo, E., Ramon, G., Diaz, S. G., D'Amato, M., & Annesi-Maesano, I. (2020). The effects of climate change on respiratory allergy and asthma induced by pollen and mold allergens. *Allergy*, *75*(9), 2219–2228. doi:10.1111/all.14476 PMID:32589303

D'Haens, G. R., & Jobin, C. (2019). Fecal microbial transplantation for diseases beyond recurrent clostridium difficile infection. *Gastroenterology*, *157*(3), 624–636. doi:10.1053/j.gastro.2019.04.053 PMID:31220424

D'Ombrain, M. C., Voss, T. S., Maier, A. G., Pearce, J. A., Hansen, D. S., Cowman, A. F., & Schofield, L. (2007). Plasmodium falciparum erythrocyte membrane protein-1 specifically suppresses early production of host interferon-γ. *Cell Host & Microbe*, *2*(2), 130–138. doi:10.1016/j.chom.2007.06.012 PMID:18005727

Dabaro, D., Birhanu, Z., Negash, A., Hawaria, D., & Yewhalaw, D. (2021). Effects of rainfall, temperature and topography on malaria incidence in elimination targeted district of Ethiopia. *Malaria Journal*, *20*(1), 104. Advance online publication. doi:10.118612936-021-03641-1 PMID:33608004

Dada, N., Jumas-Bilak, E., Manguin, S., Seidu, R., Stenström, T.-A., & Overgaard, H. J. (2014). Comparative assessment of the bacterial communities associated with *Aedes aegypti* larvae and water from domestic water storage containers. *Parasites & Vectors*, *7*(1), 391. doi:10.1186/1756-3305-7-391 PMID:25151134

Dahiri, B., Martín-Reina, J., Carbonero-Aguilar, P., Aguilera-Velázquez, J. R., Bautista, J., & Moreno, I. (2021). Impact of pesticide exposure among rural and urban female population. An overview. *International Journal of Environmental Research and Public Health*, *18*(18), 9907. doi:10.3390/ijerph18189907 PMID:34574830

Dai, J., Yang, X., Yuan, Y., Jia, Y., Liu, G., Lin, N., Xiao, H., Zhang, L., & Chen, J. (2020). Toxicity, gut microbiota and metabolome effects after copper exposure during early life in SD rats. *Toxicology*, *433*, 152395. doi:10.1016/j.tox.2020.152395 PMID:32027963

Dakal, T., Kumar, A., Majumdar, R., & Yadav, V. (2016). Mechanistic Basis of Antimicrobial Actions of Silver Nanoparticles. *Frontiers in Microbiology*, *7*. Advance online publication. doi:10.3389/fmicb.2016.01831 PMID:27899918

Dar, B., & Lone, S. A. (2021). Pre and post Vulnerability of Floods to Mental Health Among the Residents of Srinagar City, J&K-India. *Sustainability, Agri. Food and Environmental Research*, *11*(1), 1–12. doi:10.7770afer-V11N1-art2393

Daszak, P., Plowright, R. K., Epstein, J. H., Pulliam, J., Rahman, S. A., Field, H. E., Jamaluddin, A., Sharifah, S. H., Smith, C. S., Olival, K. J., Luby, S., Halpin, K., Hyatt, A. D., & Cunningham, A. A. (2006). Disease Ecology: Community Structure and Pathogen Dynamics. In S. Collinge & C. Ray (Eds.), *The emergence of Nipah and Hendra virus: pathogen dynamics across a wildlife-livestock-human continuum* (pp. 186–201). Oxford University Press.

Davies, J., & Davies, D. (2010). Origins and Evolution of Antibiotic Resistance. *Microbiology and Molecular Biology Reviews*, *74*(3), 417–433. doi:10.1128/MMBR.00016-10 PMID:20805405

Davis, R. S., Peterson, R. K., & Macedo, P. A. (2007). An ecological risk assessment for insecticides used in adult mosquito management. *Integrated Environmental Assessment and Management*, *3*(3), 373–382. doi:10.1002/ieam.5630030308 PMID:17695110

Davis, S., Calvet, E., & Leirs, H. (2005). Fluctuating rodent populations and risk to humans from rodent-borne zoonoses. *Vector Borne and Zoonotic Diseases (Larchmont, N.Y.)*, *5*(4), 305–314. doi:10.1089/vbz.2005.5.305 PMID:16417426

Day, J. F. (2016). Mosquito oviposition behavior and vector control. *Insects*, *7*(4), 65. doi:10.3390/insects7040065 PMID:27869724

de Kraker, M., Stewardson, A., & Harbarth, S. (2016). Will 10 Million People Die a Year due to Antimicrobial Resistance by 2050? *PLoS Medicine*, *13*(11), e1002184. doi:10.1371/journal.pmed.1002184 PMID:27898664 de Silva, N. R., Brooker, S., Hotez, P. J., Montresor, A., Engels, D., & Savioli, L. (2003). Soil-transmitted helminth infections: Updating the global picture. *Trends in Parasitology*, *19*(12), 547–551. doi:10.1016/j.pt.2003.10.002 PMID:14642761

DeGruttola, A. K., Low, D., Mizoguchi, A., & Mizoguchi, E. (2016). Current understanding of dysbiosis in disease in human and animal models. *Inflammatory Bowel Diseases*, *22*(5), 1137–1150. doi:10.1097/MIB.0000000000000750 PMID:27070911

Deichstetter, P. (2017). The effect of climate change on mosquito-borne diseases. *The American Biology Teacher*, *79*(3), 169–173. doi:10.1525/abt.2017.79.3.169

del Portillo, H. A., Fernandez-Becerra, C., Bowman, S., Oliver, K., Preuss, M., Sanchez, C. P., Schneider, N. K., Villalobos, J. M., Rajandream, M. A., Harris, D., da Silva, L. H. P., Barrell, B., & Lanzer, M. (2001). A superfamily of variant genes encoded in the subtelomeric region of Plasmodium vivax. *Nature*, *410*(6830), 839–842. doi:10.1038/35071118 PMID:11298455

Deng, Y., Zhang, Y., Lemos, B., & Ren, H. (2017). Tissue accumulation of microplastics in mice and biomarker responses suggest widespread health risks of exposure. *Scientific Reports*, *7*(1), 46687. doi:10.1038rep46687 PMID:28436478

Deplazes, P., Hegglin, D., Gloor, S., & Romig, T. (2004). Wilderness in the city: The urbanization of Echinococcus multilocularis. *Trends in Parasitology*, *20*(2), 77–84. doi:10.1016/j.pt.2003.11.011 PMID:14747021

Desjeux, P. (2001). The increase risk factors for Leishmaniasis worldwide. *Transactions of the Royal Society of Tropical Medicine and Hygiene*, *95*(3), 239–243. doi:10.1016/S0035-9203(01)90223-8 PMID:11490989

Despommier, D., Ellis, B. R., & Wilcox, B. A. (2006). The role of ecotones in emerging infectious diseases. *EcoHealth*, *3*(4), 281–289. doi:10.100710393-006-0063-3

Dever, L. (1991). Mechanisms of bacterial resistance to antibiotics. *Archives of Internal Medicine*, *151*(5), 886–895. doi:10.1001/archinte.1991.00400050040010 PMID:2025137

Dev, V., & Sharma, V. P. (2013). The dominant mosquito vectors of human malaria in India. In *Anopheles mosquitoes-New insights into malaria vectors*. IntechOpen. doi:10.5772/55215

Dheer, R., Patterson, J., Dudash, M., Stachler, E. N., Bibby, K. J., Stolz, D. B., Shiva, S., Wang, Z., Hazen, S. L., Barchowsky, A., & Stolz, J. F. (2015). Arsenic induces structural and compositional colonic microbiome change and promotes host nitrogen and amino acid metabolism. *Toxicology and Applied Pharmacology*, *289*(3), 397–408. doi:10.1016/j.taap.2015.10.020 PMID:26529668

Dhingra, N., Jha, P., Sharma, V. P., Cohen, A. A., Jotkar, R. M., Rodriguez, P. S., Bassani, D. G., Suraweera, W., Laxminarayan, R., & Pet, R. (2010). Adult and child malaria mortality in India: A nationally representative mortality study. *Lancet*, *376*(9754), 1768–1774. doi:10.1016/S0140-6736(10)60831-8 PMID:20970179

Ding, J., Zhang, S., Razanajatovo, R. M., Zou, H., & Zhu, W. (2018). Accumulation, tissue distribution, and biochemical effects of polystyrene microplastics in the freshwater fish red tilapia (*Oreochromis niloticus*). *Environmental Pollution*, *238*, 1–9. doi:10.1016/j.envpol.2018.03.001 PMID:29529477

Ditzen, M., Pellegrino, M., & Vosshall, L. B. (2008). Insect odorant receptors are molecular targets of the insect repellent DEET. *Science*, *319*(5871), 1838–1842. doi:10.1126cience.1153121 PMID:18339904

do Sul, J. A. I., & Costa, M. F. (2014). The present and future of microplastic pollution in the marine environment. *Environmental Pollution*, *185*, 352–364. doi:10.1016/j.envpol.2013.10.036 PMID:24275078

Dobson, A. (2004). Population dynamics of pathogens with multiple host species. *American Naturalist*, *164*(S5), 64–78. doi:10.1086/424681 PMID:15540143

Domalaon, R., Idowu, T., Zhanel, G., & Schweizer, F. (2018). Antibiotic Hybrids: The Next Generation of Agents and Adjuvants against Gram-Negative Pathogens? *Clinical Microbiology Reviews*, *31*(2), e00077-17. Advance online publication. doi:10.1128/CMR.00077-17 PMID:29540434

Dondorp, A. M., Kager, P. A., Vreeken, J., & White, N. J. (2000). Abnormal blood flow and red blood cell deformability in severe malaria. *Parasitology Today (Personal Ed.)*, *16*(6), 228–232. doi:10.1016/S0169-4758(00)01666-5 PMID:10827427

Dong, Y.-L., Zhou, P.-J., Jiang, S.-Y., Pan, X.-W., & Zhao, X.-H. (2009). Induction of oxidative stress and apoptosis by pentachlorophenol in primary cultures of Carassius carassius hepatocytes. *Comparative Biochemistry and Physiology. Toxicology & Pharmacology : CBP*, *150*(2), 179–185. doi:10.1016/j.cbpc.2009.04.010 PMID:19410655

Dreibelbis, R., Kroeger, A., Hossain, K., Venkatesh, M., & Ram, P. K. (2016). Behavior change without behavior change communication: Nudging handwashing among primary school students in Bangladesh. *International Journal of Environmental and Residential Public Health, 13*.

Dubb, S. S. (2020). Coronavirus pandemic: Applying a whole-of-society model for the whole-of-the world. *British Journal of Oral & Maxillofacial Surgery*, *58*(7), 838–842. doi:10.1016/j.bjoms.2020.05.009 PMID:32631753

Duffin, R., Mills, N. L., & Donaldson, K. (2007). Nanoparticles-a thoracic toxicology perspective. *Yonsei Medical Journal*, *48*(4), 561–572. doi:10.3349/ymj.2007.48.4.561 PMID:17722227

Duguma, D., Hall, M. W., Rugman-Jones, P., Stouthamer, R., Terenius, O., Neufeld, J. D., & Walton, W. E. (2015). Developmental succession of the microbiome of *Culex* mosquitoes. *BMC Microbiology*, *15*(1), 140. doi:10.118612866-015-0475-8 PMID:26205080

Duncan, S. H., Belenguer, A., Holtrop, G., Johnstone, A. M., Flint, H. J., & Lobley, G. E. (2007). Reduced dietary intake of carbohydrates by obese subjects results in decreased concentrations of butyrate and butyrate-producing bacteria in feces. *Applied and Environmental Microbiology, 73*(4), 1073–1078. doi:10.1128/AEM.02340-06 PMID:17189447

Duran-Reynals. (1946). *The Fever Bark Tree*. Academic Press.

Ebi, K. L., Exuzides, K. A., Lau, E., Kelsh, M., & Barnston, A. (2001). Association of normal weather periods and El Niño events with hospitalization for viral pneumonia in females: California, 1983–1998. *American Journal of Public Health, 91*(8), 1200–1208. doi:10.2105/AJPH.91.8.1200 PMID:11499104

Eddleston, M. (2020). Poisoning by pesticides. *Medicine, 48*(3), 214–217. doi:10.1016/j.mpmed.2019.12.019

EEA. (2008). *Impact of Europe's Changing Climate—2008 Indicator-based Assessment. Joint EEA-JRC-WHO report*. European Environment Agency.

Eerkes-Medrano, D., Thompson, R. C., & Aldridge, D. C. (2015). Microplastics in freshwater systems: A review of the emerging threats, identification of knowledge gaps and prioritisation of research needs. *Water Research, 75*, 63–82. doi:10.1016/j.watres.2015.02.012 PMID:25746963

Eguiluz-Gracia, I., Mathioudakis, A. G., Bartel, S., Vijverberg, S. J. H., Fuertes, E., Comberiati, P., Cai, Y. S., Tomazic, P. V., Diamant, Z., Vestbo, J., Galan, C., & Hoffmann, B. (2020). The need for clean air: The way air pollution and climate change affect allergic rhinitis and asthma. *Allergy, 75*(9), 2170–2184. doi:10.1111/all.14177 PMID:31916265

Eisenberg, J. N. S., Desai, M. N., Levy, K., Bates, S. J., Liang, S., Naumoff, K., & Scott, J. C. (2007). Environmental determinants of infectious diseases: A framework for tracking causal links and guiding public health research. *Environmental Health Perspectives, 115*(8), 1216–1223. doi:10.1289/ehp.9806 PMID:17687450

Eisen, R. J., Glass, G. E., Eisen, L., Cheek, J., Enscore, R. E., Ettestad, P., & Gage, K. L. (2007). A spatial model of shared risk for plague and hantavirus pulmonary syndrome in the southwestern United States. *The American Journal of Tropical Medicine and Hygiene, 77*(6), 999–1004. doi:10.4269/ajtmh.2007.77.999 PMID:18165511

Ellis, E. C., & Ramankutty, N. (2008). Putting people on the map: Anthropogenic biomes of the world. *Frontiers in Ecology and the Environment, 6*(8), 439–447. doi:10.1890/070062

El-Mansoury, B., Smimih, K., El Khiat, A., El Fatimi, H., Aimrane, A., El Baz, S., Elmourid, A., Saad, F. E.-Z., El Ghachi, H., & Khanouchi, M. (2023). Potential Role of Probiotics and Fecal Microbiota Transplantation in the Treatment of Parkinson's Disease. In *Experimental and Clinical Evidence of the Neuropathology of Parkinson's Disease* (pp. 296–317). IGI Global. doi:10.4018/978-1-6684-5156-4.ch017

Engin-Demir, C. (2009). Factors influencing the academic achievement of the Turkish urban poor. *International Journal of Educational Development, 29*(1), 17–29. doi:10.1016/j.ijedudev.2008.03.003

Epstein, P. R. (1999). Climate and health. *Science, 285*(5426), 347–348. doi:10.1126cience.285.5426.347 PMID:10438299

Epstein, P. R. (2000). Is global warming harmful to health? *Scientific American, 283*(2), 50–57. doi:10.1038cientificamerican0800-50 PMID:10914399

Epstein, P. R. (2001). Climate change and emerging infectious diseases. *Microbes and Infection, 3*(9), 747–754. doi:10.1016/S1286-4579(01)01429-0 PMID:11489423

Epstein, P. R., Diaz, H. F., Elias, S., Grabherr, G., Graham, N. E., Martens, W. J., Mosley Thompson, E., & Susskind, J. (1998). Biological and physical signs of climate change: Focus on mosquito-borne diseases. *Bulletin of the American Meteorological Society, 79*(3), 409–417. doi:10.1175/1520-0477(1998)079<0409:BAPSOC>2.0.CO;2

Erlanger, T. E., Keiser, J., Caldas De Castro, M., Bos, R., Singer, B. H., Tanner, M., & Utzinger, J. (2005). Effect of water resource development and management on lymphatic filariasis, and estimates of population at risk. *The American Journal of Tropical Medicine and Hygiene, 73*(3), 523–533. doi:10.4269/ajtmh.2005.73.523 PMID:16172476

Estrada-Peña, A., Ostfeld, R. S., Peterson, A. T., Poulin, R., & Fuente, J. D. L. (2014). Effects of environmental change on zoonotic disease risk: An ecological primer. *Trends in Parasitology, 30*(4), 205–214. doi:10.1016/j.pt.2014.02.003 PMID:24636356

Etkin, N. L. (2003). The Co-evolution of People, Plants, and Parasites: Biological and Cultural Adaptations to Malaria. *The Proceedings of the Nutrition Society, 62*(2), 311–317. doi:10.1079/pns2003244 PMID:14506878

European Centre for Disease Control and Prevention. (n.d.). *Tick-borne encephalitis in Europe.* Fact Sheet.

Evangelista, K. V., & Coburn, J. (2010). Leptospira as an emerging pathogen: A review of its biology, pathogenesis, and host immune responses. *Future Microbiology, 5*(9), 1413–1425. doi:10.2217/fmb.10.102 PMID:20860485

Evariste, L., Barret, M., Mottier, A., Mouchet, F., Gauthier, L., & Pinelli, E. (2019). Gut microbiota of aquatic organisms: A key endpoint for ecotoxicological studies. *Environmental Pollution, 248*, 989–999. doi:10.1016/j.envpol.2019.02.101 PMID:31091643

Ewing, D., Purse, B., Cobbold, C., & White, S. (2021). A novel approach for predicting risk of vector-borne disease establishment in marginal temperate environments under climate change: West Nile virus in the UK. *Journal of the Royal Society, Interface, 18*(178), 20210049. Advance online publication. doi:10.1098/rsif.2021.0049 PMID:34034529

Fair, J. M., Stokes, M. M., Pennington, D., & Mendenhall, I. H. (2016). Scientific collaborations: How do we measure the return on relationships? *Frontiers in Public Health*, *4*, 9. doi:10.3389/fpubh.2016.00009 PMID:26913278

Falkow, S. (2004). Molecular Koch's postulates applied to bacterial pathogenicity—A personal recollection 15 years later. *Nature Reviews. Microbiology*, *2*(1), 67–72. doi:10.1038/nrmicro799 PMID:15035010

Fang, J. (2010). Ecology: A world without mosquitoes. *Nature*, *466*(7305), 432–434. doi:10.1038/466432a PMID:20651669

Fan, J., & Malik, A. B. (2003). Toll-like receptor-4 (TLR4) signaling augments chemokine-induced neutrophil migration by modulating cell surface expression of chemokine receptors. *Nature Medicine*, *9*(3), 315–321. doi:10.1038/nm832 PMID:12592402

Fazeli, M., Hassanzadeh, P., & Alaei, S. (2011). Cadmium chloride exhibits a profound toxic effect on bacterial microflora of the mice gastrointestinal tract. *Human and Experimental Toxicology*, *30*(2), 152–159. doi:10.1177/0960327110369821 PMID:20488844

Ferguson, K. T., Cassells, R. C., MacAllister, J. W., & Evans, G. W. (2013). The physical environment and child development: An international review. *International Journal of Psychology*, *48*(4), 437–468. doi:10.1080/00207594.2013.804190 PMID:23808797

Ferguson, K. T., Kim, P., Dunn, J. R., & Evans, G. W. (2009). An ecological model of urban child health. In N. Freudenberg, S. Klitzman, & S. Saegert (Eds.), *Urban health and society: Interdisciplinary approaches to research and practice*. Jossey Bass.

Field, C. B., Barros, V. R., & Dokken, D. J. (2014). *Climate Change 2014: Impacts, Adaptation, and Vulnerability. Part A: Global and Sectoral Aspects. Working Group II Contribution to the Fifth Assessment Report of the Intergovernmental Panel on Climate Change*. Cambridge University Press.

Filho, W. L., Scheday, S., Boenecke, J., Gogoi, A., Maharaj, A., & Korovou, S. (2019). Climate Change, Health and Mosquito-Borne Diseases: Trends and Implications to the Pacific Region. *International Journal of Environmental Research and Public Health*, *16*(24), 5114. doi:10.3390/ijerph16245114 PMID:31847373

Fletcher, I. K., Hernández-Villena, J., Moreno, J. E., Drakeley, C., Jones, K., Grillet, M. E., & Lowe, R. (2021). The effect of environmental degradation and land use change on malaria re-emergence in south Venezuela: A spatiotemporal modelling study. *The Lancet. Planetary Health*, *5*, S13. doi:10.1016/S2542-5196(21)00097-8

Franklin, B. S., Parroche, P., Ataíde, M. A., Lauw, F., Ropert, C., de Oliveira, R. B., Pereira, D., Tada, M. S., Nogueira, P., da Silva, L. H. P., Bjorkbacka, H., Golenbock, D. T., & Gazzinelli, R. T. (2009). Malaria primes the innate immune response due to interferon-γ induced enhancement of toll-like receptor expression and function. *Proceedings of the National Academy of Sciences of the United States of America*, *106*(14), 5789–5794. doi:10.1073/pnas.0809742106 PMID:19297619

Freeman, M. C., Clasen, T., Brooker, S. J., Akoko, D. O., & Rheingans, R. (2013b). The impact of a school-based hygiene, water quality and sanitation intervention on soil-transmitted helminth reinfection: A cluster-randomized trial. *The American Journal of Tropical Medicine and Hygiene*, *89*(5), 875–883. doi:10.4269/ajtmh.13-0237 PMID:24019429

Freeman, M. C., Clasen, T., Dreibelbis, R., Saboori, S., Greene, L. E., Brumback, B., & Rheingans, R. (2013a). The impact of a school-based water supply and treatment, hygiene, and sanitation programme on pupil diarrhea: A cluster-randomized trial. *Epidemiology and Infection*, *142*(2), 340–351. doi:10.1017/S0950268813001118 PMID:23702047

Gabrieli, P., Caccia, S., Varotto-Boccazzi, I., Arnoldi, I., Barbieri, G., Comandatore, F., & Epis, S. (2021). Mosquito Trilogy: Microbiota, Immunity and Pathogens, and Their Implications for the Control of Disease Transmission. *Frontiers in Microbiology*, *12*, 630438. doi:10.3389/fmicb.2021.630438 PMID:33889137

Gandon, S. (2004). Evolution of multi-host parasites. *Evolution; International Journal of Organic Evolution*, *58*(3), 455–469. PMID:15119430

Gardner, M. J., Hall, N., Fung, E., White, O., Berriman, M., Hyman, R. W., Carlton, J. M., Pain, A., Nelson, K. E., Bowman, S., Paulsen, I. T., James, K., Eisen, J. A., Rutherford, K., Salzberg, S. L., Craig, A., Kyes, S., Chan, M.-S., Nene, V., ... Barrell, B. (2002). Genome sequence of the human malaria parasite Plasmodium falciparum. *Nature*, *419*(6906), 498–511. doi:10.1038/nature01097 PMID:12368864

Gascon, J., Bern, C., & Pizano, M. J. (2010). Chagas disease in Spain, the United States, and other non-endemic countries. *Acta Tropica*, *115*(1-2), 22–27. doi:10.1016/j.actatropica.2009.07.019 PMID:19646412

Gascon, M., Morales, E., Sunyer, J., & Vrijheid, M. (2013). Effects of persistent organic pollutants on the developing respiratory and immune systems: A systematic review. *Environment International*, *52*, 51–65. doi:10.1016/j.envint.2012.11.005 PMID:23291098

Genton, B., Al-Yaman, F., Mgone, C. S., Alexander, N., Paniu, M. M., & Alpers, M. P. (1995). Ovalocytosis and cerebral malaria. *Nature*, *378*(6557), 564–565. doi:10.1038/378564a0 PMID:8524388

Gerba, C. P. (1999). Virus survival and transplant in groundwater. *Journal of Industrial Microbiology & Biotechnology*, *22*(4-5), 535–539. doi:10.1038j.jim.2900663

Ghayempour, S., & Montazer, M. (2016). Micro/nanoencapsulation of essential oils and fragrances: Focus on perfumed, antimicrobial, mosquito-repellent and medical textiles. *Journal of Microencapsulation*, *33*(6), 1–14. doi:10.1080/02652048.2016.1216187 PMID:27701985

Ghendon, Y. (1994). Introduction to pandemic influenza through history. *European Journal of Epidemiology*, *10*(4), 451–453. doi:10.1007/BF01719673 PMID:7843353

Ghosh, S., Kabir, M. R., Khan, M. M. I., Shill, L. C., & Alam, S. S. (2020). Knowledge, attitude, and practice (KAP) regarding personal hygiene among primary school going children in Sadar Upazilla, Noakhali District, Bangladesh. *Indian Journal of Public Health Research & Development, 11*(6), 1464–1469.

Gibson, G. R., Probert, H. M., Van Loo, J., Rastall, R. A., & Roberfroid, M. B. (2004). Dietary modulation of the human colonic microbiota: Updating the concept of prebiotics. *Nutrition Research Reviews, 17*(2), 259–275. doi:10.1079/NRR200479 PMID:19079930

Giesen, C., Roche, J., Redondo-Bravo, L., Ruiz-Huerta, C., Gomez-Barroso, D., Benito, A., & Herrador, Z. (2020). The impact of climate change on mosquito-borne diseases in Africa. *Pathogens and Global Health, 114*(6), 287–301. doi:10.1080/20477724.2020.1783865 PMID:32584659

Gimnig, J. E., Ombok, M., Kamau, L., & Hawley, W. A. (2001). Characteristics of larval anopheline (Diptera: Culicidae) habitats in Western Kenya. *Journal of Medical Entomology, 38*(2), 282–288. doi:10.1603/0022-2585-38.2.282 PMID:11296836

Girard, M., Martin, E., Vallon, L., Raquin, V., Bellet, C., Rozier, Y., Desouhant, E., Hay, A.-E., Luis, P., Moro, C. V., & Minard, G. (2021). Microorganisms associated with mosquito oviposition sites: Implications for habitat selection and insect life histories. *Microorganisms, 9*(8), 1589. doi:10.3390/microorganisms9081589 PMID:34442667

Glatter, K. A., & Finkelman, P. (2021). History of the plague: An ancient pandemic for the age of COVID-19. *The American Journal of Medicine, 134*(2), 176–181. doi:10.1016/j.amjmed.2020.08.019 PMID:32979306

Global Burden of Disease: GBD cause and risk summaries. (2019). https://www.thelancet.com/gbd/summaries

Goldberg, T. L., Gillespie, T. R., Rwego, I. B., Estoff, E. L., & Chapman, C. A. (2008). Forest fragmentation as the cause of bacterial transmission among non-human primates, humans, and livestock, Uganda. *Emerging Infectious Diseases, 14*(9), 1375–1382. doi:10.3201/eid1409.071196 PMID:18760003

Goldin, I., & Mariathasan, M. (2014). *The Butterfly Defect - How Globalization Creates Systemic Risks, and What to Do about It.* Princeton University Press.

Gomes, B., Augusto, M., Felício, M., Hollmann, A., Franco, O., Gonçalves, S., & Santos, N. (2018). Designing improved active peptides for therapeutic approaches against infectious diseases. *Biotechnology Advances, 36*(2), 415–429. doi:10.1016/j.biotechadv.2018.01.004 PMID:29330093

Gordillo Altamirano, F., & Barr, J. (2019). Phage Therapy in the Postantibiotic Era. *Clinical Microbiology Reviews, 32*(2), e00066-18. Advance online publication. doi:10.1128/CMR.00066-18 PMID:30651225

Gouagna, L. C., Poueme, R. S., Dabiré, K. R., Ouédraogo, J. B., Fontenille, D., & Simard, F. (2010). Patterns of sugar feeding and host plant preferences in adult males of *An. gambiae* (Diptera: Culicidae). *Journal of Vector Ecology*, *35*(2), 267–276. doi:10.1111/j.1948-7134.2010.00082.x PMID:21175931

Guan, P., Huang, D., He, M., Shen, T., Guo, J., & Zhou, B. (2009). Investigating the effects of climatic variables and reservoir on the incidence of hemorrhagic fever with renal syndrome in Huludao City, China: a17-year data analysis based on structural equation model. *BMC Infectious Diseases*, *9*(109), 109. doi:10.1186/1471-2334-9-109 PMID:19583875

Guan, Y., Zheng, B. J., He, Y. Q., Liu, X. L., Zhuang, Z. X., Cheung, C. L., Luo, S. W., Li, P. H., Zhang, L. J., Guan, Y. J., Butt, K. M., Wong, K. L., Chan, K. W., Lim, W., Shortridge, K. F., Yuen, K. Y., Peiris, J. S., & Poon, L. L. (2003). Isolation and characterization of viruses related to the SARS coronavirus from animals in southern China. *Science*, *302*(5643), 276–278. doi:10.1126cience.1087139 PMID:12958366

Gubler, D. J. (2011). Dengue, urbanization, and globalization: The unholy trinity of the 21st century. *Tropical Medicine and Health*, *39*(4), 3–11. doi:10.2149/tmh.2011-S05 PMID:22500131

Guill, C. K., & Shandera, W. X. (2001). The effects of Hurricane Mitch on a community in northern Honduras. *Prehospital and Disaster Medicine*, *16*(3), 166–171. doi:10.1017/S1049023X00025929 PMID:11875801

Gupta, S., Allen-Vercoe, E., & Petrof, E. (2015). Fecal microbiota transplantation: In perspective. *Therapeutic Advances in Gastroenterology*, *9*(2), 229–239. doi:10.1177/1756283X15607414 PMID:26929784

Gurevitz, J. M., Antman, J. G., Laneri, K., & Morales, J. M. (2021). Temperature, traveling, slums, and housing drive dengue transmission in a non-endemic metropolis. *PLoS Neglected Tropical Diseases*, *15*(6), e0009465. doi:10.1371/journal.pntd.0009465 PMID:34115753

Gutierrez-Arellano, C., & Mulligan, M. (2018). A review of regulation ecosystem services and disservices from faunal populations and potential impacts of agriculturalisation on their provision, globally. *Nature Conservation*, *30*, 1–39. doi:10.3897/natureconservation.30.26989

Hadrup, N., & Lam, H. R. (2014). Oral toxicity of silver ions, silver nanoparticles and colloidal silver–a review. *Regulatory Toxicology and Pharmacology*, *68*(1), 1–7. doi:10.1016/j.yrtph.2013.11.002 PMID:24231525

Hahn, M. B., Gurley, E. S., Epstein, J. H., Islam, M. S., Patz, J. A., Daszak, P., & Luby, S. P. (2014). The role of landscape composition and configuration on Pteropus giganteus roosting ecology and Nipah virus spillover risk in Bangladesh. *The American Journal of Tropical Medicine and Hygiene*, *90*(2), 247–255. doi:10.4269/ajtmh.13-0256 PMID:24323516

Ha, J., & Yeo, M.-K. (2018). The environmental effects of microplastics on aquatic ecosystems. *Molecular & Cellular Toxicology*, *14*(4), 353–359. doi:10.100713273-018-0039-8

Haldane, J. B. S. (1948). The rate of mutation of human genes. *Hereditas*, *35*(S1), 267–273. doi:10.1111/j.1601-5223.1949.tb03339.x

Haley, R. W. (2012). Controlling urban epidemics of West Nile virus infection. *Journal of the American Medical Association*, *308*(13), 1325–1326. doi:10.1001/2012.jama.11930 PMID:22922679

Halliday, J. E. B., Knobel, D. L., & Allan, K. J., de C. Bronsvoort, B. M., Handel, I., Agwanda, B., Cutler, S., Olack, B., Ahmed, A., Hartskeerl, R. A., Njenga, M. K., Cleaveland, S. & Breiman, R. F. (2013). Urban leptospirosis in Africa: Across- sectional survey of Leptospira infection in rodents in the Kibera urban settlement, Nairobi, Kenya. *The American Journal of Tropical Medicine and Hygiene*, *89*(6), 1095–1102. doi:10.4269/ajtmh.13-0415 PMID:24080637

Hall, S. R., Duffy, M. A., & Caceres, C. E. (2005). Selective predation and productivity jointly drive complex behavior in host-parasite systems. *American Naturalist*, *165*(1), 70–81. doi:10.1086/426601 PMID:15744671

Halstead, S. B. (1996). Human factors in emerging infectious disease. *Eastern Mediterranean Health Journal*, *2*(1), 21–29. doi:10.26719/1996.2.1.21

Hamblin, M. T., & Di Rienzo, A. (2000). Detection of the signature of natural selection in humans: Evidence from the Duffy blood group locus. *American Journal of Human Genetics*, *66*(5), 1669–1679. doi:10.1086/302879 PMID:10762551

Haque, F., Sturm-Ramirez, K., Homaira, N., Gurley, E. S., Hossain, M. J., Hasan, S. M. M., Chowdhury, S., Sarkar, S., Khan, A. K. M. D., Rahman, M., Rahman, M., & Luby, S. P. (2016). Influenza B virus outbreak at a religious residential school for boys in northern Bangladesh, 2011. *Influenza and Other Respiratory Viruses*, *11*(2), 165–169. doi:10.1111/irv.12430 PMID:27603154

Harrison, M. (1996). Medicine and the Culture of Command: The Case of Malaria Control in the British Army during the Two World Wars. *Medical History*, *40*(4), 437–452. doi:10.1017/S0025727300061688 PMID:8952289

Harvell, C. D., Mitchell, C. E., Ward, J. R., Altizer, S., Dobson, A. P., Ostfeld, R. S., & Samuel, M. D. (2002). Climate warming and disease risks for terrestrial and marine biota. *Science*, *296*(5576), 2158–2162. doi:10.1126cience.1063699 PMID:12077394

Hasan, M., Barman, S. D., Islam, S., & Reza, A. W. (2019). Skin Cancer Detection Using Convolutional Neural Network. In *Proceedings of the 2019 5th International Conference on Computing and Artificial Intelligence-ICCAI 19*. Association for Computing Machinery. 10.1145/3330482.3330525

Hassaan, M. A., & El Nemr, A. (2020). Pesticides pollution: Classifications, human health impact, extraction and treatment techniques. *Egyptian Journal of Aquatic Research*, *46*(3), 207–220. doi:10.1016/j.ejar.2020.08.007

Hassell, J. M., Begon, M., Ward, M. J., & Fèvre, E. M. (2017). Urbanization and Disease Emergence: Dynamics at the Wildlife-Livestock-Human Interface. *Trends in Ecology & Evolution*, *32*(1), 55–67. doi:10.1016/j.tree.2016.09.012 PMID:28029378

Hayes, N., O'Toole, L., & Halpenny, A. M. (2017). *Introducing Bronfenbrenner: A guide for practitioners and students in early years education*. Taylor & Francis. doi:10.4324/9781315646206

Hayman, D. T. S., Bowen, R. A., Cryan, P. M., McCracken, G. F., O'Shea, T. J., Peel, A. J., Gilbert, A., Webb, C. T., & Wood, J. L. N. (2013). Ecology of zoonotic infectious diseases in bats: Current knowledge and future directions. *Zoonoses and Public Health*, *60*(1), 2–21. doi:10.1111/zph.12000 PMID:22958281

He, J., Christakos, G., Zhang, W., & Wang, Y. (2017). A Space-Time Study of Hemorrhagic Fever with Renal Syndrome (HFRS) and Its Climatic Associations in Heilongjiang Province, China. *Frontiers in Applied Mathematics and Statistics*, *0*, 16. doi:10.3389/fams.2017.00016

Hennig, B., Petriello, M. C., Gamble, M. V., Surh, Y. J., Kresty, L. A., Frank, N., Rangkadilok, N., Ruchirawat, M., & Suk, W. A. (2018). The role of nutrition in influencing mechanisms involved in environmentally mediated diseases. *Reviews on Environmental Health*, *33*(1), 87–97. doi:10.1515/reveh-2017-0038 PMID:29381475

Hery, L., Guidez, A., Durand, A. A., Delannay, C., Normandeau-Guimond, J., Reynaud, Y., Issaly, J., Goindin, D., Legrave, G., Gustave, J., Raffestin, S., Breurec, S., Constant, P., Dusfour, I., Guertin, C., & Vega-Rúa, A. (2021). Natural variation in physicochemical profiles and bacterial communities associated with *Aedes aegypti* breeding sites and larvae on Guadeloupe and French Guiana. *Microbial Ecology*, *81*(1), 93–109. doi:10.100700248-020-01544-3 PMID:32621210

Heyneman, S. P., & Loxley, W. A. (1983). The effect of primary-school quality on academic achievement across twenty-nine high- and low-income countries. *American Journal of Sociology*, *88*(6), 1162–1194. doi:10.1086/227799 PMID:6614303

Hickson, M., D'Souza, A. L., Muthu, N., Rogers, T. R., Want, S., Rajkumar, C., & Bulpitt, C. J. (2007). Use of probiotic Lactobacillus preparation to prevent diarrhoea associated with antibiotics: Randomised double blind placebo controlled trial. *BMJ (Clinical Research Ed.)*, *335*(7610), 80. doi:10.1136/bmj.39231.599815.55 PMID:17604300

Hien, T. T., & White, N. J. (1993). Qinghaosu. *Lancet*, *341*(8845), 603–608. doi:10.1016/0140-6736(93)90362-K PMID:8094838

Hill, C., Guarner, F., Reid, G., Gibson, G. R., Merenstein, D. J., Pot, B., Morelli, L., Canani, R. B., Flint, H. J., Salminen, S., Calder, P. C., & Sanders, M. E. (2014). The International Scientific Association for Probiotics and Prebiotics consensus statement on the scope and appropriate use of the term probiotic. *Nature Reviews. Gastroenterology & Hepatology*, *11*(8), 506–514. doi:10.1038/nrgastro.2014.66 PMID:24912386

Himsworth, C. G., Parsons, K. L., Jardine, C., & Patrick, D. M. (2013). Rats, cities, people, and pathogens: A systemic review and narrative synthesis of literature regarding the ecology of rat-associated zoonoses in urban centers. *Vector Borne and Zoonotic Diseases (Larchmont, N.Y.)*, *13*(6), 349–359. doi:10.1089/vbz.2012.1195 PMID:23590323

Hoffman, J. B., Petriello, M. C., Morris, A. J., Mottaleb, M. A., Sui, Y., Zhou, C., Deng, P., Wang, C., & Hennig, B. (2020). Prebiotic inulin consumption reduces dioxin-like PCB 126-mediated hepatotoxicity and gut dysbiosis in hyperlipidemic Ldlr deficient mice. *Environmental Pollution*, *261*, 114183. doi:10.1016/j.envpol.2020.114183 PMID:32105967

Honda, K., & Littman, D. R. (2016). The microbiota in adaptive immune homeostasis and disease. *Nature*, *535*(7610), 75–84. doi:10.1038/nature18848 PMID:27383982

Horton, A. A., Walton, A., Spurgeon, D. J., Lahive, E., & Svendsen, C. (2017). Microplastics in freshwater and terrestrial environments: Evaluating the current understanding to identify the knowledge gaps and future research priorities. *The Science of the Total Environment*, *586*, 127–141. doi:10.1016/j.scitotenv.2017.01.190 PMID:28169032

Hoshen, M. B., & Morse, A. P. (2004). A weather-driven model of malaria transmission. *Malaria Journal*, *3*(1), 32. doi:10.1186/1475-2875-3-32 PMID:15350206

Hossain, M. A., & Uddin, M. E. (2021). Education and health: Mediating role of socioeconomic status and health lifestyle among Muslim, Hindu, and Santal men in rural Bangladesh. *The Journal of School Health*, *4*(2), 13–29.

Huerta, E., Mendoza Vega, J., Quej, V. K., Chi, J. de los A., Sanchez del Cid, L., Quijano, C., Escalona-Segura, G., Gertsen, H., Salánki, T., & van der Ploeg, M. (2017). Bioaccumulation of microplastics in the terrestrial food chain: An example from home gardens in SE Mexico. *EGU General Assembly Conference Abstracts*, 15847.

Hughes, M. F., Beck, B. D., Chen, Y., Lewis, A. S., & Thomas, D. J. (2011). Arsenic exposure and toxicology: A historical perspective. *Toxicological Sciences*, *123*(2), 305–332. doi:10.1093/toxsci/kfr184 PMID:21750349

Hugo, G. (2010). *Climate change-induced mobility and the existing migration regime in Asia and the Pacific*. Hart Publishing.

Hunter, P. R., Risebro, H., Yen, M., Lefebvre, H., Lo, C., Hartemann, P., Longuet, C., & Jaquenoud, F. (2014). Impact of the provision of safe drinking water on school absence rates in Cambodia: A quasi-experimental study. *PLoS One*, *9*(3), e91847. doi:10.1371/journal.pone.0091847 PMID:24632573

Institute of Medicine. (2004). Saving Lives, Buying Time: Economics of Malaria Drugs in an Age of Resistance. National Academies Press.

IPCC. (2001). Climate Change 2001: Synthesis Report. In A Contribution of Working Groups I, II, and III to the Third Assessment Report of the Integovernmental Panel on Climate Change. Cambridge University Press.

IPCC. (2007). Fourth Assessment Report. *Climatic Change*.

IPCC. (2012). *Managing the Risks of Extreme Events and Disasters to Advance Climate Change Adaptation. Special Report of the Intergovernmental Panel on Climate Change*. Cambridge University Press.

IPCC. (2014a). Regional Context. In *Climate Change 2014: Impacts, Adaptation, and Vulnerability. Part B: Regional Aspects. Contribution of Working Group II to the Fifth Assessment Report of the Intergovernmental Panel on Climate Change*. Cambridge University Press.

IPCC. (2014b). Adaptation Needs and Options. In *Climate Change 2014: Impacts, Adaptation, and Vulnerability. Part A: Global and Sectoral Aspects. Contribution of Working Group II to the Fifth Assessment Report of the Intergovernmental Panel on Climate Change*. Cambridge University Press.

IPCC. (2014c). Summary for Policymakers. In *Climate Change 2014: Impacts, Adaptation, and Vulnerability. Part A: Global and Sectoral Aspects. Contribution of Working Group II to the Fifth Assessment Report of the Intergovernmental Panel on Climate Change*. Cambridge University Press.

Islam, M. S., Sharker, M. A. Y., Rheman, S., Hossain, S., Mahmud, Z. H., Islam, M. S., Uddin, A. M. K., Yunus, M., Osman, M. S., Ernst, R., Rector, I., Larson, C. P., Luby, S. P., Endtz, H. P., & Cravioto, A. (2009). Effects of local climate variability on transmission dynamics of cholera in Matlab, Bangladesh. *Transactions of the Royal Society of Tropical Medicine and Hygiene*, *103*(11), 1165–1170. doi:10.1016/j.trstmh.2009.04.016 PMID:19477477

Ismail, S. A., Saliba, V., Bernal, J. L., Ramsay, M. E., & Ladhani, S. N. (2021). SARS-CoV-2 infection and transmission in educational settings: A prospective, cross-sectional analysis of infection clusters and outbreaks in England. *The Lancet. Infectious Diseases*, *21*(3), 344–353. doi:10.1016/S1473-3099(20)30882-3 PMID:33306981

IWGCCH. (2010). A human health perspective on climate change. In K. T. Tart (Ed.), *A Report Outlining the Research Needs on the Human Health Effects of Climate Change: Environmental Health Perspectives and the National Institute of Environmental Health Sciences*. Author.

Jabr, F. (2022). *The Age of Infection*. Foreign Policy. Retrieved on 3 June 2022, from https://foreignpolicy.com/2015/09/30/the-age-of-infection-antibiotics-microbes-germ-ichip/

Jamal, F., Fletcher, A., Harden, A., Wells, H., Thomas, J., & Bonell, C. (2013). The school environment and student health: A systematic review and meta-ethnography of qualitative research. *BMC Public Health*, *13*(1), 1–11. doi:10.1186/1471-2458-13-798 PMID:24007211

Jandhyala, S. M., Talukdar, R., Subramanyam, C., Vuyyuru, H., Sasikala, M., & Reddy, D. N. (2015). Role of the normal gut microbiota. *World Journal of Gastroenterology: WJG*, *21*(29), 8787. doi:10.3748/wjg.v21.i29.8787 PMID:26269668

Jemal, Y., & Al-Thukair, A. A. (2018). Combining GIS application and climatic factors for mosquito control in Eastern Province. *Saudi Arabia. Saudi Journal of Biological Sciences*, *25*(8), 1593–1602. doi:10.1016/j.sjbs.2016.04.001 PMID:30591775

Jia, P. P., Junaid, M., Wen, P. P., Yang, Y. F., Li, W. G., Yang, X. G., & Pei, D. S. (2021). Role of germ-free animal models in understanding interactions of gut microbiota to host and environmental health: A special reference to zebrafish. *Environmental Pollution*, *279*, 116925. doi:10.1016/j.envpol.2021.116925 PMID:33744636

Jia, Z.-Q., Zhang, Y.-C., Huang, Q.-T., Jones, A. K., Han, Z.-J., & Zhao, C.-Q. (2020). Acute toxicity, bioconcentration, elimination, action mode and detoxification metabolism of broflanilide in zebrafish, Danio rerio. *Journal of Hazardous Materials*, *394*, 122521. doi:10.1016/j.jhazmat.2020.122521 PMID:32279005

Jin, Y., Chen, R., Liu, W., & Fu, Z. (2010). Effect of endocrine disrupting chemicals on the transcription of genes related to the innate immune system in the early developmental stage of zebrafish (*Danio rerio*). *Fish & Shellfish Immunology*, *28*(5–6), 854–861. doi:10.1016/j.fsi.2010.02.009 PMID:20153439

Jin, Y., Lu, L., Tu, W., Luo, T., & Fu, Z. (2019). Impacts of polystyrene microplastic on the gut barrier, microbiota and metabolism of mice. *The Science of the Total Environment*, *649*, 308–317. doi:10.1016/j.scitotenv.2018.08.353 PMID:30176444

Jin, Y., Wu, S., Zeng, Z., & Fu, Z. (2017). Effects of environmental pollutants on gut microbiota. *Environmental Pollution*, *222*, 1–9. doi:10.1016/j.envpol.2016.11.045 PMID:28086130

Jin, Y., Xia, J., Pan, Z., Yang, J., Wang, W., & Fu, Z. (2018). Polystyrene microplastics induce microbiota dysbiosis and inflammation in the gut of adult zebrafish. *Environmental Pollution*, *235*, 322–329. doi:10.1016/j.envpol.2017.12.088 PMID:29304465

Jin, Y., Zeng, Z., Wu, Y., Zhang, S., & Fu, Z. (2015). Oral exposure of mice to carbendazim induces hepatic lipid metabolism disorder and gut microbiota dysbiosis. *Toxicological Sciences*, *147*(1), 116–126. doi:10.1093/toxsci/kfv115 PMID:26071454

Jin, Y., Zhu, Z., Wang, Y., Yang, E., Feng, X., & Fu, Z. (2016). The fungicide imazalil induces developmental abnormalities and alters locomotor activity during early developmental stages in zebrafish. *Chemosphere*, *153*, 455–461. doi:10.1016/j.chemosphere.2016.03.085 PMID:27035382

Jo, A., Ding, T., & Ahn, J. (2016). Synergistic antimicrobial activity of bacteriophages and antibiotics against Staphylococcus aureus. *Food Science and Biotechnology*, *25*(3), 935–940. doi:10.100710068-016-0153-0 PMID:30263357

Jobbins, S. E., Sanderson, C. E., & Alexander, K. A. (2014). Leptospira interrogans at the human-wildlife interface in northern Botswana: A newly identified public health threat. *Zoonoses and Public Health*, *61*(2), 113–123. doi:10.1111/zph.12052 PMID:23672285

Jofre, J., Blanch, A. R., & Lucena, F. (2010). Water-borne infectious disease outbreaks associated with water scarcity and rainfall events. In S. Sabater & D. Barcelo (Eds.), *Water Scarcity in the Mediterranean: Perspectives Under Global Change*. Springer.

Johnson, C. K., Hitchens, P. L., Evans, T. S., Goldstein, T., Thomas, K., Clements, A., Joly, D. O., Wolfe, N. D., Daszak, P., Karesh, W. B., & Mazet, J. K. (2015). Spillover and pandemic properties of zoonotic viruses with high host plasticity. *Scientific Reports*, *5*, 1–8. PMID:26445169

Jones, B. A., Grace, D., Kock, R., Alonso, S., Rushton, J., Said, M. Y., McKeever, D., Mutua, F., Young, J., McDermott, J., & Pfeiffer, D. U. (2013). Zoonosis emergence is linked to agricultural intensification and environmental change. *Proceedings of the National Academy of Sciences of the United States of America*, *110*(21), 8399–8404. doi:10.1073/pnas.1208059110 PMID:23671097

Jones, F. K., Wamala, J. F., Rumunu, J., Mawien, P. N., Kol, M. T., Wohl, S., Deng, L., Pezzoli, L., Omar, L. H., Lessler, J., Quilici, M. L., Luquero, F. J., & Azman, A. S. (2020). Successive epidemic waves of cholera in South Sudan between 2014 and 2017: A descriptive epidemiological study. *The Lancet. Planetary Health*, *4*(12), e577–e587. doi:10.1016/S2542-5196(20)30255-2 PMID:33278375

Jones, K. E., Patel, N. J., Levy, M. A., Storeygard, A., Balk, D., Gittleman, J. L., & Daszak, P. (2008). Global trends in emerging infectious diseases. *Nature*, *451*(7181), 990–993. doi:10.1038/nature06536 PMID:18288193

Jones, S., Axelrad, R., & Wattigney, W. (2007). Healthy and safe school environment, part II, physical school environment: Results from the school health policies and programs study 2006. *The Journal of School Health*, *77*(8), 544–556. doi:10.1111/j.1746-1561.2007.00234.x PMID:17908107

Kagan, L. J., Aiello, A. E., & Larson, E. (2002). The role of the home environment in the transmission of infectious diseases. *Journal of Community Health*, *27*(4), 247–267. doi:10.1023/A:1016378226861 PMID:12190054

Kala, S., Naik, S. N., Patanjali, P. K., & Sogan, N. (2019). Neem oil water dispersible tablet as effective larvicide, ovicide and oviposition deterrent against *Anopheles culicifacies*. *South African Journal of Botany*, *123*, 387–392. doi:10.1016/j.sajb.2019.03.033

Kalouche, S. (2016). *Vision-Based Classification of Skin Cancer Using Deep Learning*. Available online:https://www.semanticscholar.org/paper/Vision-Based-Classification-of-Skin-Cancer-using-Kalouche/b57ba909756462d812dc20fca157b3972bc1f533

Kamat, V. (2000). Resurgence of Malaria in Bombay. *Parasitologia*, *42*, 135–148. PMID:11234324

Kan, H., Zhao, F., Zhang, X.-X., Ren, H., & Gao, S. (2015). Correlations of gut microbial community shift with hepatic damage and growth inhibition of Carassius auratus induced by pentachlorophenol exposure. *Environmental Science & Technology*, *49*(19), 11894–11902. doi:10.1021/acs.est.5b02990 PMID:26378342

Kanimozhi, T., & Murthi, D. A. (2016). Computer-Aided Melanoma Skin Cancer Detection Using Artificial Neural Network Classifier. *J. Sel. Areas Microelectron.*, *8*, 35–42.

Kapoor, G., Saigal, S., & Elongavan, A. (2021). *Action and resistance mechanisms of antibiotics: A guide for clinicians.* Academic Press.

Karesh, W. B., & Cook, R. A. (2005). The human-animal link. *Foreign Affairs, 84*(4), 38. doi:10.2307/20034419

Karesh, W. B., Dobson, A., Lloyd-Smith, J. O., Lubroth, J., Dixon, M. A., Bennett, M., Aldrich, S., Harrington, T., Formenty, P., Loh, E. H., Machalaba, C. C., Thomas, M. J., & Heymann, D. L. (2012). Ecology of zoonoses: Natural and unnatural histories. *Lancet, 380*(9857), 1936–1945. doi:10.1016/S0140-6736(12)61678-X PMID:23200502

Karim, M. R., Mondal, M. N. I., Haque, M. J., Aik, S., Chakrabarty, S., & Hossain, M. G. (2019). Preventive knowledge on seven infectious diseases in rural Rajshahi District of Bangladesh: A cross-sectional study. *Human Biology Review, 8*(1), 9–24.

Karlekar, S., & Andrew, R. (2016). Behavioral resilience of *Culex quinquefasciatus* Say 1823 at Nagpur district of Maharashtra. *International Journal of Mosquito Research, 3*(5), 25–30.

Katz, L., & Baltz, R. (2016). Natural product discovery: Past, present, and future. *Journal of Industrial Microbiology & Biotechnology, 43*(2-3), 155–176. doi:10.100710295-015-1723-5 PMID:26739136

Kaur, I. (2016). Novel Strategies to Combat Antimicrobial Resistance. *Journal of Infectious Disease and Therapy, 4*(4). Advance online publication. doi:10.4172/2332-0877.1000292

Kedzierski, L. (2011). Leishmaniasis. *Human Vaccines, 7*(11), 1204–1214. doi:10.4161/hv.7.11.17752 PMID:22048116

Keesing, F., Holt, R. D., & Ostfeld, R. S. (2006). Effects of species diversity on disease risk. *Ecology Letters, 9*(4), 485–498. doi:10.1111/j.1461-0248.2006.00885.x PMID:16623733

Kelly, M., & Coughlan, B. (2019). A theory of youth mental health recovery from a parental perspective. *Child and Adolescent Mental Health, 24*(2), 161–169. doi:10.1111/camh.12300 PMID:32677184

Kennedy, K., & Collignon, P. (2010). Colonisation with Escherichia coli resistant to "critically important" antibiotics: A high risk for international travellers. *European Journal of Clinical Microbiology & Infectious Diseases, 29*(12), 1501–1506. doi:10.100710096-010-1031-y PMID:20835879

Kermack, W. O., & McKendrick, A. G. (1922). Contribution to the mathematical theory to epidemics. *Proceedings of the Royal Society of London. Series A, Containing Papers of a Mathematical and Physical Character, 115*, 100–121.

Khan, S., Haque, M. N., & Hossen, M. S. (2021). Nutritional status of autistic school children: An evidence from Rajshahi city in Bangladesh. *JSM Nutritional Disorders, 3*(1). Advance online publication. doi:10.47739/2578-3203/1010 PMID:29881374

Kim, K.-H., Kabir, E., & Jahan, S. A. (2017). Exposure to pesticides and the associated human health effects. *The Science of the Total Environment, 575*, 525–535. doi:10.1016/j.scitotenv.2016.09.009 PMID:27614863

Klein, H. S. (1999). The Atlantic Slave Trade. Cambridge University Press.

Ko, A. I., Galvao Reis, M., Ribeiro Dourado, C. M., Johnson, W. D. Jr, & Riley, L. W. (1999). Urban epidemic of severe leptospirosis in Brazil. Salvador Leptospirosis Study Group. *Lancet*, *354*(9181), 820–825. doi:10.1016/S0140-6736(99)80012-9 PMID:10485724

Koenraadt, C. J. M. (2021). Larval source management for malaria control: Prospects for new technologies and community involvement. In C. J. M. Koenraadt, J. Spitzen, & W. Takken (Eds.), *Innovative strategies for vector control: Progress in the global vector control response, Ecology and Control of Vector-borne Diseases* (Vol. 6, pp. 155–167). Wageningen Academic Publishers. doi:10.3920/978-90-8686-895-7_8

Kolimenakis, A., Heinz, S., Wilson, M. L., Winkler, V., Yakob, L., Michaelakis, A., Papachristos, D., Richardson, C., & Horstick, O. (2021). The role of urbanisation in the spread of *Aedes* mosquitoes and the diseases they transmit—A systematic review. *PLoS Neglected Tropical Diseases*, *15*(9), e0009631. doi:10.1371/journal.pntd.0009631 PMID:34499653

Konturek, P. C., Koziel, J., Dieterich, W., Haziri, D., Wirtz, S., Glowczyk, I., Konturek, K., Neurath, M. F., & Zopf, Y. (2016). Successful therapy of Clostridium difficile infection with fecal microbiota transplantation. *Journal of Physiology and Pharmacology*, *67*(6), 859–866. PMID:28195066

Korzeniewska-Kosela, M. (2013). Tuberculosis in Poland in 2011. *Przegląd Epidemiologiczny*, *67*(2), 277–281, 375–378. PMID:24040733

Koual, M., Cano-Sancho, G., Bats, A.-S., Tomkiewicz, C., Kaddouch-Amar, Y., Douay-Hauser, N., Ngo, C., Bonsang, H., Deloménie, M., Lecuru, F., Le Bizec, B., Marchand, P., Botton, J., Barouki, R., Antignac, J.-P., & Coumoul, X. (2019). Associations between persistent organic pollutants and risk of breast cancer metastasis. *Environment International*, *132*, 105028. doi:10.1016/j.envint.2019.105028 PMID:31382183

Kovats, R. S., Menne, B., McMichael, A. J., Corvalan, C., & Bertollini, R. (2000). *Climate Change and Human Health: Impact and Adaptation*. World Health Organization.

Krief, S., Escalante, A. A., Pacheco, M. A., Mugisha, L., André, C., Halbwax, M., Fischer, A., Krief, J. M., Kasenene, J. M., Crandfield, M., Cornejo, O. E., Chavatte, J.-M., Lin, C., Letourneur, F., Grüner, A. C., McCutchan, T. F., Rénia, L., & Snounou, G. (2010). On the diversity of malaria parasites in African apes and the origin of Plasmodium falciparum from Bonobos. *PLoS Pathogens*, *6*(2), e1000765. doi:10.1371/journal.ppat.1000765 PMID:20169187

Krishnamoorthy, K., Jambulingam, P., Natarajan, R., Shriram, A. N., Das, P. K., & Sehgal, S. C. (2005). Altered environment and risk of malaria outbreak in South Andaman, Andaman & Nicobar Islands, India affected by tsunami disaster. *Malaria Journal*, *4*(1), 32. doi:10.1186/1475-2875-4-32 PMID:16029514

Krupnick, J. G., & Benovic, J. L. (1998). The role of receptor kinases and arrestins in G protein-coupled receptor regulation. *Annual Review of Pharmacology and Toxicology*, *38*(1), 289–319. doi:10.1146/annurev.pharmtox.38.1.289 PMID:9597157

Kugeler, K. J., Schwartz, A. M., Delorey, M. J., Mead, P. S., & Hinckley, A. F. (2021). Estimating the Frequency of Lyme Disease Diagnoses, United States, 2010–2018 - Volume 27, Number 2—February 2021 - Emerging Infectious Diseases journal - CDC. *Emerging Infectious Diseases, 27*(2), 616–619. doi:10.3201/eid2702.202731 PMID:33496229

Kuhn, K., Campbell-Lendrum, D., Haines, A., & Cox, J. (2005). *Using Climate to Predict Infectious Disease Epidemics*. World Health Organization.

Kumar, S., Warikoo, R., Mishra, M., Samal, R. R., Shrankhla, Panmei, K., Dagar, V. S., & Sharma, A. (2017). Impact of *Ocimum basilicum* leaf essential oil on the survival and behaviour of an Indian strain of dengue vector, *Aedes aegypti* (L.). *Vector Biology Journal, 2*(2), 2. doi:10.4172/2473-4810.1000122

Kutsyuruba, B., Klinger, D. A., & Hussain, A. (2015). Relationships among school climate, school safety, and student achievement and well-being: A review of literature. *Review of Education, 3*(2), 103–135. doi:10.1002/rev3.3043

Kwan, J. A., Novak, M. G., Hyles, T. S., & Niemela, M. K. (2009). Mortality of nontarget arthropods from an aerial application of pyrethrins. *Journal of the American Mosquito Control Association, 25*(2), 218–220. doi:10.2987/08-5858.1 PMID:19653509

Kyte, R. (2014). *Climate Change is a Challenge for Sustainable Development*. The World Bank.

Landy, J., Al-Hassi, H. O., Mclaughlin, S. D., Walker, A. W., Ciclitira, P. J., Nicholls, R. J., Clark, S. K., & Hart, A. L. (2011). faecal transplantation therapy for gastrointestinal disease. *Alimentary Pharmacology & Therapeutics, 34*(4), 409–415. doi:10.1111/j.1365-2036.2011.04737.x PMID:21682755

Langford, R., Bonell, C. P., Jones, H. E., Pouliou, T., Murphy, S. M., Waters, E., Komro, A. A., Gibbs, L. F., Magnus, D., & Campbell, R. (2014). The WHO health promoting school framework for improving the health and well-being of students and their academic achievement. *Cochrane Database of Systematic Reviews*, 4. doi:10.1002/14651858.CD008958.pub2 PMID:24737131

Langkabel, P., Zwirner, J., & Oppermann, M. (1999). Ligand-induced phosphorylation of anaphylatoxin receptors C3aR and C5aR is mediated by G protein-coupled receptor kinases. *European Journal of Immunology, 29*(9), 3035–3046. doi:10.1002/(SICI)1521-4141(199909)29:09<3035::AID-IMMU3035>3.0.CO;2-Z PMID:10508278

Latz, S., Wahida, A., Arif, A., Häfner, H., Hoß, M., Ritter, K., & Horz, H. (2016). A preliminary survey of local bacteriophages with lytic activity against multi-drug resistant bacteria. *Journal of Basic Microbiology, 56*(10), 1117–1123. doi:10.1002/jobm.201600108 PMID:27194637

Lawson, B., de Pinna, E., Horton, R. A., Macgregor, S. K., John, S. K., Chantrey, J., Duff, J. P., Kirkwood, J. K., Simpson, V. R., Robinson, R. A., Wain, J., & Cunningham, A. A. (2014). Epidemiological evidence that garden birds are a source of human salmonellosis in England and Wales. *PLoS One, 9*(2), e88968. doi:10.1371/journal.pone.0088968 PMID:24586464

Lederberg, J. (2000). Infectious History. *Science*, *288*(5464), 287–293. doi:10.1126cience.288.5464.287 PMID:10777411

Leech, J. H., Barnwell, J. W., Miller, L. H., & Howard, R. J. (1984). Identification of a strain-specific malarial antigen exposed on the surface of *Plasmodium falciparum*-infected erythrocytes. *The Journal of Experimental Medicine*, *159*(6), 1567–1575. doi:10.1084/jem.159.6.1567 PMID:6374009

Lee, S. A., Economou, T., Barcellos, C., Catão, R., Carvalho, M. S., & Lowe, R. (2021). Effect of climate change, connectivity, and socioeconomic factors on the expansion of the dengue virus transmission zone in 21st century Brazil: An ecological modelling study. *The Lancet. Planetary Health*, *5*, S14. doi:10.1016/S2542-5196(21)00098-X

Lefever, D. E., Xu, J., Chen, Y., Huang, G., Tamas, N., & Guo, T. L. (2016). TCDD modulation of gut microbiome correlated with liver and immune toxicity in streptozotocin (STZ)-induced hyperglycemic mice. *Toxicology and Applied Pharmacology*, *304*, 48–58. doi:10.1016/j.taap.2016.05.016 PMID:27221631

Leoratti, F. M. D. S., Trevelin, S. C., Cunha, F. Q., Rocha, B. C., Costa, P. A. C., Gravina, H. D., Tada, M. S., Pereira, D. B., Golenbock, D. T., do Valle Antonelli, L. R., & Gazzinelli, R. T. (2012). Neutrophil paralysis in Plasmodium vivax malaria. *PLoS Neglected Tropical Diseases*, *6*(6), e1710. doi:10.1371/journal.pntd.0001710 PMID:22745844

Li, L., Jiang, C., Murtugudde, R., Liang, X.-Z., & Sapkota, A. (2021). Global Population Exposed to Extreme Events in the 150 Most Populated Cities of the World: Implications for Public Health. *International Journal of Environmental Research and Public Health, 18*(3), 1293. doi:10.3390/ijerph18031293

Libertucci, J., & Young, V. B. (2019). The role of the microbiota in infectious diseases. *Nature Microbiology*, *4*(1), 35–45. doi:10.103841564-018-0278-4 PMID:30546094

Li, J., Liu, H., & Chen, J. P. (2018). Microplastics in freshwater systems: A review on occurrence, environmental effects, and methods for microplastics detection. *Water Research*, *137*, 362–374. doi:10.1016/j.watres.2017.12.056 PMID:29580559

Lim, J., Nam, C., Yang, J., Rha, K. H., Lim, K.-M., & Jee, S. H. (2017). Serum persistent organic pollutants (POPs) and prostate cancer risk: A case-cohort study. *International Journal of Hygiene and Environmental Health*, *220*(5), 849–856. doi:10.1016/j.ijheh.2017.03.014 PMID:28420543

Li, N., Ulucak, R., & Danish. (2022). Turning points for environmental sustainability: The potential role of income inequality, human capital, and globalization. *Environmental Science and Pollution Research International*, *29*(27), 40878–40892. doi:10.100711356-021-18223-6 PMID:35083689

Linkon, M. R., Prodhan, U. K., Liton, M., Islam, M. K., Lisa, L. A., & Paul, D. K. (2015). Prevalence of contagious diseases of school going children in Bangladesh. *Journal of Biosciences and Medicines*, *3*(7), 17–24. doi:10.4236/jbm.2015.37003

Lin, L., & Zhang, J. (2017). Role of intestinal microbiota and metabolites on gut homeostasis and human diseases. *BMC Immunology*, *18*(1), 1–25. doi:10.118612865-016-0187-3 PMID:28061847

Lippard, C. N., La Paro, K. M., Rouse, H. L., & Crosby, D. A. (2018). A closer look at teacher–child relationships and classroom emotional context in preschool. *Child and Youth Care Forum*, *47*(1), 1–21. doi:10.100710566-017-9414-1

Li, R., Jiang, Z. B., & Xu, B. (2014). Global spatiotemporal and genetic footprint of the H5N1 avian influenza virus. *International Journal of Health Geographics*, *13*(1), 14. doi:10.1186/1476-072X-13-14 PMID:24885233

Liu, C., Pitts, R. J., Bohbot, J. D., Jones, P. L., Wang, G., & Zwiebel, L. J. (2010). Distinct olfactory signaling mechanisms in the malaria vector mosquito *Anopheles gambiae*. *PLoS Biology*, *8*(8), e1000467. doi:10.1371/journal.pbio.1000467 PMID:20824161

Liu-Helmersson, J., Brännström, Å., Sewe, M. O., Semenza, J. C., & Rocklöv, J. (2019). Estimating past, present, and future trends in the global distribution and abundance of the arbovirus vector *Aedes aegypti* under climate change scenarios. *Frontiers in Public Health*, *7*, 148. doi:10.3389/fpubh.2019.00148 PMID:31249824

Liu, Y., Li, Y., Xia, Y., Liu, K., Ren, L., & Ji, Y. (2020). The dysbiosis of gut microbiota caused by low-dose cadmium aggravate the injury of mice liver through increasing intestinal permeability. *Microorganisms*, *8*(2), 211. doi:10.3390/microorganisms8020211 PMID:32033263

Liu, Z., Wang, Y., Zhu, Z., Yang, E., Feng, X., Fu, Z., & Jin, Y. (2016). Atrazine and its main metabolites alter the locomotor activity of larval zebrafish (Danio rerio). *Chemosphere*, *148*, 163–170. doi:10.1016/j.chemosphere.2016.01.007 PMID:26803580

Liu, Z., Zhang, Z., Lai, Z., Zhou, T., Jia, Z., Gu, J., Wu, K., & Chen, X.-G. (2017). Temperature Increase Enhances Aedes albopictus Competence to Transmit Dengue Virus. *Frontiers in Microbiology*, *0*(DEC), 2337. doi:10.3389/fmicb.2017.02337 PMID:29250045

Li, W., Shi, Z., Yu, M., Ren, W., Smith, C., Epstein, J. H., Wang, H., Crameri, G., Hu, Z., Zhang, H., Zhang, J., McEachern, J., Field, H., Daszak, P., Eaton, B. T., Zhang, S., & Wang, L. F. (2005). Bats are natural reservoirs of SARS-like coronaviruses. *Science*, *310*(5748), 676–679. doi:10.1126cience.1118391 PMID:16195424

Lloyd, S. J., Kovats, R. S., & Armstrong, B. G. (2007). Global diarrhoea morbidity, weather and climate. *Climate Research*, *34*, 119–127. doi:10.3354/cr034119

Lloyd-Smith, J. O., George, D., Pepin, K. M., Pitzer, V. E., Pulliam, J. R. C., Dobson, A. P., Hudson, P. J., & Grenfell, B. T. (2009). Epidemic dynamics at the human-animal interface. *Science*, *326*(5958), 1362–1367. doi:10.1126cience.1177345 PMID:19965751

Lofgren, E., Fefferman, N. H., Naumov, Y. N., Gorski, J., & Naumova, E. N. (2007). Influenza seasonality: Underlying causes and modeling theories. *Journal of Virology*, *81*(11), 5429–5436. doi:10.1128/JVI.01680-06 PMID:17182688

Longdon, B., Hadfield, J. D., Webster, C. L., Obbard, D. J., & Jiggins, F. M. (2011). Host phylogeny determines viral persistence and replication in novel hosts. *PLoS Pathogens*, *7*(9), e1002260. Advance online publication. doi:10.1371/journal.ppat.1002260 PMID:21966271

Lowy, F. (2003). Antimicrobial resistance: The example of Staphylococcus aureus. *The Journal of Clinical Investigation*, *111*(9), 1265–1273. doi:10.1172/JCI18535 PMID:12727914

Loy, D. E., Liu, W., Li, Y., Learn, G. H., Plenderleith, L. J., Sundararaman, S. A., Sharp, P. M., & Hahn, B. H. (2017). Out of Africa: Origins and evolution of the human malaria parasites *Plasmodium falciparum* and *Plasmodium vivax*. *International Journal for Parasitology*, *47*(2-3), 87–97. doi:10.1016/j.ijpara.2016.05.008 PMID:27381764

Lucas, R. M., Yazar, S., Young, A. R., Norval, M., De Gruijl, F. R., Takizawa, Y., Rhodes, L. E., Sinclair, C. A., & Neale, R. E. (2019). Human health in relation to exposure to solar ultraviolet radiation under changing stratospheric ozone and climate. *Photochemical & Photobiological Sciences*, *18*(3), 641–680. doi:10.1039/c8pp90060d PMID:30810559

Lucius, R., Loos-Frank, B., Lane, R., Poulin, R., Roberts, C., & Grencis. R. (2017). The Biology of Parasites. Wiley.

Lu, K., Abo, R. P., Schlieper, K. A., Graffam, M. E., Levine, S., Wishnok, J. S., Swenberg, J. A., Tannenbaum, S. R., & Fox, J. G. (2014). Arsenic exposure perturbs the gut microbiome and its metabolic profile in mice: An integrated metagenomics and metabolomics analysis. *Environmental Health Perspectives*, *122*(3), 284–291. doi:10.1289/ehp.1307429 PMID:24413286

Lu, L., Luo, T., Zhao, Y., Cai, C., Fu, Z., & Jin, Y. (2019). Interaction between microplastics and microorganism as well as gut microbiota: A consideration on environmental animal and human health. *The Science of the Total Environment*, *667*, 94–100. doi:10.1016/j.scitotenv.2019.02.380 PMID:30826685

Lu, L., Wan, Z., Luo, T., Fu, Z., & Jin, Y. (2018). Polystyrene microplastics induce gut microbiota dysbiosis and hepatic lipid metabolism disorder in mice. *The Science of the Total Environment*, *631*, 449–458. doi:10.1016/j.scitotenv.2018.03.051 PMID:29529433

Lumpkin, W. P., Stirek, K. R., & Dyer, L. A. (2020). Macrophyte diversity and complexity reduce larval mosquito abundance. *Journal of Medical Entomology*, *57*(4), 1041–1048. doi:10.1093/jme/tjaa012 PMID:32006421

Lutz, E. K., Lahondère, C., Vinauger, C., & Riffell, J. A. (2017). Olfactory learning and chemical ecology of olfaction in disease vector mosquitoes: A life history perspective. *Current Opinion in Insect Science*, *20*, 75–83. doi:10.1016/j.cois.2017.03.002 PMID:28602240

Lv, L.-X., Jiang, H.-Y., Yan, R., & Li, L. (2019). Interactions Between Gut Microbiota and Hosts and Their Role in Infectious Diseases. *Infectious Microbes and Diseases*, *1*(1), 3–9. doi:10.1097/IM9.0000000000000001

Macfarlane, G. T., & Macfarlane, S. (2011). Fermentation in the human large intestine: Its physiologic consequences and the potential contribution of prebiotics. *Journal of Clinical Gastroenterology*, *45*, S120–S127. doi:10.1097/MCG.0b013e31822fecfe PMID:21992950

Maciel-Fiuza, M. F., Muller, G. C., Campos, D. M. S., do Socorro Silva Costa, P., Peruzzo, J., Bonamigo, R. R., Veit, T., & Vianna, F. S. L. (2023). Role of gut microbiota in infectious and inflammatory diseases. *Frontiers in Microbiology*, *14*(March), 1098386. Advance online publication. doi:10.3389/fmicb.2023.1098386 PMID:37051522

Maddalon, A., Galbiati, V., Colosio, C., Mandić-Rajčević, S., & Corsini, E. (2021). Glyphosate-based herbicides: Evidence of immune-endocrine alteration. *Toxicology*, *459*, 152851. doi:10.1016/j.tox.2021.152851 PMID:34246717

Mahlapuu, M., Håkansson, J., Ringstad, L., & Björn, C. (2016). Antimicrobial Peptides: An Emerging Category of Therapeutic Agents. *Frontiers in Cellular and Infection Microbiology*, *6*. Advance online publication. doi:10.3389/fcimb.2016.00194 PMID:28083516

Mallhi, T. H., Khan, Y. H., Khan, A. H., Tanveer, N., & Qadir, M. I. (2017). First chikungunya outbreak in Pakistan: A trail of viral attacks. *New Microbes and New Infections*, *19*, 13–14. doi:10.1016/j.nmni.2017.05.008 PMID:28663798

Mandal, S., Sarkar, R. R., & Sinha, S. (2011). Mathematical models of malaria -a review. *Malaria Journal*, *10*(202), 1–19. PMID:21777413

Manimunda, S. P., Sugunan, A. P., Sha, W. A., Singh, S. S., Shriram, A. N., & Vijayachari, P. (2011). Tsunami, post-tsunami malaria situation in Nancowry group of islands, Nicobar district, Andaman and Nicobar Islands. *The Indian Journal of Medical Research*, *133*(1), 76–82. PMID:21321423

Mara, D., Lane, J., Scott, B., & Trouba, D. (2010). Sanitation and health. *PLoS Medicine*, *7*(11), 1–7. doi:10.1371/journal.pmed.1000363 PMID:21125018

Marchesi, J. R., Adams, D. H., Fava, F., Hermes, G. D. A., Hirschfield, G. M., Hold, G., Quraishi, M. N., Kinross, J., Smidt, H., Tuohy, K. M., Thomas, L. V., Zoetendal, E. G., & Hart, A. (2016). The gut microbiota and host health: A new clinical frontier. *Gut*, *65*(2), 330–339. doi:10.1136/gutjnl-2015-309990 PMID:26338727

Markowiak, P., & Śliżewska, K. (2017). Effects of probiotics, prebiotics, and synbiotics on human health. *Nutrients*, *9*(9), 1021. doi:10.3390/nu9091021 PMID:28914794

Marten, G. G. (2007). Larvicidal algae. *Journal of the American Mosquito Control Association*, *23*(sp2), 177–183. doi:10.2987/8756-971X(2007)23[177:LA]2.0.CO;2 PMID:17855939

Martens, E., & Demain, A. (2017). The antibiotic resistance crisis, with a focus on the United States. *The Journal of Antibiotics*, *70*(5), 520–526. doi:10.1038/ja.2017.30 PMID:28246379

Martens, P., & McMichael, A. J. (Eds.). (2009). *Environmental change, climate and health: issues and research methods*. Cambridge University Press.

McFadden, G. I., & Yeh, E. (2017). The apicoplast: Now you see it, now you don't. *International Journal for Parasitology, 47*(2–3), 137–144. doi:10.1016/j.ijpara.2016.08.005 PMID:27773518

McLaren, L., & Hawe, P. (2005). Ecological perspectives in health research. *Journal of Epidemiology and Community Health, 59*(1), 6–14. doi:10.1136/jech.2003.018044 PMID:15598720

McMichael, C. (2019). Water, sanitation, and hygiene (WASH) in schools in low-income countries: A review of evidence of impact. *International Journal of Environmental Research and Public Health, 16*(3), 1–21. doi:10.3390/ijerph16030359 PMID:30696023

Meentemeyer, R. K., Haas, S. E., & Vaclavik, T. (2012). Landscape epidemiology of emerging infectious diseases in natural and human-altered eco- systems. *Annual Review of Phytopathology, 50*(1), 379–402. doi:10.1146/annurev-phyto-081211-172938 PMID:22681449

Mehand, M. S., Al-Shorbaji, F., Millett, P., & Murgue, B. (2018). The WHO R&D Blueprint: 2018 review of emerging infectious diseases requiring urgent research and development efforts. *Antiviral Research, 159*, 63–67. doi:10.1016/j.antiviral.2018.09.009 PMID:30261226

Mellor, P. S., & Leake, C. J. (2000). Climatic and geographic influences on arboviral infections and vectors. *Revue Scientifique et Technique (International Office of Epizootics), 19*(1), 41–54. doi:10.20506/rst.19.1.1211 PMID:11189725

Meng, X. J., Lindsay, D. S., & Sriranganathan, N. (2009). Wild boars as sources of infectious diseases in livestock and humans. *Philosophical Transactions of the Royal Society of London. Series B, Biological Sciences, 364*(1530), 2697–2707. doi:10.1098/rstb.2009.0086 PMID:19687039

Merisa, D., Soeyoko, S., & Sutomo, A. H. (2014). Sanitation of house and school, personal hygiene and infection of soil transmitted helminths among elementary school students. *International Journal of Public Health Sciences, 3*(1), 43–50.

Merritt, R. W., Dadd, R. H., & Walker, E. D. (1992). Feeding behavior, natural food, and nutritional relationships of larval mosquitoes. *Annual Review of Entomology, 37*(1), 349–376. doi:10.1146/annurev.en.37.010192.002025 PMID:1347208

Michael, C. A., Dominey-Howes, D., & Labbate, M. (2014). The antimicrobial resistance crisis: Causes, consequences, and management. *Frontiers in Public Health, 2*. Advance online publication. doi:10.3389/fpubh.2014.00145 PMID:25279369

Mikolajkow, A., & Małyszczak, K. (2021). Stress level and general mental state in Polish pregnant women during COVID-19 pandemic. *Journal of Reproductive and Infant Psychology, 2023 Apr;41(2):165-182,* 1–18. doi:10.1080/02646838.2021.1976402 PMID:34555951

Miller, L. H., Mason, S. J., Clyde, D. F., & McGinniss, M. H. (1976). The resistance factor to *Plasmodium vivax* in blacks: The Duffy blood- group genotype, FyFy. *The New England Journal of Medicine, 295*(6), 302–304. doi:10.1056/NEJM197608052950602 PMID:778616

Mohanty, S., Mishra, S., Jena, P., Jacob, B., Sarkar, B., & Sonawane, A. (2012). An investigation on the antibacterial, cytotoxic, and antibiofilm efficacy of starch-stabilized silver nanoparticles. *Nanomedicine; Nanotechnology, Biology, and Medicine*, *8*(6), 916–924. doi:10.1016/j.nano.2011.11.007 PMID:22115597

Möhler, J., Sim, W., Blaskovich, M., Cooper, M., & Ziora, Z. (2018). Silver bullets: A new luster on an old antimicrobial agent. *Biotechnology Advances*, *36*(5), 1391–1411. doi:10.1016/j.biotechadv.2018.05.004 PMID:29847770

Moore, C. J. (2008). Synthetic polymers in the marine environment: A rapidly increasing, long-term threat. *Environmental Research*, *108*(2), 131–139. doi:10.1016/j.envres.2008.07.025 PMID:18949831

Moossavi, S., & Bishehsari, F. (2019). Microbes: Possible link between modern lifestyle transition and the rise of metabolic syndrome. *Obesity Reviews*, *20*(3), 407–419. doi:10.1111/obr.12784 PMID:30548384

Mora, C., Dousset, B., Caldwell, I. R., Powell, F. E., Geronimo, R. C., Bielecki, C. R., Counsell, C. W. W., Dietrich, B. S., Johnston, E. T., Louis, L. V., Lucas, M. P., McKenzie, M. M., Shea, A. G., Tseng, H., Giambelluca, T. W., Leon, L. R., Hawkins, E., & Trauernicht, C. (2017). Global risk of deadly heat. *Nature Climate Change 7*, *7*(7), 501–506. doi:10.1038/nclimate3322

Mordecai, E. A., Ryan, S. J., Caldwell, J. M., Shah, M. M., & LaBeaud, A. D. (2020). Climate change could shift disease burden from malaria to arboviruses in Africa. *The Lancet. Planetary Health*, *4*(9), e416–e423. doi:10.1016/S2542-5196(20)30178-9 PMID:32918887

Moreno, M., Marinotti, O., Krzywinski, J., Tadei, W. P., James, A. A., Achee, N. L., & Conn, J. E. (2010). Complete mtDNA genomes of *Anopheles darlingi* and an approach to anopheline divergence time. *Malaria Journal*, *9*(1), 127. doi:10.1186/1475-2875-9-127 PMID:20470395

Morse, S. S. (1995). Factors in the emergence of infectious diseases. *Emerging Infectious Diseases*, *1*(1), 7–15. doi:10.3201/eid0101.950102 PMID:8903148

Mulani, M. S., Kamble, E. E., Kumkar, S. N., Tawre, M. S., & Pardesi, K. R. (2019). Emerging Strategies to Combat ESKAPE Pathogens in the Era of Antimicrobial Resistance: A Review. *Frontiers in Microbiology*, *10*, 539. Advance online publication. doi:10.3389/fmicb.2019.00539 PMID:30988669

Müller, G. C., Beier, J. C., Traore, S. F., Toure, M. B., Traore, M. M., Bah, S., Doumbia, S., & Schlein, Y. (2010). Successful field trial of attractive toxic sugar bait (ATSB) plant-spraying methods against malaria vectors in the *Anopheles gambiae* complex in Mali, West Africa. *Malaria Journal*, *9*(1), 210. doi:10.1186/1475-2875-9-210 PMID:20663142

Mullish, B. H., Quraishi, M. N., Segal, J. P., McCune, V. L., Baxter, M., Marsden, G. L., Moore, D. J., Colville, A., Bhala, N., Iqbal, T. H., Settle, C., Kontkowski, G., Hart, A. L., Hawkey, P. M., Goldenberg, S. D., & Williams, H. R. T. (2018). The use of faecal microbiota transplant as treatment for recurrent or refractory *Clostridium difficile* infection and other potential indications: Joint British Society of Gastroenterology (BSG) and Healthcare Infection Society (HIS) guidelines. *Gut, 67*(11), 1920–1941. doi:10.1136/gutjnl-2018-316818 PMID:30154172

Nagpal, B. N., & Sharma, V. P. (1995). *Indian Anophelines*. Oxford & IBH Publishing Co. Pvt. LTD.

Nah, K., Bede-Fazekas, Á., Trájer, A. J., & Wu, J. (2020). The potential impact of climate change on the transmission risk of tick-borne encephalitis in Hungary. *BMC Infectious Diseases, 20*(1), 1–10. doi:10.1186/s12879-019-4734-4

Napier, L. E., & Campbell, H. (1932). Observations on a Plasmodium Infection Which Causes Hæmoglobinuria in Certain Species of Monkey. *Indian Medical Gazette, 67*(5), 246–249. PMID:29011028

Narain, J. P. (2008). Malaria in the South-East Asia: Myth & the reality. *The Indian Journal of Medical Research, 128*, 1–3. PMID:18820349

Nasr-Esfahani, E., Samavi, S., Karimi, N., Soroushmehr, S. M. R., Jafari, M. H., Ward, K., & Najarian, K. (2016). Melanoma Detection by Analysis of Clinical Images Using Convolutional Neural Network. *Proceedings of the 2016 38th Annual International Conference of the IEEE Engineering in Medicine and Biology Society (EMBC)*, 1373. 10.1109/EMBC.2016.7590963

National Cancer Institute. (2022). *Photodynamic Therapy to Treat Cancer*. Retrieved 14 June 2022, from https://www.cancer.gov/about-cancer/treatment/types/photodynamic-therapy

National Vector Borne Disease Control Programme. (n.d.). Government of India. http://www.nvbdcp.gov.in

Newton, P. N., Fernández, F. M., Plançon, A., Mildenhall, D. C., Green, M. D., Ziyong, L., Christophel, E. M., Phanouvong, S., Howells, S., McIntosh, E., Laurin, P., Blum, N., Hampton, C. Y., Faure, K., Nyadong, L., Soong, C. W. R., Santoso, B., Zhiguang, W., Newton, J., & Palmer, K. (2008). A collaborative epidemiological investigation into the criminal fake artesunate trade in South East Asia. *PLoS Medicine, 5*(2), e32. doi:10.1371/journal.pmed.0050032 PMID:18271620

Nicholson, J. K., Holmes, E., Kinross, J., Burcelin, R., Gibson, G., Jia, W., & Pettersson, S. (2012). Host-gut microbiota metabolic interactions. *Science, 336*(6086), 1262–1267. doi:10.1126cience.1223813 PMID:22674330

Nichols, R. G., Zhang, J., Cai, J., Murray, I. A., Koo, I., Smith, P. B., Perdew, G. H., & Patterson, A. D. (2019). Metatranscriptomic analysis of the mouse gut microbiome response to the persistent organic pollutant 2, 3, 7, 8-tetrachlorodibenzofuran. *Metabolites, 10*(1), 1. doi:10.3390/metabo10010001 PMID:31861317

Nilsson, L. K. J., Sharma, A., Bhatnagar, R. K., Bertilsson, S., & Terenius, O. (2018). Presence of *Aedes* and *Anopheles* mosquito larvae is correlated to bacteria found in domestic water-storage containers. *FEMS Microbiology Ecology*, *94*(6), fiy058. doi:10.1093/femsec/fiy058 PMID:29617987

Njambi, E., Magu, D., Masaku, J., Okoyo, C., & Njenga, S. M. (2020). Prevalence of intestinal parasitic infections and associated water, sanitation, and hygiene risk factors among school children in Mwea irrigation scheme, Kirinyaga County, Kenya. *Journal of Tropical Medicine*, *2020*, 1–9. doi:10.1155/2020/3974156 PMID:32454837

Nosrat, C., Altamirano, J., Anyamba, A., Caldwell, J. M., Damoah, R., Mutuku, F., Ndenga, B., & LaBeaud, A. D. (2021). Impact of recent climate extremes on mosquito-borne disease transmission in Kenya. *PLoS Neglected Tropical Diseases*, *15*(3), e0009182. doi:10.1371/journal.pntd.0009182 PMID:33735293

O'Neill, J. (2016). *Tackling Drug-Resistant Infections Globally: Final Report and Recommendations*. Review on Antimicrobial Resistance. Welcome Trust and UK Government. https://amr-review.org/sites/default/files/160518_Final%20paper_with%20cover.pdf

Obbard, R. W., Sadri, S., Wong, Y. Q., Khitun, A. A., Baker, I., & Thompson, R. C. (2014). Global warming releases microplastic legacy frozen in Arctic Sea ice. *Earth's Future*, *2*(6), 315–320. doi:10.1002/2014EF000240

Oberdörster, G. (1995). Lung particle overload: Implications for occupational exposures to particles. *Regulatory Toxicology and Pharmacology*, *21*(1), 123–135. doi:10.1006/rtph.1995.1017 PMID:7784625

Ogunfowokan, A. A., Esan, D. T., Salau, O. R., Bamidele, Y. R., & Ojo, D. E. (2017). Assessment of the physical school environment of selected private high schools in Ile-Ife, Nigeria. *British Journal of School Nursing*, *11*(10), 498–505.

Olson, D. H., Aanensen, D. M., Ronnenberg, K. L., Powell, C. I., Walker, S. F., Bielby, J., Garner, T. W. J., Weaver, G., Group, B. M., & Fisher, M. C. (2013). Mapping the Global Emergence of *Batrachochytrium dendrobatidis*, the Amphibian Chytrid Fungus. *PLoS One*, *8*(2), 1–13. doi:10.1371/journal.pone.0056802 PMID:23463502

Omar, E. H., Balzano, T., Abdelmonaim, E. H. M., & Jayakumar, A. R. (2022). The gut microbiota and brain interactions in healthy and pathological circumstances: From understanding to therapy. *Frontiers in Neuroscience*, *16*, 1052350. doi:10.3389/fnins.2022.1052350 PMID:36325481

Ostfeld, R. S., & Brunner, J. L. (2015). Climate change and Ixodes tick-borne diseases of humans. *Philosophical Transactions of the Royal Society of London. Series B: Biological Science*, *370*(20).

Ostholm-Balkhed, A., Ta¨rnberg, M., Nilsson, M., Nilsson, L. E., Hanberger, H., & Ha¨llgren, A. (2013). Travel-associated fecal colonization with ESBL-producing Enterobacteriaceae: Incidence and risk factors. *The Journal of Antimicrobial Chemotherapy*, *68*(9), 2144–2153. doi:10.1093/jac/dkt167 PMID:23674762

Owolabi, D. O., & Bagbe, A. S. (2019). Assessment of physico-chemical and ecological variables in selected natural breeding sites of mosquitoes in Ibadan, Oyo State, Nigeria. Archives of Pharmacy and Pharmacology Research, 1(5).

Paat, Y. F. (2013). Working with immigrant children and their families: An application of Bronfenbrenner's ecological systems theory. *Journal of Human Behavior in the Social Environment*, *23*(8), 954–966. doi:10.1080/10911359.2013.800007

Padgett, D. A., & Glaser, R. (2003). How stress influences the immune response. *Trends in Immunology*, *24*(8), 444–448. doi:10.1016/S1471-4906(03)00173-X PMID:12909458

PAHO. (1998). Impact of Hurricane Mitch on Central America. *Epidemiological Bulletin*, *19*, 1–14. PMID:10330785

Palinkas, L. A. (1991). Effects of physical and social environments on the health and well-being of Antarctic Winter-over personnel. *Environment and Behavior*, *23*(6), 782–799. doi:10.1177/0013916591236008

Pampana, E. (1969). *A Textbook of Malaria Eradication* (2nd ed.). Oxford University Press.

Pandey, K. R., Naik, S. R., & Vakil, B. V. (2015). Probiotics, prebiotics and synbiotics-a review. *Journal of Food Science and Technology*, *52*(12), 7577–7587. doi:10.100713197-015-1921-1 PMID:26604335

Panigrahi, S. K., Barik, T. K., Mohanty, S., & Tripathy, N. K. (2014). Laboratory evaluation of oviposition behavior of field collected *Aedes* mosquitoes. *Journal of Insects*, *207489*, 1–8. doi:10.1155/2014/207489

Parekh, F. K., Yeh, K. B., Olinger, G., & Ribeiro, F. A. (2020). Infectious disease risks and vulnerabilities in the aftermath of an environmental disaster in Minas Gerais, Brazil. *Vector Borne and Zoonotic Diseases (Larchmont, N.Y.)*, *20*(5), 387–389. doi:10.1089/vbz.2019.2501 PMID:31944914

Paterson, D. L., Wright, H., & Harris, P. N. A. (2018). Health Risks of Flood Disasters. *Clinical Infectious Diseases*, *67*(9), 1450–1454. doi:10.1093/cid/ciy227 PMID:30986298

Patil, S. S., Patil, S. R., Durgawale, P. M., & Patil, A. G. (2013). A study of the outbreak of Chikungunya fever. *Journal of Clinical and Diagnostic Research : JCDR*, *7*(6), 1059. PMID:23905103

Patz, J. A., Epstein, P. R., Burke, T. A., & Balbus, J. M. (1996). Global climate change and emerging infectious diseases. *Journal of the American Medical Association*, *275*(3), 217–223. doi:10.1001/jama.1996.03530270057032 PMID:8604175

Patz, J. A., Graczyk, T. K., Geller, N., & Vittor, A. Y. (2000). Effects of environmental change on emerging parasitic diseases. *International Journal for Parasitology*, *30*(12-13), 1395–1405. doi:10.1016/S0020-7519(00)00141-7 PMID:11113264

Paveley, N. D., Sylvester-Bradley, R., Scott, R. K., Craigon, J., & Day, W. (2001). Steps in predicting the relationship of yield on fungicide dose. *Phytopathology, 91*(7), 708–716. doi:10.1094/PHYTO.2001.91.7.708 PMID:18943002

Pavlovich, S. G., & Rockett, C. L. (2000). Color, bacteria, and mosquito eggs as ovipositional atiractants for *Aedes aegypti* and *Aedes albopictus* (Diptera: Culicidae). *Great Lakes Entomologist, 33*(2), 7, 141–153.

Paz, S. (2015). Climate change impacts on West Nile virus transmission in a global context. *Philosophical Transactions of the Royal Society B: Biological Sciences, 370*(1665), 20130561.

Peña-García, V. H., Triana-Chávez, O., & Arboleda-Sánchez, S. (2017). Estimating Effects of Temperature on Dengue Transmission in Colombian Cities. *Annals of Global Health, 83*(3–4), 509–518. doi:10.1016/j.aogh.2017.10.011 PMID:29221523

Penrose, K., de Castro, M. C., Werema, J., & Ryan, E. T. (2010). Informal urban settlements and cholera risk in Dar es Salaam, Tanzania. *PLoS Neglected Tropical Diseases, 4*(3), 1–11. doi:10.1371/journal.pntd.0000631 PMID:20300569

Pepin, K. M., Lass, S., Pulliam, J. R. C., Read, A. F., & Llyod-Smith, J. O. (2010). Identifying genetic markers of adaptation for surveillance of viral host jumps. *Nature Reviews. Microbiology, 8*(11), 802–813. doi:10.1038/nrmicro2440 PMID:20938453

Phillips, D. R. (1993). Urbanization and human health. *Parasitology, 106*(S1), 93–107. doi:10.1017/S0031182000086145 PMID:8488075

Piccioni, A., de Cunzo, T., Valletta, F., Covino, M., Rinninella, E., Raoul, P., Zanza, C., Mele, M. C., & Franceschi, F. (2021). Gut microbiota and environment in coronary artery disease. *International Journal of Environmental Research and Public Health, 18*(8), 4242. Advance online publication. doi:10.3390/ijerph18084242 PMID:33923612

Pickering, A. J., Davis, J., Blum, A. G., Scalmanini, J., Oyier, B., ... Ram, P. K. (2013). Access to waterless hand sanitizer improves student hand hygiene behavior in primary schools in Nairobi, Kenya. *The American Journal of Tropical Medicine and Hygiene, 89*(3), 411–418. doi:10.4269/ajtmh.13-0008 PMID:23836575

Pickett, S. T. A., Cadenasso, M. L., Grove, J. M., Boone, C. G., Groffman, P. M., Irwin, E., Kaushal, S. S., Marshall, V., McGrath, B. P., Nilon, C. H., Pouyat, R. V., Szlavecz, K., Troy, A., & Warren, P. (2011). Urban ecological systems: Scientific foundations and a decade of progress. *Journal of Environmental Management, 92*(3), 331–362. doi:10.1016/j.jenvman.2010.08.022 PMID:20965643

Pilosof, S., Fortuna, M. A., Cosson, J. F., Galan, M., Kittipong, C., Ribas, A., Segal, E., Krasnov, B. R., Morand, S., & Bascompte, J. (2014). The host-parasite network structure is associated with community-level immunogenetic diversity. *Nature Communications, 5*(1), 1–9. doi:10.1038/ncomms6172 PMID:25312328

Pinter-Wollman, N., Jelic, A., & Wells, N. M. (2018). The impact of the built environment on health behaviors and disease transmission in social systems. *Philosophical Transaction of the Royal Society B, 373*, 20170245. doi:10.1098/rstb.2017.0245

Pirnay, J., Verbeken, G., Ceyssens, P., Huys, I., De Vos, D., Ameloot, C., & Fauconnier, A. (2018). The Magistral Phage. *Viruses, 10*(2), 64. doi:10.3390/v10020064 PMID:29415431

Plowright, R. K., Eby, P., Hudson, P. J., Smith, I. L., Westcott, D., Bryden, W. L, Middleton, D., Reid, P. A., McFarlane, R. A., Martin, G., Tabor, G. M., Skerratt, L. F., Anderson, D. L., Crameri, G., Quammen, D., Jordan, D., Freeman, P., Wang, L. F., Epstein, J. H., Marsh, G. A., Kung, N. Y., & McCallum, H. (2015). Ecological dynamics of emerging bat virus spillover. *Proceedings. Biological Sciences, 282*(1798).

Plowright, R. K., Foley, P., Field, H. E., Dobson, A. P., Foley, J. E., Eby, P., & Daszak, P. (2011). Urban habituation, ecological connectivity, and epidemic dampening: the emergence of Hendra virus from flying foxes (Pteropus spp.). *Proceedings of the Royal Society B: Biological Sciences, 278*(1725), 3703-3712.

Plowright, R. K., Reaser, J. K., Locke, H., Woodley, S. J., Patz, J. A., Becker, D. J., Oppler, G., Hudson, P. J., & Tabor, G. M. (2021). Land use-induced spillover: A call to action to safeguard environmental, animal, and human health. *The Lancet. Planetary Health, 5*(4), e237–e245. doi:10.1016/S2542-5196(21)00031-0 PMID:33684341

Poerwanto, S. H., Ahmad, S., Giyantolin, & Windyaraini, D. H. (2020). Characterization of mosquitoes breeding site (Diptera: Culicidae) in Bogowonto Lagoon Ecosystem, Kulonprogo Districts, special region of Yogyakarta, Indonesia. AIP Conference Proceedings, 2260.

Ponting, J., Kelly, T. J., Verhoef, A., Watts, M. J., & Sizmur, T. (2021). The impact of increased flooding occurrence on the mobility of potentially toxic elements in floodplain soil - A review. *The Science of the Total Environment, 754*, 142040. doi:10.1016/j.scitotenv.2020.142040 PMID:32916489

Popli, S., Badgujar, P. C., Agarwal, T., Bhushan, B., & Mishra, V. (2022). Persistent organic pollutants in foods, their interplay with gut microbiota and resultant toxicity. *The Science of the Total Environment, 832*, 155084. doi:10.1016/j.scitotenv.2022.155084 PMID:35395291

Poulin, R. (2004). Macroecological patterns of species richness in parasite assemblages. *Basic and Applied Ecology, 5*(5), 423–434. doi:10.1016/j.baae.2004.08.003

Powers, A. M., & Logue, C. H. (2007). Changing patterns of chikungunya virus: Re-emergence of a zoonotic arbovirus. *The Journal of General Virology, 88*(9), 2363–2377. doi:10.1099/vir.0.82858-0 PMID:17698645

Preidis, G. A., & Versalovic, J. (2009). Targeting the human microbiome with antibiotics, probiotics, and prebiotics: Gastroenterology enters the metagenomics era. *Gastroenterology, 136*(6), 2015–2031. doi:10.1053/j.gastro.2009.01.072 PMID:19462507

Pulliam, J. R. C., Epstein, J. H., Dushoff, J., Rahman, S. A., Bunning, M., Jamaluddin, A. A., Hyatt, A. D., Field, H. E., Dobson, A. P., & Daszak, P. (2012). Agricultural intensification, priming for persistence and the emergence of Nipah virus: A lethal bat-borne zoonosis. *Journal of the Royal Society, Interface*, *9*(66), 89–101. doi:10.1098/rsif.2011.0223 PMID:21632614

Qayyum, S., Oves, M., & Khan, A. (2017). Obliteration of bacterial growth and biofilm through ROS generation by facilely synthesized green silver nanoparticles. *PLoS One*, *12*(8), e0181363. doi:10.1371/journal.pone.0181363 PMID:28771501

Qin, D., Manning, M., Chen, Z., Marquis, M., Averyt, K. B., & Tignor, M. (Eds.). (2007). *IPCC Fourth Assessment Report: Climate Change 2007: The Physical Science Basis*. Cambridge University Press.

Qinghaosu Antimalaria Coordinating Research Group. (1979). Antimalaria Studies on Qinghaosu. *Chinese Medical Journal*, *92*, 811–816. PMID:117984

Qiu, Y.-W., Zeng, E. Y., Qiu, H., Yu, K., & Cai, S. (2017). Bioconcentration of polybrominated diphenyl ethers and organochlorine pesticides in algae is an important contaminant route to higher trophic levels. *The Science of the Total Environment*, *579*, 1885–1893. doi:10.1016/j.scitotenv.2016.11.192 PMID:27939082

Qualls, W. A., Müller, G. C., Traore, S. F., Traore, M. M., Arheart, K. L., Doumbia, S., Schlein, Y., Kravchenko, V. D., Xue, R. D., & Beier, J. C. (2015). Indoor use of attractive toxic sugar bait (ATSB) to effectively control malaria vectors in Mali, West Africa. *Malaria Journal*, *14*(1), 301. doi:10.118612936-015-0819-8 PMID:26242186

Quammen, D. (2013). *Spillover: Animal Infections and the Next Human Pandemic*. Vintage.

Radi, M. F. M., Hashim, J. H., Jaafar, M. H., Hod, R., Ahmad, N., Nawi, A. M., Baloch, G. M., Ismail, R., & Ayub, N. I. F. (2018). Leptospirosis Outbreak After the 2014 Major Flooding Event in Kelantan, Malaysia: A Spatial-Temporal Analysis. *The American Journal of Tropical Medicine and Hygiene*, *98*(5), 1281–1295. doi:10.4269/ajtmh.16-0922 PMID:29532771

Ragavendran, C., & Natarajan, D. (2018). Larval toxicity of entomopathogenic bacteria *Bacillus cereus* against *Aedes aegypti, Anopheles stephensi*, and *Culex quinquefasciatus* mosquitoes. In B. K. Tyagi & D. Dhanasekaran (Eds.), *Microbial Control of Vector-Borne Diseases* (pp. 113–135). CRC Press. doi:10.1201/b22203-7

Rahman, M. M., Alil, M. A., Parvez, A., Shahriar, M., & Rahman, T., & Shameem, D. (2019). Knowledge and practice of personal hygiene among primary school students in Dhaka, Bangladesh. *Journal of Pharmaceutical Sciences & Research*, *11*(9), 3140–3144.

Raj, S., Roodbar, S., Brinkley, C., & Wolfe, D. W. (2022). Food Security and Climate Change: Differences in impacts and adaptation strategies for rural communities in the Global South and North. *Frontiers in Sustainable Food Systems*, *5*(1), 691191. Advance online publication. doi:10.3389/fsufs.2021.691191

Rana, M., Singh, S. S. J., & Yadav, S. (2017). Effect of microencapsulated plant extracts on mosquito repellency. *Journal of Applied and Natural Science*, *9*(4), 2127–2131. doi:10.31018/jans.v9i4.1498

Ranasinghe, H. A. K., & Amarasinghe, L. D. (2020). Naturally occurring microbiota associated with mosquito breeding habitats and their effects on mosquito larvae. *BioMed Research International*, *4065315*, 1–11. doi:10.1155/2020/4065315 PMID:33381553

Rassi, A. Jr, Rassi, A., & Marin-Neto, J. A. (2010). Chagas disease. *Lancet*, *375*(9723), 1388–1402. doi:10.1016/S0140-6736(10)60061-X PMID:20399979

Reinhardt, A., & Neundorf, I. (2016). Design and Application of Antimicrobial Peptide Conjugates. *International Journal of Molecular Sciences*, *17*(5), 701. doi:10.3390/ijms17050701 PMID:27187357

Reisen, W. K. (2010). Landscape epidemiology of vector-borne diseases. *Annual Review of Entomology*, *55*(1), 461–483. doi:10.1146/annurev-ento-112408-085419 PMID:19737082

Rejmánková, E., Grieco, J., Achee, N., & Roberts, D. R. (2013). Ecology of larval habitats. In S. Manguin (Ed.), *Anopheles mosquitoes: New insights into malaria vectors* (pp. 397–446). IntechOpen. doi:10.5772/55229

Reperant, L. A. (2010). Applying the theory of island biogeography to emerging pathogens: Toward predicting the sources of future emerging zoonotic and vector-borne diseases. *Vector Borne and Zoonotic Diseases (Larchmont, N.Y.)*, *10*(2), 105–110. doi:10.1089/vbz.2008.0208 PMID:19589061

Rezania, S., Park, J., Din, M. F. M., Taib, S. M., Talaiekhozani, A., Yadav, K. K., & Kamyab, H. (2018). Microplastics pollution in different aquatic environments and biota: A review of recent studies. *Marine Pollution Bulletin*, *133*, 191–208. doi:10.1016/j.marpolbul.2018.05.022 PMID:30041307

Ribeiro Marques, M., Henrique Pereira, J., Oliveira, J. S., Augusto Basso, L., Filgueira de Azevedo, W., Santiago Santos, D., & Sergio Palma, M. (2007). The inhibition of 5-enolpyruvylshikimate-3-phosphate synthase as a model for development of novel antimicrobials. *Current Drug Targets*, *8*(3), 445–457. doi:10.2174/138945007780058951 PMID:17348837

Rich, S. M., Leendertz, F. H., Xu, G., LeBreton, M., Djoko, C. F., Aminake, M. N., Takang, E. E., Diffo, J. L., Pike, B. L., Rosenthal, B. M., Formenty, P., Boesch, C., Ayala, F. J., & Wolfe, N. D. (2009). The origin of malignant malaria. *Proceedings of the National Academy of Sciences of the United States of America*, *106*(35), 14902–14907. doi:10.1073/pnas.0907740106 PMID:19666593

Ridenhour, B. J., Braun, A., Teyrasse, T., & Goldsman, D. (2011). Controlling the spread of disease in schools. *PLoS One*, *6*(12), e29640. doi:10.1371/journal.pone.0029640 PMID:22242138

Rios, A., Moutinho, C., Pinto, F., Del Fiol, F., Jozala, A., Chaud, M., Vila, M. M. D. C., Teixeira, J. A., & Balcão, V. M. (2016). Alternatives to overcoming bacterial resistances: State-of-the-art. *Microbiological Research*, *191*, 51–80. doi:10.1016/j.micres.2016.04.008 PMID:27524653

Roberfroid, M. (2007). Prebiotics: The concept revisited. *The Journal of Nutrition, 137*(3), 830S–837S. doi:10.1093/jn/137.3.830S PMID:17311983

Roberts, D. (2014). Mosquito larvae change their feeding behavior in response to kairomones from some predators. *Journal of Medical Entomology, 51*(2), 368–374. doi:10.1603/ME13129 PMID:24724285

Rocco, F. (2003). The Miraculous Fever-Tree: Malaria and the Quest for a Cure That Changed the World. New York: HarperCollins.

Rochlin, I., White, G., Reissen, N., Martheswaran, T., & Faraji, A. (2022). Effects of aerial adulticiding for mosquito management on nontarget insects: A Bayesian and community ecology approach. *Ecosphere, 13*(1), e3896. doi:10.1002/ecs2.3896

Rohde, C., Resch, G., Pirnay, J., Blasdel, B., Debarbieux, L., Gelman, D., Górski, A., Hazan, R., Huys, I., Kakabadze, E., Łobocka, M., Maestri, A., Almeida, G., Makalatia, K., Malik, D., Mašlaňová, I., Merabishvili, M., Pantucek, R., Rose, T., ... Chanishvili, N. (2018). Expert Opinion on Three Phage Therapy Related Topics: Bacterial Phage Resistance, Phage Training and Prophages in Bacterial Production Strains. *Viruses, 10*(4), 178. doi:10.3390/v10040178 PMID:29621199

Rohmah, F. A., Setiawan, R., Adriyani, R., & Sham, S. (2022). Personal hygiene as a risk factor of helminthiasis among primary school students in Asia and Africa: A literature review. *Journal of Environmental Health, 14*(3), 139–152. Advance online publication. doi:10.20473/jkl.v14i3.2022.139-152

Romeo-Aznar, V., Paul, R., Telle, O., & Pascual, M. (2018). Mosquito-borne transmission in urban landscapes: the missing link between vector abundance and human density. *Proceedings. Biological Sciences, 285*(1884), 20180826.

Rose, N. H., Sylla, M., Badolo, A., Lutomiah, J., Ayala, D., Aribodor, O. B., Ibe, N., Akorli, J., Otoo, S., Mutebi, J.-P., Kriete, A. L., Ewing, E. G., Sang, R., Gloria-Soria, A., Powell, J. R., Baker, R. E., White, B. J., Crawford, J. E., & McBride, C. S. (2020). Climate and urbanization drive mosquito preference for humans. *Current Biology, 30*(18), 3570–3579. doi:10.1016/j.cub.2020.06.092 PMID:32707056

Rosen, L., Manor, O., Engelhard, D., Brody, D., Rosen, B., Peleg, H., Meir, M., & Zucker, D. (2006). Can a handwashing intervention make a difference? Results from a randomized controlled trial in Jerusalem preschools. *Preventive Medicine, 42*(1), 27–32. doi:10.1016/j.ypmed.2005.09.012 PMID:16300823

Rosselló, J., Gallego, M. S., & Awan, W. (2017). Infectious disease risk and international tourism demand. *Health Policy and Planning, 32*(4), 538–548. doi:10.1093/heapol/czw177 PMID:28104695

Ross, R. (1915). Some a priori pathometric equations. *British Medical Journal, 1*(2830), 546–447. doi:10.1136/bmj.1.2830.546 PMID:20767555

Rothhammer, V., & Quintana, F. J. (2019). The aryl hydrocarbon receptor: An environmental sensor integrating immune responses in health and disease. *Nature Reviews. Immunology, 19*(3), 184–197. doi:10.103841577-019-0125-8 PMID:30718831

Rothschild, D., Weissbrod, O., Barkan, E., Kurilshikov, A., Korem, T., Zeevi, D., Costea, P. I., Godneva, A., Kalka, I. N., Bar, N., Shilo, S., Lador, D., Vila, A. V., Zmora, N., Pevsner-Fischer, M., Israeli, D., Kosower, N., Malka, G., Wolf, B. C., ... Segal, E. (2018). Environment dominates over host genetics in shaping human gut microbiota. *Nature, 555*(7695), 210–215. doi:10.1038/nature25973 PMID:29489753

Round, J. L., O'Connell, R. M., & Mazmanian, S. K. (2010). Coordination of tolerogenic immune responses by the commensal microbiota. *Journal of Autoimmunity, 34*(3), J220–J225. doi:10.1016/j.jaut.2009.11.007 PMID:19963349

Rueda-Ruzafa, L., Cruz, F., Roman, P., & Cardona, D. (2019). Gut microbiota and neurological effects of glyphosate. *Neurotoxicology, 75*, 1–8. doi:10.1016/j.neuro.2019.08.006 PMID:31442459

Rutter, M. (1983). School effects on pupil progress: Research findings and policy implications. *Child Development, 54*(1), 1–29. doi:10.2307/1129857

Sadique, M. Z., Adams, E. J., & Edmunds, W. J. (2008). Estimating the costs of school closure for mitigating an influenza pandemic. *BMC Public Health, 8*(1), 8. doi:10.1186/1471-2458-8-135 PMID:18435855

SagarA.DheebaJ. (2020). Convolutional Neural Networks for Classifying Melanoma Images. bioRxiv. doi:10.1101/2020.05.22.110973

Samoli, E., Touloumi, G., Schwartz, J., Anderson, H. R., Schindler, C., Forsberg, B., Vigotti, M. A., Vonk, J., Košnik, M., Skorkovsky, J., & Katsouyanni, K. (2007). Short-Term Effects of Carbon Monoxide on Mortality: An Analysis within the APHEA Project. *Environmental Health Perspectives, 115*(11), 1578–1583. doi:10.1289/ehp.10375 PMID:18007988

Sanders, E. J., Rigau-Perez, J. G., Smits, H. L., Deseda, C. C., Vorndam, V. A., Aye, T., Spiegel, R. A., Weyant, R. S., & Bragg, S. L. (1999). Increase of leptospirosis in dengue-negative patients after a hurricane in Puerto Rico in 1996. *The American Journal of Tropical Medicine and Hygiene, 61*(3), 399–404. doi:10.4269/ajtmh.1999.61.399 PMID:10497979

Santos, J., & Meneses, B. M. (2017). An integrated approach for the assessment of the *Aedes aegypti* and *Aedes albopictus* global spatial distribution, and determination of the zones susceptible to the development of Zika virus. *Acta Tropica, 168*, 80–90. doi:10.1016/j.actatropica.2017.01.015 PMID:28111132

Savage, H. M., Anderson, M., Gordon, E., McMillen, L., Colton, L., Delorey, M., Sutherland, G., Aspen, S., Charnetzky, D., Burkhalter, K., & Godsey, M. (2008). Host-seeking heights, host-seeking activity patterns, and West Nile virus infection rates for members of the Culex pipiens complex at different habitat types within the hybrid zone, Shelby County, TN, 2002 (Diptera: Culicidae). *Journal of Medical Entomology, 45*(2), 276–288. doi:10.1093/jmedent/45.2.276 PMID:18402144

Schraufnagel, D. E., Balmes, J. R., Cowl, C. T., De Matteis, S., Jung, S. H., Mortimer, K., Perez-Padilla, R., Rice, M. B., Riojas-Rodriguez, H., Sood, A., Thurston, G. D., To, T., Vanker, A., & Wuebbles, D. J. (2019). Air Pollution and Noncommunicable Diseases: A Review by the Forum of International Respiratory Societies' Environmental Committee, Part 1: The Damaging Effects of Air Pollution. *Chest, 155*(2), 409–416. doi:10.1016/j.chest.2018.10.042 PMID:30419235

Sekhsaria, P. (2021, Mar. 20). NITI Ayog vision for Great Nicobar ignores tribal, ecological concerns. *The Hindu.*

Semenza, J. C., Lindgren, E., Balkanyi, L., Espinosa, L., Almqvist, M. S., Penttinen, P., & Rocklöv, J. (2016). Determinants and drivers of infectious disease threat events in Europe. *Emerging Infectious Diseases*, *22*(4), 581–589. doi:10.3201/eid2204.151073 PMID:26982104

Seo, M., Won, H., Kim, J., Mishig-Ochir, T., & Lee, B. (2012). Antimicrobial Peptides for Therapeutic Applications: A Review. *Molecules (Basel, Switzerland)*, *17*(10), 12276–12286. doi:10.3390/molecules171012276 PMID:23079498

Shackleton, C. M., Ruwanza, S., Sinasson Sanni, G. K., Bennett, S., De Lacy, P., Modipa, R., Mtati, N., Sachikonye, M., & Thondhlana, G. (2016). Unpacking pandora's box: Understanding and categorising ecosystem disservices for environmental management and human wellbeing. *Ecosystems (New York, N.Y.)*, *19*(4), 587–600. doi:10.100710021-015-9952-z

Shah, S. (2010). *The Fever: How Malaria has ruled humankind for 500,000 years*. Academic Press.

Shaman, J., & Kohn, M. (2009). Absolute humidity modulates influenza survival, transmission, and seasonality. *Proceedings of the National Academy of Sciences of the United States of America*, *106*(9), 3243–3248. doi:10.1073/pnas.0806852106 PMID:19204283

Shankar, P. (2016). Book review: Tackling drug-resistant infections globally. *Archives of Pharmacy Practice (Mumbai)*, *7*(3), 110. doi:10.4103/2045-080X.186181

Shapiro, E. D., & Gerber, M. A. (2000). Lyme Disease. *Clinical Infectious Diseases*, *31*(2), 533–542. doi:10.1086/313982 PMID:10987718

Sharma, P., Sharma, S., Maurya, R. K., De, T. D., Thomas, T., Lata, S., Singh, N., Pandey, K. C., Valecha, N., & Dixit, R. (2014). Salivary glands harbor more diverse microbial communities than gut in *Anopheles culicifacies. Parasites & Vectors*, *7*(1), 235. doi:10.1186/1756-3305-7-235 PMID:24886293

Sharma, S. N., Srivastava, P. K., Singh, S., Sharma, R. S., Sonal, G. S., & Dhariwal, A. C. (2014). Urban malaria scheme—Past, present and future. *The Journal of Communicable Diseases*, *46*(2), 77–84.

Sharma, S. N., Subbarao, S. K., Choudhury, D. S., & Pandey, K. C. (1993). Role of *An. culicifacies* and *An. stephensi* in malaria transmission in urban Delhi. *Indian Journal of Malariology*, *30*, 155–168. PMID:8131883

Sharma, S., & Chatterjee, S. (2017). Microplastic pollution, a threat to marine ecosystem and human health: A short review. *Environmental Science and Pollution Research International, 24*(27), 21530–21547. doi:10.100711356-017-9910-8 PMID:28815367

Sharma, S., Chattopadhyay, R., Srivastava, V. K., Das, B. S., Pati, S. S., Mahanty, S., Tyagi, P. K., Chitnis, C., Misra, S. K., Adak, T., & Chakrabarti, K. (2004). Epidemiology of Malaria Transmission and Development of Natural Immunity in a Malaria endemic Village, San Dulakudra, in Orissa State, India. *The American Journal of Tropical Medicine and Hygiene, 71*(4), 457–465. doi:10.4269/ajtmh.2004.71.457

Shivalli, S., Pai, S., Akshaya, K. M., & D'Souza, N. (2016). Construction site workers' malaria knowledge and treatment-seeking pattern in a highly endemic urban area of India. *Malaria Journal, 15*(1), 1–10. PMID:27530110

Shi, Z., Lei, H., Chen, G., Yuan, P., Cao, Z., Ser, H.-L., Zhu, X., Wu, F., Liu, C., Dong, M., Song, Y., Guo, Y., Chen, C., Hu, K., Zhu, Y., Zeng, X., Zhou, J., Lu, Y., Patterson, A. D., & Zhang, L. (2021). Impaired intestinal *Akkermansia muciniphila* and aryl hydrocarbon receptor ligands contribute to nonalcoholic fatty liver disease in mice. *mSystems, 6*(1), 10–1128. doi:10.1128/mSystems.00985-20 PMID:33622853

Shoieb, D. A., Aly, W. M., & Youssef, S. M. (2016). Basal Cell Carcinoma Detection in Full-Field OCT Images Using Convolutional Neural Networks. *JOIG, 4*, 122–129. doi:10.18178/joig.4.2.122-129

Shukla, M. A., Woc-Colburn, L., & Weatherhead, J. E. (2018). Infectious Diseases in the Aftermath of Hurricanes in the United States. *Current Tropical Medicine Reports, 5*(4), 217–223. doi:10.100740475-018-0162-6

Silberbush, A., Markman, S., Lewinsohn, E., Bar, E., Cohen, J. E., & Blaustein, L. (2010). Predator-released hydrocarbons repel oviposition by a mosquito. *Ecology Letters, 13*(9), 1129–1138. doi:10.1111/j.1461-0248.2010.01501.x PMID:20618841

Silk, D. B. A., Davis, A., Vulevic, J., Tzortzis, G., & Gibson, G. R. (2009). Clinical trial: The effects of a trans-galactooligosaccharide prebiotic on faecal microbiota and symptoms in irritable bowel syndrome. *Alimentary Pharmacology & Therapeutics, 29*(5), 508–518. doi:10.1111/j.1365-2036.2008.03911.x PMID:19053980

Simonsen, P. E., & Mwakitalu, M. E. (2013). Urban lymphatic filariasis. *Parasitology Research, 112*(1), 35–44. doi:10.100700436-012-3226-x PMID:23239094

Sims, L. D., Domenech, J., Benigno, C., Kahn, S., Kamata, A., Lubroth, J., Martin, V., & Roeder, P. (2005). Origin and evolution of highly pathogenic H5N1 avian influenza in Asia. *The Veterinary Record, 157*(6), 159–164. doi:10.1136/vr.157.6.159 PMID:16085721

Singh, B., Lee, K. S., Asmad, M., Anand, R., Sunita, S., & Shamsul, G. (2004). A Large Focus of Naturally Acquired *Plasmodium knowlesi* Infections in Human Beings. *Lancet, 363*(9414), 1017–1024. doi:10.1016/S0140-6736(04)15836-4 PMID:15051281

Singh, M. M., & Mullin, G. E. (2017). Diet, environmental chemicals, and the gut microbiome. In *Integrative Environmental Medicine* (pp. 115–140). Oxford University Press New York.

Singh, R. K., Mittal, P. K., Gourshettiwar, M. P., Pande, S. J., & Dhiman, R. C. (2012). Susceptibility of malaria vectors to insecticides in Gadchiroli district (Maharashtra), India. *Journal of Vector Borne Diseases*, *49*, 42–44. PMID:22585244

Singh, V., & Nwogu, I. (2018). Analysing Skin Lesions in Dermoscopy Images Using Convolutional Neural Networks. *Proceedings of the 2018 IEEE International Conference on Systems, Man, and Cybernetics (SMC),* 4035–4040. 10.1109/SMC.2018.00684

Skretkowicz, V. (1992). Florence Nightingale's notes on nursing. London: Scutari Press.

Smith, A. R., Macfarlane, S., Furrie, E., Ahmed, S., Bahrami, B., Reynolds, N., & Macfarlane, G. T. (2011). Microbiological and immunological effects of enteral feeding on the upper gastrointestinal tract. *Journal of Medical Microbiology*, *60*(3), 359–365. doi:10.1099/jmm.0.026401-0 PMID:21088127

Snedeker, S. M., & Hay, A. G. (2012). Do interactions between gut ecology and environmental chemicals contribute to obesity and diabetes? *Environmental Health Perspectives*, *120*(3), 332–339. doi:10.1289/ehp.1104204 PMID:22042266

Song, X., Jiang, L., Zhang, D., Wang, X., Ma, Y., Hu, Y., Tang, J., Li, X., Huang, W., Meng, Y., Shi, A., Feng, Y., & Zhang, Y. (2021). Impact of short-term exposure to extreme temperatures on diabetes mellitus morbidity and mortality? A systematic review and meta-analysis. *Environmental Science and Pollution Research International*, *2021*(41), 1–15. doi:10.100711356-021-14568-0 PMID:34105073

Souza, R. S., Virginio, F., Riback, T. I. S., Suesdek, L., Barufi, J. B., & Genta, F. A. (2019). Microorganism-based larval diets affect mosquito development, size and nutritional reserves in the yellow fever mosquito *Aedes aegypti* (Diptera: Culicidae). *Frontiers in Physiology*, *10*, 152. doi:10.3389/fphys.2019.00152 PMID:31024326

Souza, T. B., Morais, M. B., Tahan, S., Melli, L. C., Rodrigues, M. S., & Scaletsky, I. C. (2009). High prevalence of antimicrobial drug-resistant diarrheagenic Escherichia coli in asymptomatic children living in an urban slum. *The Journal of Infection*, *59*(4), 247–251. doi:10.1016/j.jinf.2009.08.007 PMID:19706305

Stach, J., & Good, L. (2011). Synthetic RNA Silencing in Bacteria – Antimicrobial Discovery and Resistance Breaking. *Frontiers in Microbiology*, *2*, 185. doi:10.3389/fmicb.2011.00185 PMID:21941522

Stanczyk, N. M., Brookfield, J. F., Ignell, R., Logan, J. G., & Field, L. M. (2010). Behavioral insensitivity to DEET in Aedes aegypti is a genetically determined trait residing in changes in sensillum function. *Proceedings of the National Academy of Sciences of the United States of America*, *107*(19), 8575–8580. doi:10.1073/pnas.1001313107 PMID:20439757

Stewart, G. J., Nelson, B. S., Acton, W. J. F., Vaughan, A. R., Hopkins, J. R., Yunus, S. S. M., Hewitt, C. N., Wild, O., Nemitz, E., Gadi, R., Sahu, L. K., Mandal, T. K., Gurjar, B. R., Rickard, A. R., Lee, J. D., & Hamilton, J. F. (2021). Emission estimates and inventories of non-methane volatile organic compounds from anthropogenic burning sources in India. *Atmospheric Environment: X*, *11*, 100115. doi:10.1016/j.aeaoa.2021.100115

Stocks, M. E., Ogden, S., Haddad, D., Addiss, D. G., McGuire, C., & Freeman, M. C. (2014). Effect of water, sanitation, and hygiene on the prevention of trachoma: A systematic review and meta-analysis. *PLoS Medicine*, *11*(2), e1001605. doi:10.1371/journal.pmed.1001605 PMID:24586120

Streicher, L. M. (2021). Exploring the future of infectious disease treatment in a post-antibiotic era: A Comparative Review of Alternative Therapeutics. *Journal of Global Antimicrobial Resistance*, *24*, 285–295. doi:10.1016/j.jgar.2020.12.025 PMID:33484895

Strunz, E. C., Addiss, D. G., Stocks, M. E., Ogden, S., Utzinger, J., & Freeman, M. C. (2014). Water, sanitation, hygiene, and soil-transmitted helminth infection: A systematic review and meta-analysis. *PLoS Medicine*, *11*(3), e1001620. doi:10.1371/journal.pmed.1001620 PMID:24667810

Stumm, S., von, Smith-Woolley, E., & Cheesman, R. (2021). School quality ratings are weak predictors of students' achievement and well-being. *Journal of Child Psychology and Psychiatry, and Allied Disciplines*, *62*(3), 339–348. doi:10.1111/jcpp.13276 PMID:32488912

Subbarao, S. K. (1998). Anopheline Species Complexes in South-East Asia. *World Health Organization. Technical Publication, SEARO*, *18*, 82.

Subbarao, S. K., Vasantha, K., & Sharma, V. P. (1998). Response of *Anopheles culicifacies* sibling species A and B to DDT and HCH in India: Implications in malaria control. *Medical and Veterinary Entomology*, *2*(3), 219–223. doi:10.1111/j.1365-2915.1988.tb00186.x PMID:2485170

Sueoka, K., Chikama, T., Latief, M., Ko, J., Kiuchi, Y., Sakaguchi, T., & Obana, A. (2018). Time-dependent antimicrobial effect of photodynamic therapy with TONS 504 on Pseudomonas aeruginosa. *Lasers in Medical Science*, *33*(7), 1455–1460. doi:10.100710103-018-2490-0 PMID:29589177

Suez, J., Korem, T., Zeevi, D., Zilberman-Schapira, G., Thaiss, C. A., Maza, O., Israeli, D., Zmora, N., Gilad, S., Weinberger, A., Kuperman, Y., Harmelin, A., Kolodkin-Gal, I., Shapiro, H., Halpern, Z., Segal, E., & Elinav, E. (2014). Artificial sweeteners induce glucose intolerance by altering the gut microbiota. *Nature*, *514*(7521), 181–186. doi:10.1038/nature13793 PMID:25231862

Sultana, F., Nizame, F. A., Southern, D. L., Unicomb, L., Winch, P. J., & Luby, S. P. (2017). Pilot of an elementary school cough etiquette intervention: Acceptability, feasibility, and potential for sustainability. *The American Journal of Tropical Medicine and Hygiene*, *97*(6), 1876–1885. doi:10.4269/ajtmh.16-0914 PMID:29141703

Sunish, I. P., Shriram, A. N., De, A., & Vijayachari, P. (2015). Malaria in the Andaman and Nicobar Islands: Challenges and opportunities for elimination. *Asian Pacific Journal of Tropical Disease*, *5*(10), 837–840. doi:10.1016/S2222-1808(15)60942-2

Swirski, A. L., Pearl, D. L., Williams, M. L., Homan, H. J., Linz, G. M., Cernicchiaro, N., & LeJeune, J. T. (2014). Spatial epidemiology of Escherichia coli O157:H7 in dairy cattle about night roosts of Sturnus vulgaris (European starling) in Ohio, USA (2007-2009). *Zoonoses and Public Health*, *61*(6), 427–435. doi:10.1111/zph.12092 PMID:24279810

Ta¨ngde'n, T., Cars, O., Melhus, A., & Lo¨wdin, E. (2010). Foreign travel is a major risk factor for colonization with Escherichia coli producing CTX-M-type extended-spectrum beta-lactamases: A prospective study with Swedish volunteers. *Antimicrobial Agents and Chemotherapy*, *54*(9), 3564–3568. doi:10.1128/AAC.00220-10 PMID:20547788

Talaat, M., Afiffi, S., Dueger, E., El-Ashry, N., Marfin, A., ... El-Sayed, N. (2011). Effects of hand hygiene campaigns on incidence of laboratory-confirmed influenza and absenteeism in school children, Cairo, Egypt. *Emerging Infectious Diseases*, *17*(4), 619–625. doi:10.3201/eid1704.101353 PMID:21470450

Tall, J. A., & Gatton, M. L. (2020). Flooding and Arboviral Disease: Predicting Ross River Virus Disease Outbreaks Across Inland Regions of South-Eastern Australia. *Journal of Medical Entomology*, *57*(1), 241–251. doi:10.1093/jme/tjz120 PMID:31310648

Tamagawa, E., Bai, N., Morimoto, K., Gray, C., Mui, T., Yatera, K., Zhang, X., Xing, L., Li, Y., Laher, I., Sin, D. D., Man, S. F. P., & van Eeden, S. F. (2008). Particulate matter exposure induces persistent lung inflammation and endothelial dysfunction. *American Journal of Physiology. Lung Cellular and Molecular Physiology*, *295*(1), L79–L85. Advance online publication. doi:10.1152/ajplung.00048.2007 PMID:18469117

Tariq, R., Pardi, D. S., Bartlett, M. G., & Khanna, S. (2019). Low cure rates in controlled trials of fecal microbiota transplantation for recurrent Clostridium difficile infection: A systematic review and meta-analysis. *Clinical Infectious Diseases*, *68*(8), 1351–1358. doi:10.1093/cid/ciy721 PMID:30957161

Taylor, T. N., Caudle, W. M., & Miller, G. W. (2011). *VMAT2-Deficient Mice Display Nigral and Extranigral Pathology and Motor and Nonmotor Symptoms of Parkinson's Disease.* doi:10.4061/2011/124165

Tchioffo, M. T., Boissière, A., Abate, L., Nsango, S. E., Bayibéki, A. N., Awono-Ambéné, P. H., Christen, R., Gimonneau, G., & Morlais, I. (2016). Dynamics of Bacterial Community Composition in the malaria mosquito's epithelia. *Frontiers in Microbiology*, *6*, 1500. doi:10.3389/fmicb.2015.01500 PMID:26779155

Telfer, S., Lambin, X., Birtles, R., Beldomenico, P., Burthe, S., Paterson, S., & Begon, M. (2010). Species interactions in a parasite community drive infection risk in a wildlife population. *Science*, *330*(6001), 243–246. doi:10.1126cience.1190333 PMID:20929776

Tenywa G. (2007). *Chimps Eat Herbs to Cure Malaria.* AllAfrica.com

Thaiss, C. A., Zmora, N., Levy, M., & Elinav, E. (2016). The microbiome and innate immunity. *Nature*, *535*(7610), 65–74. doi:10.1038/nature18847 PMID:27383981

The Potsdam Institute for Climate Impact Research and Climate Analytics. (2012). Turn-down the Heat – Why a 4 degree Warmer World Must be Avoided. International Bank for Reconstruction and Development and World Bank.

Thompson, R. C., Olsen, Y., Mitchell, R. P., Davis, A., Rowland, S. J., John, A. W. G., McGonigle, D., & Russell, A. E. (2004). Lost at sea: Where is all the plastic? *Science, 304*(5672), 838. doi:10.1126cience.1094559 PMID:15131299

Thundiyil, J. G., Stober, J., Besbelli, N., & Pronczuk, J. (2008). Acute pesticide poisoning: A proposed classification tool. *Bulletin of the World Health Organization, 86*(3), 205–209. doi:10.2471/BLT.08.041814 PMID:18368207

Tian, H. Y., Zhou, S., Dong, L., Van Boeckel, T. P., Cui, Y. J., Wu, Y. R., Cazelles, B., Huang, S. Q., Yang, R. F., Grenfell, B. T., & Xu, B. (2015). Avian influenza H5N1 viral and bird migration networks in Asia. *Proceedings of the National Academy of Sciences of the United States of America, 112*(1), 172–177. doi:10.1073/pnas.1405216112 PMID:25535385

Tian, X., Yu, Z., Feng, P., Ye, Z., Li, R., Liu, J., Hu, J., Kakade, A., Liu, P., & Li, X. (2019). Lactobacillus plantarum TW1-1 alleviates diethylhexylphthalate-induced testicular damage in mice by modulating gut microbiota and decreasing inflammation. *Frontiers in Cellular and Infection Microbiology, 9*, 221. doi:10.3389/fcimb.2019.00221 PMID:31297340

Tian, Y., Gui, W., Rimal, B., Koo, I., Smith, P. B., Nichols, R. G., Cai, J., Liu, Q., & Patterson, A. D. (2020). Metabolic impact of persistent organic pollutants on gut microbiota. *Gut Microbes, 12*(1), 1–16. doi:10.1080/19490976.2020.1848209 PMID:33295235

Tilg, H., & Moschen, A. R. (2014). Microbiota and diabetes: An evolving relationship. *Gut, 63*(9), 1513–1521. doi:10.1136/gutjnl-2014-306928 PMID:24833634

Tishkoff, S. A., Varkonyi, R., Cahinhinan, N., Abbes, S., Argyropoulos, G., Destro-Bisol, G., Drousiotou, A., Dangerfield, B., Lefranc, G., Loiselet, J., Piro, A., Stoneking, M., Tagarelli, A., Tagarelli, G., Touma, E. H., Williams, S. M., & Clark, A. G. (2001). Haplotype diversity and linkage disequilibrium at human *G6PD*: Recent origin of alleles that confer malarial resistance. *Science, 293*(5529), 455–462. doi:10.1126cience.1061573 PMID:11423617

Tjaden, N. B., Suk, J. E., Fischer, D., Thomas, S. M., Beierkuhnlein, C., & Semenza, J. C. (2017). Modelling the effects of global climate change on Chikungunya transmission in the 21 st century. *Scientific Reports, 7*(1), 1–11. doi:10.1038/s41598-017-03566-3

Tomb, R., Maclean, M., Coia, J., MacGregor, S., & Anderson, J. (2017). Assessment of the potential for resistance to antimicrobial violet-blue light in Staphylococcus aureus. *Antimicrobial Resistance and Infection Control, 6*(1), 100. Advance online publication. doi:10.118613756-017-0261-5 PMID:29046782

Tomczuk, K., Grzybek, M., Szczepaniak, K., Studzinska, M., Demkowska-Kutrzepa, M., Roczen-Karczmarz, M., & Klockiewicz, M. (2015). Analysis of intrinsic and extrinsic factors influencing the dynamics of bovine Eimeria spp. from central– eastern Poland. *Veterinary Parasitology, 214*(1-2), 22–28. doi:10.1016/j.vetpar.2015.09.027 PMID:26455571

Toole, M. J. (1997). Communicable Diseases and Disease Control. In E. Noji (Ed.), *The Public Health Consequences of Disasters*. Oxford University Press.

Toyinbo, O., Matilainen, M., Turunen, M., Putus, T., Shaughnessy, R., & Haverinen-Shaughnessy, U. (2016). Modeling associations between principals' reported indoor environmental quality and students' self-reported respiratory health outcomes using GLMM and ZIP models. *International Journal of Environmental Research and Public Health*, *13*(4), 1–12. doi:10.3390/ijerph13040385 PMID:27043595

Trinies, V., Garn, J. V., Chang, H. H., & Freeman, M. C. (2016). The impact of a school-based water, sanitation, and hygiene program on absenteeism, diarrhea, and respiratory infection: A matched-control trial in Mali. *The American Journal of Tropical Medicine and Hygiene*, *94*(6), 1418–1425. doi:10.4269/ajtmh.15-0757 PMID:27114292

Tristan, A., Bes, M., Meugnier, H., Lina, G., Bozdogan, B., Courvalin, P., Reverdy, M. E., Enright, M. C., Vandenesch, F., & Etienne, J. (2007). Global distribution of Panton-Valentine leucocidin - positive methicillin-resistant Staphylococcus aureus, 2006. *Emerging Infectious Diseases*, *13*(4), 594–600. doi:10.3201/eid1304.061316 PMID:17553275

Turnipseed, R. K., Moran, P. J., & Allan, S. A. (2018). Behavioral responses of gravid *Culex quinquefasciatus*, *Aedes aegypti*, and *Anopheles quadrimaculatus* mosquitoes to aquatic macrophyte volatiles. *Journal of Vector Ecology*, *43*(2), 252–260. doi:10.1111/jvec.12309 PMID:30408300

Tyagi, R. K., Das, M. K., Singh, S. S., & Sharma, Y. D. (2013). Discordance in drug resistance-associated mutation patterns in marker genes of *Plasmodium falciparum* and *Plasmodium knowlesi* during coinfections. *The Journal of Antimicrobial Chemotherapy*, *68*(5), 1081–1088. doi:10.1093/jac/dks508 PMID:23292346

Uddin, M. E. (2017a). Ecological framework for primary school attainment in the tri-ethnic communities in rural Bangladesh. *Child Indicators Research*, *10*(3), 693–713. doi:10.100712187-016-9401-3

Uddin, M. E. (2017b). Family demographic mechanisms linking of socioeconomic status to subjective physical health in rural Bangladesh. *Social Indicators Research*, *130*(3), 1263–1279. doi:10.100711205-015-1209-x

Uddin, M. E. (2022). Parental investment mediates associations between lower family income and primary math achievement. *Journal of Applied Developmental Psychology*, *81*(4), 1–11. doi:10.1016/j.appdev.2022.101422

Uddin, M. E., & Ferdous, Z. (2009). Gender of birth control methods practice in sexual behavior patterns in Bangladesh: A literatures review. *Antrocom: Online Journal of Anthropology*, *5*(2), 87–91.

Uddin, M. E., & Ferdous, Z. (2015). Men's condom use or non-use and health risk: A comparative study on premarital, marital and extramarital men at Rajshahi city, Bangladesh. *Advances in Applied Psychology*, *1*(2), 128–134.

UNFCCC (United Nations Framework Convention on Climate Change). (2016). *Task Force on Displacement*. UNFCCC.

UNFCCC. (2011). *Report of the Conference of the Parties: The Cancun Agreements, Outcome of the work of the Ad Hoc Working Group on the Long-term Cooperative Action under the Convention*. FCCC/CP/2010/7/Add.1, Part Two: Action taken by the Conference of the Parties at its sixteenth session. Bonn: UNFCCC, http://unfccc.int/resource/ docs/2010/cop16/eng/07a01.pdf#page=4

United Nations, Department of Economic and Social Affairs. (2014). World urbanization prospects: 2014 revision (ST/ESA/SER.A/366). Population Division.

Urban, K., Mehrmal, S., Uppal, P., Giesey, R. L., & Delost, G. R. (2021). The global burden of skin cancer: A longitudinal analysis from the Global Burden of Disease Study, 1990–2017. *JAAD International*, *2*, 98–108. doi:10.1016/j.jdin.2020.10.013 PMID:34409358

Usher, K., Durkin, J., & Bhullar, N. (2019). Eco-anxiety: How thinking about climate change-related environmental decline is affecting our mental health. International Journal of Mental Health Nursing, 28(6), 1233–1234. doi:10.1111/inm.12673

Van Cauwenberghe, L., & Janssen, C. R. (2014). Microplastics in bivalves cultured for human consumption. *Environmental Pollution*, *193*, 65–70. doi:10.1016/j.envpol.2014.06.010 PMID:25005888

Van de Wiele, T., Gallawa, C. M., Kubachk, K. M., Creed, J. T., Basta, N., Dayton, E. A., Whitacre, S., Du Laing, G., & Bradham, K. (2010). Arsenic metabolism by human gut microbiota upon in vitro digestion of contaminated soils. *Environmental Health Perspectives*, *118*(7), 1004–1009. doi:10.1289/ehp.0901794 PMID:20603239

Van Den Elsen, L. W. J., Poyntz, H. C., Weyrich, L. S., Young, W., & Forbes-Blom, E. E. (2017). Embracing the gut microbiota: The new frontier for inflammatory and infectious diseases. *Clinical & Translational Immunology*, *6*(1), e125. Advance online publication. doi:10.1038/cti.2016.91 PMID:28197336

van Seventer, J. M., & Hochberg, N. S. (2017). Principles of infectious diseases: Transmission, diagnosis, prevention, and control. International Encyclopedia of Public Health, 6, 22-39. doi:10.1016/B978-0-12-803678-5.00516-6

Vardakou, M., Palop, C. N., Christakopoulos, P., Faulds, C. B., Gasson, M. A., & Narbad, A. (2008). Evaluation of the prebiotic properties of wheat arabinoxylan fractions and induction of hydrolase activity in gut microflora. *International Journal of Food Microbiology*, *123*(1–2), 166–170. doi:10.1016/j.ijfoodmicro.2007.11.007 PMID:18279987

Ventola, C.L. (2015). The antibiotic resistance crisis: part 1: causes and threats. *P & T: A Peer-Reviewed Journal for Formulary Management*, *40*(4), 277-83.

Ventura-Garcia, L., Roura, M., Pell, C., Posada, E., Gascón, J., Aldasoro, E., Munoz, J., & Pool, R. (2013). Socio-cultural aspects of Chagas disease: A systematic review of qualitative research. *PLoS Neglected Tropical Diseases*, *7*(9), 1–8. doi:10.1371/journal.pntd.0002410 PMID:24069473

Verhoeven, J. I., Allach, Y., Vaartjes, I. C. H., Klijn, C. J. M., & de Leeuw, F. E. (2021). Ambient air pollution and the risk of ischaemic and haemorrhagic stroke. *The Lancet. Planetary Health, 5*(8), e542–e552. doi:10.1016/S2542-5196(21)00145-5 PMID:34390672

Vested, A., Giwercman, A., Bonde, J. P., & Toft, G. (2014). Persistent organic pollutants and male reproductive health. *Asian Journal of Andrology, 16*(1), 71. doi:10.4103/1008-682X.122345 PMID:24369135

Viana, M., Mancy, R., Biek, R., Cleaveland, S., Cross, P. C., Lloyd-Smith, J. O., & Haydon, D. T. (2014). Assembling evidence for identifying reservoirs of infection. *Trends in Ecology & Evolution, 29*(5), 270–279. doi:10.1016/j.tree.2014.03.002 PMID:24726345

Viboud, C., Pakdaman, K., Boelle, P.-Y., Wilson, M. L., Myers, M. F., Valleron, A.-J., & Flahault, A. (2004). Association of influenza epidemics with global climate variability. *European Journal of Epidemiology, 19*, 1055–1059. doi:10.100710654-004-2450-9 PMID:15648600

Vikram, K., Nagpal, B. N., Pande, V., Srivastava, A., & Gupta, S. K., Anushrita, Singh, V. P., Singh, H., Saxena, R., Tuli, N. R., Yadav, N. K., Paul, R., Valecha, N., & Telle, O. (2015). Comparison of *Ae. aegypti* breeding in localities of different socio-economic groups of Delhi, India. *International Journal of Mosquito Research, 2*(2), 83–88.

Vikram, K., Nagpal, B. N., Pande, V., Srivastava, A., Saxena, R., Singh, H., Gupta, S. K., Tuli, N. R., Yadav, N. K., Olivier, T., & Richard, P. (2015). Detection of dengue virus in individual *Aedes aegypti* mosquitoes in Delhi, India. *Journal of Vector Borne Diseases, 52*(2), 129. PMID:26119544

Wadood, M. A., Mamun, A. S. M. A., Rafi, M. A., Islam, M. K., Mohd, S., Lee, L. L., & Hossain, M. G. (2021). *Knowledge, attitude, practice and perception regarding COVID-19 among students in Bangladesh: Survey in Rajshahi University*. SciMedicine Journal. doi:10.28991/SciMedJ-2021-03-SI-9

Walsh, M. G. (2014). Rat sightings in New York City are associated with neighborhood sociodemographics, housing characteristics, and proximity to open public spaces. *PeerJ, 2*, e533. Advance online publication. doi:10.7717/peerj.533 PMID:25237595

Wamai, R. G., Kahn, J., McGloin, J., & Ziaggi, G. (2020). Visceral leishmaniasis: A global overview. *Journal of Global Health Science, 2*(1), e3. Advance online publication. doi:10.35500/jghs.2020.2.e3

Wamaket, N., Dieng, H., Komalamisra, N., Apiwathnasorn, C., Morales, R. E., Thanomsub, B. W., Srisawat, R., & Attrapadung, S. (2018). Larvicidal and adulticidal activities of castor oil against the dengue vector, *Aedes aegypti. Tropical Biomedicine, 35*(3), 610–618. PMID:33601748

Wang, G., Li, X., & Zasloff, M. (2010). A database view of naturally occurring antimicrobial peptides: nomenclature, classification, and amino acid sequence analysis. *Antimicrobial Peptides: Discovery, Design and Novel Therapeutic Strategies*, 1-21. doi:10.1079/9781845936570.0001

Wang, S., Zhang, W., Wang, H., Wang, J., & Jiang, M. J. (2021). How Does Income Inequality Influence Environmental Regulation in the Context of Corruption? A Panel Threshold Analysis Based on Chinese Provincial Data. *International Journal of Environmental Research and Public Health, 18*(15), 8050. doi:10.3390/ijerph18158050 PMID:34360341

Wang, Y., Yim, S., Yang, Y., & Morin, C. W. (2020). The effect of urbanization and climate change on the mosquito population in the Pearl River Delta region of China. *International Journal of Biometeorology, 64*(3), 501–512. doi:10.100700484-019-01837-4 PMID:31811391

Warrell and Gilles. (2002). *Essential Malariology* (4th ed.). CRC.

Watts, N., Adger, W. N., Ayeb-Karlsson, S., Bai, Y., Byass, P., Campbell-Lendrum, D., Colbourn, T., Cox, P., Davies, M., Depledge, M., Depoux, A., Dominguez-Salas, P., Drummond, P., Ekins, P., Flahault, A., Grace, D., Graham, H., Haines, A., Hamilton, I., ... Costello, A. (2017). The Lancet Countdown: Tracking progress on health and climate change. *Lancet, 389*(10074), 1151–1164. doi:10.1016/S0140-6736(16)32124-9 PMID:27856085

Watts, N., Amann, M., Ayeb-Karlsson, S., Belesova, K., Bouley, T., Boykoff, M., Byass, P., Cai, W., Campbell-Lendrum, D., Chambers, J., Cox, P. M., Daly, M., Dasandi, N., Davies, M., Depledge, M., Depoux, A., Dominguez-Salas, P., Drummond, P., Ekins, P., ... Costello, A. (2018). The Lancet Countdown on health and climate change: From 25 years of inaction to a global transformation for public health. *Lancet, 391*(10120), 581–630. doi:10.1016/S0140-6736(17)32464-9 PMID:29096948

Webb, J. L. A. (2005). Malaria and the Peopling of Early Tropical Africa. *Journal of World History, 16*(3), 269–291. doi:10.1353/jwh.2006.0011

Wei, T., Yang, S. L., Moore, J. C., Shi, P. J., Cui, X. F., Duan, Q. Y., Xu, B., Dai, Y. J., Yuan, W. P., Wei, X., Yang, Z., Wen, T., Teng, F., Gao, Y., Chou, J., Yan, X., Wei, Z., Guo, Y., Jiang, Y., ... Dong, W. (2012). Developed and developing world responsibilities for historical climate change and CO_2 mitigation. *Proceedings of the National Academy of Sciences of the United States of America, 109*(32), 12911–12915. doi:10.1073/pnas.1203282109 PMID:22826257

Wellington, E. M. H., Boxall, A. B., Cross, P., Feil, E. J., Gaze, W. H., Hawkey, P. M., Johnson-Rollings, A. S., Jones, D. L., Lee, N. M., Otten, W., Thomas, C. M., & Williams, A. P. (2013). The role of the natural environment in the emergence of antibiotic resistance in gram-negative bacteria. *The Lancet. Infectious Diseases, 13*(2), 155–165. doi:10.1016/S1473-3099(12)70317-1 PMID:23347633

Wen, T. H., Tsai, C. T., & Chin, W. C. B. (2016). Evaluating the role of disease importation in the spatiotemporal transmission of indigenous dengue outbreak. *Applied Geography (Sevenoaks, England), 76*, 137–146. doi:10.1016/j.apgeog.2016.09.020

Wigle, D. T. (2003). *Child health and the environment.* Oxford University Press.

Wilcove, D. S., McLellan, C. H., & Dobson, A. P. (1986). Habitat fragmentation in the temperate zone. In Conservation Biology: The Science of Scarcity and Diversity (pp. 237-256). Sinauer Associates, Inc. Publishers.

Wilcox, B. A., & Colwell, R. R. (2005). Emerging and reemerging infectious Diseases: Biocomplexity as an interdisciplinary paradigm. *EcoHealth*, *2*(4), 244–257. doi:10.100710393-005-8961-3

Wilcox, B. A., & Gubler, D. J. (2005). Disease ecology and the global emergence of zoonotic pathogens. *Environmental Health and Preventive Medicine*, *10*(5), 263–272. doi:10.1007/BF02897701 PMID:21432130

Wilder-Smith, A. (2012). Dengue infections in travellers. *Paediatrics and International Child Health*, *32*(sup1, s1), 28–32. doi:10.1179/2046904712Z.00000000050 PMID:22668447

Wilding, L. A., Bassis, C. M., Walacavage, K., Hashway, S., Leroueil, P. R., Morishita, M., Maynard, A. D., Philbert, M. A., & Bergin, I. L. (2016). Repeated dose (28-day) administration of silver nanoparticles of varied size and coating does not significantly alter the indigenous murine gut microbiome. *Nanotoxicology*, *10*(5), 513–520. doi:10.3109/17435390.2015.1078854 PMID:26525505

Wilke, A. B. B., Chase, C., Vasquez, C., Carvajal, A., Medina, J., Petrie, W. D., & Beier, J. C. (2019). Urbanization creates diverse aquatic habitats for immature mosquitoes in urban areas. *Scientific Reports*, *9*(1), 15335. doi:10.103841598-019-51787-5 PMID:31653914

Williams, K., Milner, J., Boudreau, M. D., Gokulan, K., Cerniglia, C. E., & Khare, S. (2015). Effects of subchronic exposure of silver nanoparticles on intestinal microbiota and gut-associated immune responses in the ileum of Sprague-Dawley rats. *Nanotoxicology*, *9*(3), 279–289. doi:10.3109/17435390.2014.921346 PMID:24877679

Willox, A. C., Stephenson, E., Allen, J., Bourque, F., Drossos, A., Elgarøy, S., Kral, M. J., Mauro, I., Moses, J., & Pearce, T. (2015). Examining relationships between climate change and mental health in the Circumpolar North. *Regional Environmental Change*, *15*(1), 169–182. doi:10.100710113-014-0630-z

Wilson, P., Atkinson, M., Hornby, G., Thompson, M., Cooper, M., Hooper, C. M., & Southall, A. (2002). *Young minds in our schools-a guide for teachers and others working in schools.* Young Minds.

Wilson, M. L. (2001). Ecology and infectious disease. In J. L. Aron & J. A. Patz (Eds.), *Ecosystem Change and Public Health: A Global Perspective* (pp. 283–324). Johns Hopkins University Press.

Woolhouse, M. E. J., & Gowtage-Sequeria, S. (2005). Host range and emerging and reemerging pathogens. *Emerging Infectious Diseases*, *11*(12), 1842–1847. doi:10.3201/eid1112.050997 PMID:16485468

Woolhouse, M. E. J., Haydon, D. T., & Antia, R. (2005). Emerging pathogens: The epidemiology and evolution of species jumps. *Trends in Ecology & Evolution*, *20*(5), 238–244. doi:10.1016/j.tree.2005.02.009 PMID:16701375

World Health Organization (WHO). (2017). *Antibacterial agents in clinical development: an analysis of the antibacterial clinical development pipeline, including tuberculosis*. Geneva, Switzerland: WHO. https://www.who.int/medicines/areas/rational_use/antibacterial_agents_clinical_development/en/

World Health Organization. (2004). *The physical school environment: An essential elements of a health promoting school*. World Health Organization.

World Health Organization. (2007). *The world health report 2007- a safer future: global public health security in the 21st century*. World Health Organization.

World Health Organization. (2008). *Our cities, our health, our future - acting on social determinants for health equity in urban settings*. World Health Organization.

World Health Organization. (2010). *Hidden cities: unmasking and overcoming health inequities in urban settings*. World Health Organization.

World Health Organization. (2013). *Sustaining the drive to overcome the global impact of neglected tropical diseases _ second WHO report on neglected diseases*. World Health Organization.

World Health Organization. (2014). *Progress on drinking water and sanitation - Joint Monitoring Programme update 2014*. World Health Organization.

World Health Organization. (2015). *School environment: policies and current status*. WHO Regional Office for Europe.

World Health Organization. (2016). *School and youth health-global school health initiative*. https://www.who.int/school_youth_health/gshi/en

World Health Organization. (2019). *Urban settings as a social determinant of health*. World Health Organization.

World Health Organization. (2020). *Novel Coronavirus (2019-nCoV): situation report, 11*. World Health Organization.

World Health Organization. (2020). *Soil-transmitted helminth infections*. World Health Organization.

World Health Organization. (2020). *World malaria report 2020*. https://www.who.int/publications/i/item/9789240015791

World Health Organization. (2021a). *Dengue and severe dengue*. Fact Sheet. https://www.who.int/news-room/fact-sheets/detail/dengue-and-severe-dengue

World Health Organization. (2021b). *Leishmaniasis*. Fact Sheet. https://www.who.int/news-room/fact-sheets/detail/leishmaniasis

World Health Organization. Report of an Informal Consultation. (2007). *Containment of Malaria Multi-drug Resistance on the Cambodia-Thailand Border*. WHO.

World malaria report 2021. (2021). Geneva: World Health Organization.

World Tourism Organization. (2014). UNWTO Tourism Highlights, 2014 Edition (eISBN 9789284416226). UNTWO Publications.

World Urban Forum. (Ed.). (2012). *State of the world's cities 2012/2013*. United Nations Human Settlements Programme (UN-HABITAT).

Wright, A. N., & Gompper, M. E. (2005). Altered parasite assemblages in raccoons in response to manipulated resource availability. *Oecologia*, *144*(1), 148–156. doi:10.100700442-005-0018-3 PMID:15891856

Wright, S. L., & Kelly, F. J. (2017). Plastic and human health: A micro issue? *Environmental Science & Technology*, *51*(12), 6634–6647. doi:10.1021/acs.est.7b00423 PMID:28531345

Wu, H., Kioumourtzoglou, M. A., Just, A. C., Kloog, I., Sanders, A., Svensson, K., McRae, N., Tamayo-Ortiz, M., Solano-González, M., Wright, R. O., Téllez-Rojo, M. M., & Baccarelli, A. A. (2020). Association of ambient PM2·5 exposure with maternal bone strength in pregnant women from Mexico City: A longitudinal cohort study. *The Lancet. Planetary Health*, *4*(11), e530–e537. doi:10.1016/S2542-5196(20)30220-5 PMID:33159880

Wu, X. X., Tian, H. Y., Zhou, S., Chen, L. F., & Xu, B. (2014). Impact of global change on transmission of human infectious diseases. *Science China. Earth Sciences*, *57*(2), 189–203. doi:10.100711430-013-4635-0 PMID:32288763

Wu, X., Lu, Y., Zhou, S., Chen, L., & Xu, B. (2016). Impact of climate change on human infectious diseases: Empirical evidence and human adaptation. *Environment International*, *86*, 14–23. doi:10.1016/j.envint.2015.09.007 PMID:26479830

Xiang, J., Hansen, A., Liu, Q., Tong, M. X., Liu, X., Sun, Y., Cameron, S., Hanson-Easey, S., Han, G. S., Williams, C., Weinstein, P., & Bi, P. (2018). Impact of meteorological factors on hemorrhagic fever with renal syndrome in 19 cities in China, 2005–2014. *The Science of the Total Environment*, *636*, 1249–1256. doi:10.1016/j.scitotenv.2018.04.407 PMID:29913587

Xia, Y., Wang, G., Buscariollo, D., Pitts, R. J., Wenger, H., & Zwiebel, L. J. (2008). The molecular and cellular basis of olfactory-driven behavior in *Anopheles gambiae* larvae. *Proceedings of the National Academy of Sciences of the United States of America*, *105*(17), 6433–6438. doi:10.1073/pnas.0801007105 PMID:18427108

Xie, F., Fan, H., Li, Y., Jiang, Z., Meng, R., & Bovik, A. (2017). Melanoma Classification on Dermoscopy Images Using a Neural Network Ensemble Model. *IEEE Transactions on Medical Imaging*, *36*(3), 849–858. doi:10.1109/TMI.2016.2633551 PMID:27913337

Xu, B., Jin, Z. Y., Jiang, Z. B., Guo, J. P., Timberlake, M., & Ma, X. L. (2014). Climatological and geographical impacts on global pandemic of influenza A (H1N1). In Q. Weng (Ed.), *Global Urban Monitoring and Assessment through Earth Observation*. Taylor & Francis/CRC Press.

Xu, C., Liu, Q., Huan, F., Qu, J., Liu, W., Gu, A., Wang, Y., & Jiang, Z. (2015). Changes in gut microbiota may be early signs of liver toxicity induced by epoxiconazole in rats. *Chemotherapy*, *60*(2), 135–142. doi:10.1159/000371837 PMID:25766299

Yadav, N. P., Rai, V. K., Mishra, N., Sinha, P., Bawankule, D. U., Pal, A., Tripathi, A. K., & Chanotiya, C. S. (2014). A novel approach for development and characterization of effective mosquito repellent cream formulation containing citronella oil. *BioMed Research International*, *786084*, 1–11. doi:10.1155/2014/786084 PMID:25379509

Yang, D., Shi, H., Li, L., Li, J., Jabeen, K., & Kolandhasamy, P. (2015). Microplastic pollution in table salts from China. *Environmental Science & Technology*, *49*(22), 13622–13627. doi:10.1021/acs.est.5b03163 PMID:26486565

Yang, J., Feng, P., Ling, Z., Khan, A., Wang, X., Chen, Y., Ali, G., Fang, Y., Salama, E.-S., Wang, X., Liu, P., & Li, X. (2023). Nickel exposure induces gut microbiome disorder and serum uric acid elevation. *Environmental Pollution*, *324*, 121349. doi:10.1016/j.envpol.2023.121349 PMID:36870597

Yee, D. A., Kesavaraju, B., & Juliano, S. A. (2004). Larval feeding behavior of three co-occurring species of container mosquitoes. *Journal of Vector Ecology*, *29*(2), 315–322. PMID:15707290

Yeh, K. B., Parekh, F. K., Mombo, I., Leimer, J., Hewson, R., Olinger, G., Fair, J. M., Sun, Y., & Hay, J. (2023). Climate change and infectious disease: A prologue on multidisciplinary cooperation and predictive analytics. *Frontiers in Public Health*, *11*, 1018293. doi:10.3389/fpubh.2023.1018293 PMID:36741948

Yeh, K. B., Parekh, F. K., Tabynov, K., Tabynov, K., Hewson, R., Fair, J. M., Essbauer, S., & Hay, J. (2021). Operationalizing cooperative research for infectious disease surveillance: Lessons learned and ways forward. *Frontiers in Public Health*, *9*, 659695. doi:10.3389/fpubh.2021.659695 PMID:34568249

Yeh, K., Fair, M., Smith, J., Martinez Torres, W., Lucas, T., Monagin, J., & … . (2020). Assessing climate change impact on ecosystems and infectious disease: Important roles for genomic sequencing and a one health perspective. *Tropical Medicine and Infectious Disease*, *5*, 90. doi:10.3390/tropicalmed5020090 PMID:32503239

Ye, J., Zhao, M., Liu, J., & Liu, W. (2010). Enantioselectivity in environmental risk assessment of modern chiral pesticides. *Environmental Pollution*, *158*(7), 2371–2383. doi:10.1016/j.envpol.2010.03.014 PMID:20378218

Yilmaz, B., Terekeci, H., Sandal, S., & Kelestimur, F. (2020). Endocrine disrupting chemicals: Exposure, effects on human health, mechanism of action, models for testing and strategies for prevention. *Reviews in Endocrine & Metabolic Disorders*, *21*(1), 127–147. doi:10.100711154-019-09521-z PMID:31792807

Yoldascan, E., Kurtaran, B., Koyunchu, M., & Koyunchu, E. (2010). Modeling the economic impact of pandemic influenza: A case study in Turkey. *Journal of Medical Systems*, *34*(2), 139–145. doi:10.100710916-008-9225-x PMID:20433052

Yoneyama, H., & Katsumata, R. (2006). Antibiotic Resistance in Bacteria and Its Future for Novel Antibiotic Development. *Bioscience, Biotechnology, and Biochemistry*, *70*(5), 1060–1075. doi:10.1271/bbb.70.1060 PMID:16717405

Yue, R. P. H., & Lee, H. F. (2020). Drought-induced spatio-temporal synchrony of plague outbreak in Europe. *The Science of the Total Environment*, *698*, 134138. doi:10.1016/j.scitotenv.2019.134138 PMID:31505345

Zahouli, J. B. Z., Koudou, B. G., Muèller, P., Malone, D., Tano, Y., & Utzinger, J. (2017). Urbanization is a main driver for the larval ecology of *Aedes mosquitoes* in arbovirus-endemic settings in south-eastern Côte d'Ivoire. *PLoS Neglected Tropical Diseases*, *11*(7), e0005751. doi:10.1371/journal.pntd.0005751 PMID:28704434

Zaman, S., Hussain, M., Nye, R., Mehta, V., Mamun, K., & Hossain, N. (2017). A Review on Antibiotic Resistance: Alarm Bells are Ringing. *Cureus*. Advance online publication. doi:10.7759/cureus.1403 PMID:28852600

Zarn, J. A., Brüschweiler, B. J., & Schlatter, J. R. (2003). Azole fungicides affect mammalian steroidogenesis by inhibiting sterol 14 alpha-demethylase and aromatase. *Environmental Health Perspectives*, *111*(3), 255–261. doi:10.1289/ehp.5785 PMID:12611652

Zell, R. (2004). Global climate change and the emergence/re-emergence of infectious diseases. *International Journal of Medical Microbiology NLM*, *293*, 16–26. doi:10.1016/S1433-1128(04)80005-6 PMID:15146981

Zeng, Z., Zhan, J., Chen, L., Chen, H., & Cheng, S. (2021). Global, regional, and national dengue burden from 1990 to 2017: A systematic analysis based on the global burden of disease study 2017. *EClinicalMedicine*, *32*, 100712. doi:10.1016/j.eclinm.2020.100712 PMID:33681736

Zhai, Q., Tian, F., Zhao, J., Zhang, H., Narbad, A., & Chen, W. (2016). Oral administration of probiotics inhibits absorption of the heavy metal cadmium by protecting the intestinal barrier. *Applied and Environmental Microbiology*, *82*(14), 4429–4440. doi:10.1128/AEM.00695-16 PMID:27208136

Zhang, F., Cui, B., He, X., Nie, Y., Wu, K., Fan, D., & Group, F. S. (2018). Microbiota transplantation: Concept, methodology and strategy for its modernization. *Protein & Cell*, *9*(5), 462–473. doi:10.100713238-018-0541-8 PMID:29691757

Zhang, F., Luo, W., Shi, Y., Fan, Z., & Ji, G. (2012). Should we standardize the 1,700-year-old fecal microbiota transplantation? [-author.]. *The American Journal of Gastroenterology*, *107*(11), 1755. doi:10.1038/ajg.2012.251 PMID:23160295

Zhang, L., Nichols, R. G., Correll, J., Murray, I. A., Tanaka, N., Smith, P. B., Hubbard, T. D., Sebastian, A., Albert, I., Hatzakis, E., Gonzalez, F. J., Perdew, G. H., & Patterson, A. D. (2015). Persistent organic pollutants modify gut microbiota–host metabolic homeostasis in mice through aryl hydrocarbon receptor activation. *Environmental Health Perspectives*, *123*(7), 679–688. doi:10.1289/ehp.1409055 PMID:25768209

Zhang, S., Jin, Y., Zeng, Z., Liu, Z., & Fu, Z. (2015). Subchronic exposure of mice to cadmium perturbs their hepatic energy metabolism and gut microbiome. *Chemical Research in Toxicology*, *28*(10), 2000–2009. doi:10.1021/acs.chemrestox.5b00237 PMID:26352046

Zhao, J., Ran, W., Teng, J., Liu, Y., Liu, H., Yin, X., Cao, R., & Wang, Q. (2018). Microplastic pollution in sediments from the Bohai Sea and the Yellow Sea, China. *The Science of the Total Environment*, *640*, 637–645. doi:10.1016/j.scitotenv.2018.05.346 PMID:29870939

Zhao, Y., Zhou, C., Wu, C., Guo, X., Hu, G., Wu, Q., Xu, Z., Li, G., Cao, H., Li, L., Latigo, V., Liu, P., Cheng, S., & Liu, P. (2020). Subchronic oral mercury caused intestinal injury and changed gut microbiota in mice. *The Science of the Total Environment*, *721*, 137639. doi:10.1016/j.scitotenv.2020.137639 PMID:32172103

Zhong, S., Yang, L., Toloo, S., Wang, Z., Tong, S., Sun, X., Crompton, D., FitzGerald, G., & Huang, C. (2018). The long-term physical and psychological health impacts of flooding: A systematic mapping. *The Science of the Total Environment*, *626*, 165–194. doi:10.1016/j.scitotenv.2018.01.041 PMID:29339262

Zhong, X., Li, J., Lu, F., Zhang, J., & Guo, L. (2022). Application of zebrafish in the study of the gut microbiome. *Animal Models and Experimental Medicine*, *5*(4), 323–336. doi:10.1002/ame2.12227 PMID:35415967

Zhou, Y. B., Zhuang, J. L., Yang, M. X., Zhang, Z. J., Wei, J. G., Peng, W. X., Zhao, G. M., Zhang, S. M., & Jiang, Q. W. (2010). Effects of low temperature on the schistosome-transmitting snail *Oncomelania hupensis* and the implications of global climate change. *Molluscan Research*, *30*, 102–108.

Zietz, B. P., & Dunkelberg, H. (2004). The history of the plague and the research on the causative agent Yersinia *pestis. International Journal of Hygiene and Environmental Health*, *207*(2), 165–178. doi:10.1078/1438-4639-00259 PMID:15031959

Zwiebel, L. J., & Takken, W. (2004). Olfactory regulation of mosquito-host interactions. *Insect Biochemistry and Molecular Biology*, *34*(7), 645–652. doi:10.1016/j.ibmb.2004.03.017 PMID:15242705

About the Contributors

P. A. Azeez is a researcher working in the field of Environmental Science/ Ecological research for about 37 years. He was the Director of the Sálim Ali Centre for Ornithology and Natural History (SACON), Coimbatore, India, for almost a decade. He is an active member of many global research organisations and has served as Principal Investigator or Chief Member for more than 50 Environmental Impact Assessments. He has guided several doctoral and post-doctoral researchers. Dr. Azeez has served as chair or member of several policy-making committees organised by governmental bodies, courts of law, universities or research institutions on environmental issues, nature conservation or academics. His current interest lies in environmental policies, environmental health, climate change and sustainable development.

P. P. Nikhil Raj is a transdisciplinary academician. He has been undertaking research and teaching activities in the core area of environment and sustainability. Dr. Raj's research question broadly lies in understanding how natural systems tolerate/check the changing scenario of climate and the pressures from various anthropogenic pressures; he is fascinated by studying the trade-off between the social and the environmental systems and the crucial roles played by money and political power in shaping them. Dr. Raj has published his research findings in many peer-reviewed journals and books. He also writes popular articles of socio-environmental importance in newspapers and magazines.

R. Mohanraj is a Professor of Environmental Sciences at Bharathidasan University, Tiruchirappalli, India. His primary research interest lies in Environmental Impact and Risk Assessment, Urban Air Pollution, Hydro-Geochemistry, Ecological Restoration, Forest Carbon Dynamics, and Corporate Environmental Management. Dr Mohanraj has published 80 research articles in peer-reviewed international and Indian journals and edited volumes. He has authored and edited seven books by Indian and International publishers. He has presented more than 80 research and review papers in international and national level seminars /symposia/conferences, including COP meetings, as an invited speaker. He had also organised/coordinated several national-level conferences/webinars/refresher courses, academic events and outreach programs.

* * *

Youssef Ait Hamdan is a PhD student in Biochemistry at Cadi Ayyad University, Morocco.

Melisa Anderson Cross is a Lecturer in the School of Allied Health and Wellness, College of Health Sciences, University of Technology.

Suresh B. Arakera is working as an Assistant Professor at Department of Applied Genetics, Karnatak University, Dharwad since 2014. Currently he is supervising six PhD scholars who are working in the areas of Molecular Medicine & Microbial Genetics. More than 25 Masters' dissertations have been completed till date under his supervision in the fields of molecular medicine and microbial genetics. Apart from this, Dr. Arakera has also been a Laboratory Nodal Officer in Covid-19 testing laboratory in DIMHANS, Dharwad where he studied the Covid-19 diversity & Immunology. He has more than 13 years of research experience and contributed in more than 13 research articles and 4 book chapters.

Sushmitha Baskar is presently working at the School of Interdisciplinary and Transdisciplinary Studies, Indira Gandhi National Open University (IGNOU), New Delhi. She earned her M.Sc (1995) (Gold medalist) and M.Phil in Environmental Science (1997) from the Department of Environmental Sciences, Bharathiar University, Coimbatore, Tamil Nadu. She obtained her Ph.D. in Environmental Science (2002) from the Department of Environmental Science and Engineering, Guru Jambheshwar University of Science and Technology, Hisar. She is awarded with various national fellowships: CSIR (SRF), CSIR (RA), CSIR- Pool Scientist (SRA), UGC Dr. DS Kothari post-doctoral fellowship. She is awarded with international fellowships: Swiss Fellowship (2002-04) to work at Swiss Federal Institute of Technology, Zurich, Switzerland, Indo-Norwegian Grant (2008-09) to work at University of Bergen, Norway and Swedish Grant (2009) at Stockholm University, Sweden where she carried out research in the field of Cave geomicrobiology. She has also visited Technical University, Munich, Germany, Czech Republic, Thailand, Malaysia and Indonesia for part of her research training and as an invited speaker in trainings and conferences. She was recently awarded the university Silver medal for "Best researcher 2019 IGNOU" by the Vice President of India, Sh. Venkaiah Naidu. Her interests include Environmental Science, Environmental Microbiology and

Geomicrobiology. She has been an invited speaker at Mahidol University, Thailand and the Summer school on Speleology, Yogyakarta, Indonesia, 2018. She has made significant contributions in peer reviewed international journals with 40 research publications and 3 books. She is a reviewer in peer reviewed journals 'Current Science' and 'Astrobiology'. She was an invited Speaker at Indian Science Congress 2014, 2016, 2020. She has written several chapters for IGNOU, chapters for E-PG pathshala and contributed for online course development at IGNOU and refresher courses (ARPIT) offered through SWAYAM (MOOC-IGNOU), MOOC-AMU, Aligarh, MOOC-GJUST, Hisar, MOOC-MLSU, Udaipur). She is the coordinator for IESO International Earth Science Olympaid in Delhi-NCR region from 2012 till present. She regularly gives lectures on Gyanvani (FM105.6) and Gyandarshan programmes at IGNOU. She is also a science journalist for Current Science, Indian Academy of Sciences, Bangalore. At IGNOU she is the Programme Coordinator for Ph.D. Environmental Science, PG Diploma in Environmental and Occupational Health and the Course Coordinator for M.Sc. Environmental Science papers in Earth processes, Disaster management and Environment and Society.

Debraj Biswal is currently employed as Assistant Professor of Zoology in a Government College in West Bengal, India. His teaching career spans for more than 13 years. His research activities for over 14 years mainly include areas in parasitology, helminthology, vector biology and microbiology. He has published a number of articles related to these topics. Additionally, he has keen interests in ecology, environmental biology and eco-toxicology with a few publications in these fields too.

Shafa Contractor is a research scholar in the Department of Applied Genetics, Karnatak University, Dharwad.

Ahmed Draoui is a PhD specializing in neuroscience, neuropharmacology, neurobehavioral, and health, my research mainly focuses on the pathophysiology of neurodegenerative diseases, particularly Parkinson's disease.

Hanane El Fatimi is a PhD student in neuroscience, Laboratory of Neuroscience and Cellular Physiology, Mohammed VI University of Health Sciences, Morocco.

Omar El Hiba is a Professor at Chouaib Doukkali University, El Jadida, Morocco.

Abdelaati El Khiat is a PhD Professor at ISPITS Ouarzazate, Ministry of Health, Morocco.

Bilal El-Mansoury is a Ph.D. in Neurosciene and Physiology, Faculty of Science, Chouaib Doukkali University, Morocco. He was graduated with a master degree in Pathophysiology and Anthropobiology from cadi Ayyad University in 2019, Morocco.

Samia Elouali is a PhD student in polymer chemistry and biochemistry at Cadi Ayyad University, Morocco.

Debojyoty Ghosh has completed his MSc in Zoology from West Bengal State University. Currently he is working as Research Assistant in Sundarban Tiger Reserve, Directorate of Forest, Government of West Bengal.

Pradipta Kumar Ghosh has completed his M.Sc. in Zoology from Barasat Government College under West Bengal State University. Currently registered under West Bengal State University as a PhD scholar. He is pursuing his doctoral research on estuarine fish juveniles in mangroves of Indian Sundarban.

Lorenzo Gordon is the Vice-Dean and Lecturer at the Caribbean School of Medical Sciences, Jamaica.

Arumugam Radhakrishnan Jayakumar is a professor at University of Miami, USA.

Banani Mandal completed her M.Phil and PhD from University of Calcutta. She has more than 30 publications. Currently she is working as an Assistant Professor and Head of the Department of Zoology, Jogesh Chandra Chaudhuri College in Kolkata.

Shelly McFarlane, a mental health epidemiologist, with a masters in Diabetology, is a member of the team of researchers within The Caribbean Institute for Health Research, formerly The Tropical Medicine Research Institute. She has a wealth of experience as project coordinator, health researcher, and project manager for a number of large and small studies, including clinical trials. She has been a member of the investigator and management team large national surveys such as the Jamaica Health and Lifestyle Surveys II and III and the Jamaica Youth Risk and Resiliency Behaviour Survey 2006. She has been member of the team of investigators for multi-country studies such as the "Development and Implementation of Culture Sensitive Peer /Lay Diabetes Education Program for Adults with Type 2 Diabetes in Six English Speaking Caribbean Countries", and "Congregations Taking Action Against NCDs". More recently, she has had to meet the rigorous standards for good clinical practices as project coordinator/manager for clinical trials in several ground breaking studies carried out under the aegis of the Caribbean Institute for

Medical Research. She has a history of successful acquisition of grant funding to carry out research. She has been with the Caribbean Institute for Health Research for more than 10 years and has served as workshop facilitator for a number of basic and advanced research skills training workshops aimed at equipping health professionals and other persons seeking development of careers in research. She is an expert trainer in Effective Project Planning and Evaluation and Implementation Research and, in addition to proving training locally, has been called to provide training for persons in the wider Caribbean region, North and Central America and in Africa. Dr McFarlane is also special advisor and a board member of the World Walk Foundation Jamaica Chapter and has also conducted trainings in foot-care management in the wider Caribbean and Nigeria. Dr McFarlane holds membership in several learned societies including, The Jamaica Statistical Society, Bioethics Society of the English-Speaking Caribbean Society for Epidemiologic Research, Society for Clinical Research Associates (SOCRA) and Association of Clinical Research Professional (ACRP).

Donovan McGrowder holds Bachelor of Science (Honours) and Doctor of Philosophy (PhD) degrees in Biochemistry and Chemistry, and Biochemistry respectively from the University of the West Indies, Mona, Kingston, Jamaica. Dr. McGrowder also holds a Master of Science degree from University of Westminster, London, UK and a Master of Arts in Psychology and Counselling from St. Stephens College, Alberta, Canada. He also has post-graduate training and experience in Clinical Biochemistry and Molecular Biology. Dr. McGrowder joined the staff of The University of the West Indies, Mona in 2001 as Research Fellow in the Department of Basic Medical Sciences – Biochemistry Section and was promoted to Lecturer in 2003 and Senior Lecturer in 2009. He is also Consultant of Chemical Pathology at the University Hospital of the West Indies. He is Fellow of the Institute of Biomedical Science, UK, Fellow of the Royal Society of Tropical Medicine and Hygiene, UK, and Member of the Institute of Health Promotion and Education, UK.

Fabian Miller is a Senior Lecturer, Researcher and the Head of Department of Physical Education at The Mico University College; he has participated in research in the Medical field and Sport Science field. He has worked on research in sport services in the CARIFORUM region which aim was to establish regional policies for sport services for the member states involved and to establish sport services as a viable industry in the region. His current research is aimed to establish Biological Passports of College/University Athletes who compete in Sprinting and Long Distance events. Mr. Miller also published peer reviewed journal articles and book chapters. He co-authored the manuscript titled "Prostate Specific Antigen Testing in a Tertiary Teaching Hospital" and co-authored the book chapter titled "Dietary

Antioxidants in the Chemoprevention of Prostate Cancer. Mr. Miller is a graduate from G.C Foster College where he completed a Diploma in Education and a Bachelors Degree in Physical Education and Sport, he went on to pursue and complete his Masters of Science in Physical Education at The University of Technology, Jamaica and he is currently a PhD student in Biotechnology, Faculty of Science and Technology, University of the West Indies, Mona. Mr. Miller is a board member of the Social and Human Sciences Advisory Committee of Jamaica National Commission for UNESCO (Period of 2016-2019).

Arunava Mukherjee is currently working as an Assistant Professor in Post Graduate Department of Zoology of Ramakrishna Mission Vivekananda Centenary College (Autonomous), Rahara, Kolkata. He has completed his MPhil and PhD from University of Calcutta. He has published more than 30 research articles in various journals, proceedings or in books as book chapters.

Magdalene Nwokocha is a consultant Haematologist/Oncologist in the University Hospital of the West Indies and an Associate Lecturer in Haematology in the University of the West Indies. She has been involved in several peer-reviewed publications ranging from the Continental distribution of blood group variants in the 1000 Genome project, to the use of natural products in mitigating the effect of heavy metal poisoning, to Hypertension awareness and screening as part of the International May Measurement Month, to Antibody response following the COVID-19 Astra Zeneca vaccine and now Acute kidney injury in relation to COVID-19 severity. Dr. Nwokocha is involved in clinical practice where she treats both Haematology and Oncology cases with a special interest in treating patients with bleeding disorders.

Venugopal N. Puluamaghatta has completed his masters in Biological Anthropology and Ph. D in Molecular Anthropology from University of Mysore, Mysore. At present, he is working as an Assistant Anthropologist, Anthropological Survey of India, Eastern Regional Centre, Kolkata. Under the national project "DNA Polymorphism and Disease" he explored the genetic causes of the Handigodu disease. The major contribution of the said study was identification of a novel 39 base pair deletion; this novel deletion is not reported elsewhere in any of the literature cited in any of the genome and mutation database. The results of his study provide an opportunity for prenatal and post natal screening to avoid the ill effects of the Handigodu syndrome and there by promoting healthy society for the future generations. For the same he has received prestigious "Young Scientist Award" from Indian Society of Human Genetics at 38th Annual Conference and International Symposium Developmental and Complex Disorders in 2012. Also, he is closely associated with National project "DNA Polymorphism in Contemporary Indian Population", which involves sequencing of mtDNA and Y chromosome to explore maternal and paternal inheritance and to know evolutionary changes in genetic structure of Indian population. Under this project for the research work "Peopling of Uttarakhand" he received Young Scientist Award from Uttarakhand State Council for Science and Technology, Ministry of Science and Technology, Government of Uttarakhand at 10thUttarakhand State Science and Technology Congress in 2016. Further, published more than 40 research papers in National and International Journals and serving as editorial member in three journals.

Supriya Ray has done her bachelor's degree in Food Science and Nutrition Management from J.D. Birla Institute under Jadavpur University, Kolkata in 2010. She did her master's degree in Food Science and Technology from Gitam University, Andhra Pradesh in 2015. After that, she pursued her career in research and development. She worked as a Junior Research Fellow in CSIR-Indian Institute of Toxicology Research till 2018. She then joined Anthropological Survey of India as a Junior Research Fellow in 2019. Besides, she also registered for Ph. D in Applied Genetics at Karnatak University, Dharwad from 2019. She did her bachelor's degree project, titled "A study on the Nutritional Status of Pregnant Women belonging to the Bengali Community" and her master's degree project, titled "Physico-Chemical Properties of Spinach Leaves with Different Pre-treatments and drying Methods". She has published 1 paper in International Journal, participated in poster presentation at IFCON in 2013 and attended several conferences. She secured 1st rank in master's degree and achieved DST INSPIRE Fellowship in 2016. She also got special appreciation for poster presentation in UGC sponsored National conference held in Visakhapatnam in 2014.

Kamal Smimih is a PhD student in Neuroscience and physiology, at Sultan Moulay Slimane University Faculty of Science and Technology of Beni-Mellal (FSTBM).

Abirami Subramanian is an Environmental Science Research Scholar in the School of Interdisciplinary and Transdisciplinary Studies, Indira Gandhi National Open University, New Delhi. She is the recipient of the Savitribai Phule Fellowship and holds a UGC-NET (LS) qualification in Environmental Science. She received her undergraduate (2017) and postgraduate (2019) degrees in Zoology from the Avinashilingam Institute for Home Science and Higher Education for Women in Coimbatore, Tamil Nadu. She also obtained her Postgraduate Certificate in Climate Change from Indira Gandhi National Open University. She was a First Proficiency holder (Gold medalist) in her M.Sc and a Third Proficiency holder in her B.Sc. She had authored/co-authored 2 book chapters and 2 research articles and also presented 3 papers at national and international conferences. Her research interest includes Environmental Science, Environmental and Applied Microbiology, and Geomicrobiology.

Nagarjuna Telagam is with the Electronics and Communication Engineering Department, GITAM University, Bangalore, India and His research interests are 5G, IoT, VLSI technologies, Virtual Instrumentation. He is Anna University rank holder (20) for Masters Degree in State Level. Certified LabVIEW Associate Developer (CLAD) from National Instruments.

Cameil Wilson-Clarke is presently a Clinical Pharmacist and the Programme Director for the Doctor of Pharmacy Programme. She also is a lecturer at the Pharmacology & Pharmacy Section, Faculty of Medical Sciences, University of West Indies, Mona Campus. She has been lecturing in the field of clinical pharmacy for more than a decade. She achieved her Doctor of Pharmacy (Pharm.D) at The Ohio State University, Columbus, Ohio, USA. She also holds a Master of Arts in Teaching, at the University of the West Indies, Mona. Her areas of focus are Sickle Cell Disease, Oncology (Prostate & Breast), Diabetes Research/Education, Medicinal Cannabis and Pharmacovigilance. Currently she is involved in clinical trials research in sickle cell disease and medicinal cannabis.

Index

A

AMR 152, 154-155, 157-158, 163-166
and Humidity 8
Antibiotic Resistance 65, 152-156, 158, 160, 164, 167-170, 172
Apicomplexa 101
Artemisinin 101, 117

B

Bangladesh 39, 50, 55, 59, 126-127, 130-133, 135-136, 140, 142-148, 150
Basic Reproduction Number (R0) 46, 66
Behavior Pattern 134-135, 138
Bridge Host 43, 49, 66

C

CDC 25, 35, 38, 154, 157, 165, 168
Climate Change 1-2, 4-5, 7-8, 10-28, 31-42, 44, 60, 68-69, 71, 81-84, 87-94, 98, 114
CNN Model 208, 216-217, 223, 226
Crisis 19, 54, 104, 115, 157, 167, 170, 172

D

Disease transmission 26, 31-32, 34-36, 54, 86, 92, 95, 126, 131-133, 137, 141, 148, 155
Duffy 59, 100, 105-106, 116, 121-122

E

Ecological Systems Theory 126, 128-129, 148
Ecopsychology 19, 30
Environment 4-5, 8-9, 14, 19, 22, 28-30, 32-33, 38, 47, 51-52, 54, 57, 65, 72, 81, 86-87, 96, 98, 110, 113, 121, 126-134, 136-143, 145-148, 150-151, 173, 176-180, 188, 190-203, 207

F

Falciparum 14, 101-107, 109, 111, 114-116, 119-122, 125
Favism 100, 104

G

Gut Dysbiosis 174-175, 186, 190, 197

H

Health 1-8, 10-12, 14-17, 19-27, 29-33, 35-42, 44-45, 48, 50-56, 58-59, 61-66, 68, 73, 84, 87, 90-94, 110-113, 118-119, 124-128, 130, 132, 142-151, 153-154, 157, 162, 164, 170, 172-180, 184-188, 190-191, 193-194, 196, 198-207
Hosts 31, 33-34, 46-49, 52, 54, 56, 61, 67, 69, 71, 81, 89, 101-103, 108-110, 129, 174, 181, 183, 186, 200

Human Diseases 31, 199
Human Health 1, 4, 7-8, 16-17, 20, 25, 31-32, 40, 62-63, 68, 87, 174, 176-177, 179, 184, 188, 190-191, 196, 199-201, 203, 206

I

Inception Model 208
Infectious Disease 29, 34-37, 39-40, 42, 45, 54, 63, 97, 126, 129-133, 135-137, 139-141, 170, 172, 186
Interface 23, 45-47, 59, 61, 63, 66, 75

L

Larval Source Management 68, 84-85, 88-89, 94
Livelihood 30

M

Maintenance Host 46, 66
Microorganisms 93, 153-154, 160, 166, 174-175, 177, 181, 186-188, 190, 200
Micropollutants 178
Morbidity 1, 6, 27, 30, 35, 38, 40, 111, 127, 130-131, 153
Mortality 1, 6, 15, 22, 27, 30, 38, 46, 50, 54, 94, 111, 119-120, 127, 130, 153, 167, 210
Mosquitoes 14, 33-34, 48, 68-83, 85-89, 91-92, 94-96, 98-99, 101, 109-110, 112-115
Multi-Drug Resistance 125, 152

N

Network Theory 53-54, 67

O

Opportunistic Breeding Behaviour 68
Ovalocytosis 105, 120

P

Parasite 11-12, 14, 21, 33, 51, 63-64, 66, 101-102, 106-110, 115-120
Pathogens 2, 31-35, 37, 43-48, 51, 53-54, 57, 59, 61, 63, 65, 82, 92-93, 101, 113, 116, 121, 129, 152-154, 157-159, 161-164, 166, 169-170, 174, 180, 185-187
Phylogenetics 67, 81
Physical School Environment 126-134, 136-142, 147-148, 151
Pollution 2, 9, 22-23, 27, 29, 36, 44, 47, 61, 174, 193-198, 202-203, 205-207
Population Genetics 54, 67, 118
Primary School Context 126-129, 132, 136, 139

Q

Quinine 116-117

R

Rainfall 2, 13-14, 23, 32-34, 40, 81-82, 112
Ronald Ross 110

S

Schizogony 102
Spillover 46, 48, 59-60, 62-63, 67, 108-109, 111, 123

T

Target Host 48, 67
Temperature 2-5, 8, 13-14, 23-26, 32-35, 38, 42, 74, 84, 91, 114
Temporary Breeding Habitats 68, 71-72, 81
Transformer Model 208, 212-213, 216, 219-220, 224, 226

U

Urban Growth 44, 68

Urbanization 1, 43-47, 50-54, 56-59, 62, 64, 68-69, 82-84, 87-89, 93, 97-99, 114

V

Vector 1, 12, 14, 21, 32, 34, 36, 40, 47, 51-52, 56-57, 59, 63, 70-71, 73, 77, 79, 82-85, 88-89, 92-95, 97-102, 109-113, 115, 123-124, 218, 222

W

Wild Life 43
Wind 4, 8, 33-35

Z

Zoonoses 45-46, 48-49, 55, 57, 59-60, 64, 100
Zoonotic 1, 11-12, 22, 26, 40, 43, 45-47, 49, 51-54, 56-60, 63, 100, 108, 111, 168

www.ingramcontent.com/pod-product-compliance
Lightning Source LLC
LaVergne TN
LVHW061550290926
849162LV00004B/963